Yun Zhang · Shenghan Gao

High Efficiency Non-isolated DC-DC Converters with Wide Voltage Gain Range for Renewable Energies

Springer

Yun Zhang
School of Electrical and Information Engineering
Tianjin University
Tianjin, China

Shenghan Gao
School of Electrical and Information Engineering
Tianjin University
Tianjin, China

ISSN 2520-8853 ISSN 2520-8861 (electronic)
CPSS Power Electronics Series
ISBN 978-981-97-0647-1 ISBN 978-981-97-0648-8 (eBook)
https://doi.org/10.1007/978-981-97-0648-8

This Springer imprint is published by the registered company Springer Nature Singapore Pte Ltd.
The registered company address is: 152 Beach Road, #21-01/04 Gateway East, Singapore 189721, Singapore

Paper in this product is recyclable.

Preface

The efficient use of the renewable energy sources presents unique challenges, including the need for effective power conversion and control techniques. This book focuses on the non-isolated DC-DC converters with wide voltage gain range, exploring a family of novel topologies and control schemes for renewable energy applications. It offers innovative approaches that enhance efficiency, reduce costs, and maximize the utilization of energy resources. Moreover, detailed explanations based on circuit diagrams, as well as mathematical models, are presented to help readers to grasp the operating principles.

Chapter 1 introduces the basic principles, modulation techniques, composition, and performance indicators of power converters. A detailed overview of conventional non-isolated DC-DC converter is provided. Furthermore, the specific application of DC-DC converters in electric vehicles and photovoltaic systems is discussed.

Chapter 2 presents two bidirectional H-bridge DC-DC converters. Both the converters can achieve wide voltage gain with non-extreme duty cycles that are close to 0.5 and exhibit doubled equivalent switching frequency. By applying synchronous rectification control scheme, the conducting losses and reverse recovery losses of diodes are reduced.

Chapter 3 shows a variety of three-level DC-DC converters, where the wide voltage gain range and capacitor voltage self-balance as well as soft-switching can be achieved.

Chapter 4 gives detailed analysis on quasi-Z-source DC-DC converters, which offer wider voltage range and improved efficiency. In addition, a novel variable switching-frequency control strategy is applied to realize soft-switching operation.

Chapter 5 presents a family of DC-DC converters adopting switched-capacitor cells. The benefits include high voltage gain, low voltage stresses, and common grounds, enhancing the application potential of the converters for renewable energies.

Chapter 6 provides practical examples of the non-isolated DC-DC converters applied in energy storage systems, fuel cell vehicles, and photovoltaic systems. The parameter design procedure and experimental results are presented to demonstrate the correctness and feasibility of the introduced topologies in practical application.

Tianjin, China

Yun Zhang
Shenghan Gao

Acknowledgments The authors would like to express their sincere appreciation to the editors at Springer for their help and support in producing the book. Additionally, gratitude is extended to the individuals who proposed and developed the converters discussed in this book: Prof. Ping Wang, Yongping Gao, Xingtao Sun, Qiangqiang Liu, Lei Zhou, Heyu Liu, Jilong Shi, Chendong Zhao, Chuanzhi Fu, and Wei Zhang. Their innovative ideas and extensive research have made significant contributions to the development of non-isolated DC-DC converters.

About This Book

The book focuses on the topologies of non-isolated DC-DC converters with a wide voltage gain range for renewable energy systems. It addresses both theoretical principles and engineering practices. Each chapter introduces a family of non-isolated DC-DC converters, including advanced control methods. The inclusion of modeling and experimental verification provides in-depth applications in fuel cell vehicles, electric vehicles with multiple sources including batteries and super-capacitors, and photovoltaic systems. The book is beneficial for researchers, engineers, and graduate students in the fields of renewable energies, as well as electrical and electronic engineering.

Contents

Chapter 1
DC–DC Converters and Applications for Renewable Energies

1.1 Introduction

Given the shortage of fossil fuels and concerns about environmental pollution, the imperative to substitute fossil fuels with renewable energies has become crucial. Clean and pollution-free energy sources, such as photovoltaic (PV) cells, fuel cells, batteries, and super-capacitors, have gained widespread adoption in diverse power systems. The ongoing enhancement of energy utilization efficiency heavily relies on the performance of DC–DC converters that interface with these energy units.

The conversion of solar energy into electrical power by PV cells has attracted considerable attention in industrial applications [1]. Nevertheless, the inherent challenges in developing efficient and environmentally friendly PV systems stem from the unpredictable and variable nature of environmental conditions. Photovoltaic cells display fluctuating and nonlinear output characteristics due to variations in irradiance and internal parameters [2, 3]. Consequently, the adoption of DC–DC converters to achieve power point tracking becomes imperative to guarantee optimal power conversion and overall system performance.

Fuel cells harness chemical energy from hydrogen to achieve efficient and clean power generation, and thus have become a viable solution for addressing concerns related to fossil fuels and air pollution [4–6]. Vehicles powered by fuel cells can attain environmentally friendly propulsion, producing zero emissions and optimizing energy utilization [7–9]. Nevertheless, the widespread adoption of fuel cells faces limitations due to their low output voltage and high output current. This necessitates the use of DC–DC converters to bridge the voltage gap and ensure a stable DC-link bus voltage [10–12].

The increasing prevalence of electric vehicles (EVs) in the automotive market has spurred advances in battery technology, motors, and electric control systems [13, 14]. EVs offer a cleaner and more environmentally friendly alternative to fossil fuel vehicles, with low noise levels and high energy conversion efficiency. To optimize the performance and longevity of EVs, a multisource energy system that

Y. Zhang and S. Gao, *High Efficiency Non-isolated DC-DC Converters with Wide Voltage Gain Range for Renewable Energies*, CPSS Power Electronics Series,
https://doi.org/10.1007/978-981-97-0648-8_1

combines batteries and super-capacitors has been introduced [15]. Bidirectional DC–DC converters play a pivotal role in this system. The batteries supply low-frequency power to maintain a stable DC bus voltage, while the super-capacitors compensate for or absorb high-frequency power during load transients.

In this chapter, the techniques related to non-isolated DC–DC converters are comprehensively explored. The operating principles, modulation techniques, composition, and performance indicators of power converters are introduced, establishing a foundation for the understanding of non-isolated DC–DC converters as applied in renewable energy.

Additionally, an overview of non-isolated DC–DC converters is provided, with a focus on Boost converter and Buck converter. Various techniques for topology derivation and expansion are presented and characterized. Furthermore, the application scenarios of DC–DC converters for renewable energies are discussed, including energy storage devices, fuel cells, and photovoltaic systems.

1.1.1 Operating Principles of Power Converters

The power converters function as power interfaces, controlling the output voltage/current of sources by adjusting the switching states of semiconductor components. Through proper topology design, these converters can facilitate power transition with minimal losses, high conversion efficiency, and compact size. As a result, they find widespread application in various fields, including communication equipment, digital controller devices, household appliances, and renewable energy systems. Power converters integrated with digital controllers offer advantages such as stable output, high reliability, and remote control. Consequently, they represent the current development trend in power supply technology.

During the operation of the power converter, the power switches are periodically turned ON and OFF by pulse driving signals, resulting in a pulse voltage derived from the input voltage. The desired output voltage is then obtained from this pulse voltage through a filter circuit. According to the area equivalence principle, the value of the output voltage can be controlled by adjusting the conducting duty cycle of the power switches. The duty cycle D is defined as follows, where t_{ON} is the turn-on interval of the power switch and T_{s} is the switching cycle.

$$D = \frac{t_{\mathrm{ON}}}{T_{\mathrm{s}}} \tag{1.1}$$

Taking the conventional Buck converter as an example, the pre-filtered output voltage U_{out} is a pulse voltage, with an amplitude equal to the input voltage U_{in}. The relationship between the input and output voltage is expressed by the following equation.

$$U_{\mathrm{out}} = U_{\mathrm{in}} D \tag{1.2}$$

1.1.2 Modulation Techniques of Power Converters

According to the above analysis, the control of power switches is crucial for achieving efficient and stable converter operation. The control techniques for power converters can be further classified into pulse-width modulation (PWM), pulse-frequency modulation (PFM), and combined pulse-width and pulse-frequency (PWPF) modulation.

PWM: The duty cycle is controlled by adjusting the turn-on interval of power switches. Due to the fixed switching cycle, the design and optimization of the output filter circuit are easily implemented. Therefore, PWM is the most commonly applied control scheme for power converters.

PFM: The turn-on interval t_{ON} remains constant while adjusting the switching frequency f_s to control the duty cycle. Due to the variable switching frequency, the design of the output filter circuit is relatively more complex.

PWPF: This scheme adjusts both the turn-on interval t_{ON} and the switching frequency f_s as variables to achieve duty cycle control.

In comparison, the PWM scheme exhibits higher voltage stabilization accuracy and conversion efficiency under heavy load, whereas the PFM scheme possesses better transient performance and higher efficiency under light load. The PWPF scheme combines PWM and PFM to obtain comprehensive advantages, employing PWM in heavy load conditions and PFM in light load conditions. This approach ensures high efficiency across the full load range.

The implementation of PWM can be classified into leading-edge modulation, trailing-edge modulation, and dual-edge modulation, as illustrated in Fig. 1.1. In this illustration, v_c represents the carrier wave, and v_m is the modulation wave generated from the voltage/current controller. V_{PWM} is the gate signal obtained from v_c and v_m, as shown below.

$$V_{PWM} = \begin{cases} 1, & v_m > v_c \\ 0, & v_m < v_c \end{cases} \tag{1.3}$$

According to (1.3), the power switch is turned ON as v_m is greater than v_c, while it is turned OFF when v_m is smaller than v_c. All three modulation techniques are based on the same approach of adjusting the modulation signal before the next switching

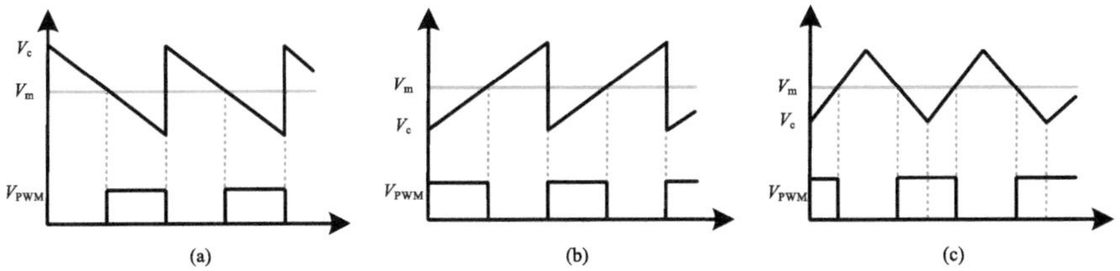

Fig. 1.1 PWM implementation with **a** leading-edge, **b** trailing-edge, and **c** dual-edge modulation

cycle, so that the error between the reference and the sampled voltage/current can be dynamically reduced.

As shown in Fig. 1.1a, under leading-edge modulation, the power switch is turned OFF at the beginning of each switching cycle and then turned ON after a time interval of DT_S. Conversely, in trailing-edge modulation, depicted in Fig. 1.1b, the power switch is turned ON at the start of each switching cycle and turned OFF after a time interval of DT_S. In dual-edge modulation, the power switch is turned ON after the beginning and before the end of the switching cycle. In other words, both the leading and trailing edges of the PWM can be adjusted, providing dual-edge modulation with better performance in dynamic response. However, implementing dual-edge modulation is more complex than implementing single-edge modulation techniques.

In addition, a mismatch between modulation techniques and the topologies or control strategies of converters could result in stability problems. Therefore, selecting the appropriate PWM modulation technique based on the specific application is necessary.

1.1.3 Composition of Power Converters

The semiconductor devices are utilized as power switches in converters to generate a stable output by being turned ON and OFF periodically. In addition to the main power circuit, which contains power switches, diodes, inductors, and capacitors, the converter also mainly consists of a driving circuit, a digital control circuit, a sensing circuit, and an auxiliary power supply circuit.

The composition of a power converter is presented in Fig. 1.2, and the functions of each stage are explained as follows.

Main power circuit: The main power circuit is the core of a converter, with various proposed topologies.

Driving circuit: The driving circuit amplifies and reshapes the PWM signals to drive the power switches. Isolated driving circuits enable electrical isolation between the control and power circuits.

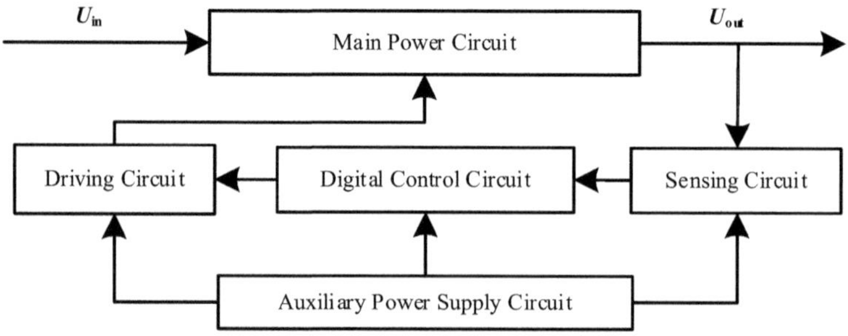

Fig. 1.2 Fundamental composition of a power converter

Digital control circuit: The digital converter generates PWM signals through sampled signal processing and control variable calculation, directly impacting system performance.

Sensing circuit: The current/voltage sensing circuit is employed for current/ voltage sensing and analog/digital signal conversion, providing feedback signals for the controller.

Auxiliary power supply circuit: All the integrated circuits are powered by the auxiliary power supply circuit.

Additionally, start-up circuits, over current/voltage protection circuits, filter circuits, and indicator circuits are often adopted to further improve the performance of converter.

1.1.4 Performances Indicators of Power Converters

The performance of power converters concerns the reliability and efficiency of the whole system. As a result, the following performance indicators of converters should be considered for specific practical applications:

Galvanic isolation: Converters with galvanic isolation capability can ensure safety for the system, especially in high power applications.

EMI issues: The EMI performance of converters is related to operational reliability and efficiency. Major solutions for EMI suppression include auxiliary filter circuits, soft-switching operations, grounded shielding enclosures, optimization of PCB design, etc.

Output polarity: The output polarity of converters can either be the same or reversed as the input.

Power level: The rated power level at which the converter can operate continuously should be greater than the maximum power of the application condition.

Dynamic response: The converter is expected to respond rapidly to sudden changes in operation conditions, such as load, input, and reference current/voltage.

Static accuracy: Converters with high static accuracy can track the reference current/voltage with minimal error during steady state.

Current/Voltage ripples: The ripples in inductor current and capacitor voltage are related to the values of inductance and capacitance, respectively. These ripples have significant influences on the converter's efficiency and output quality.

Voltage gain: Converters require a wide voltage gain to match the input and output voltages. Those with limited voltage gain may experience a narrowed operating range and extreme duty cycles.

Conversion efficiency: Conversion efficiency is a key performance metric for a converter, directly related to the rate of energy utilization. Low conversion efficiency results in high power losses, leading to potential heat dissipation problems.

Stability: Converter stability can be analyzed based on small-signal models. Optimal design of control parameters is essential to improve system stability and robustness.

1.2 Overview of Non-isolated DC–DC Converters

The DC–DC converters are circuits that convert a DC input to another DC output with different or adjustable values. They can be classified into isolated and non-isolated converters based on the galvanic isolation characteristic. The input and output of isolated DC–DC converters are isolated through the utilization of a transformer, which exhibits better safety performance. Thus, isolated converters are more applicable for high power levels than non-isolated DC–DC converters. Additionally, by adjusting the turns ratio of the transformer, high voltage gain can be achieved for isolated DC–DC converters. However, due to the complex circuit structure and a large number of components, the application of isolated DC–DC converters is usually limited by their large volume and high cost.

The conventional non-isolated DC–DC converters include the Buck, Boost, Buck-Boost, Cuk, Sepic, and Zeta converters. The topologies of these converters are presented in Fig. 1.3, where *D* represents diodes, *Q* represents power switches, R_{load} represents the output load, and *L* and *C* represent inductors and capacitors, respectively. Among these converters, the Buck and Boost converters are the most representative non-isolated DC–DC converters, which are analyzed in detail as follows.

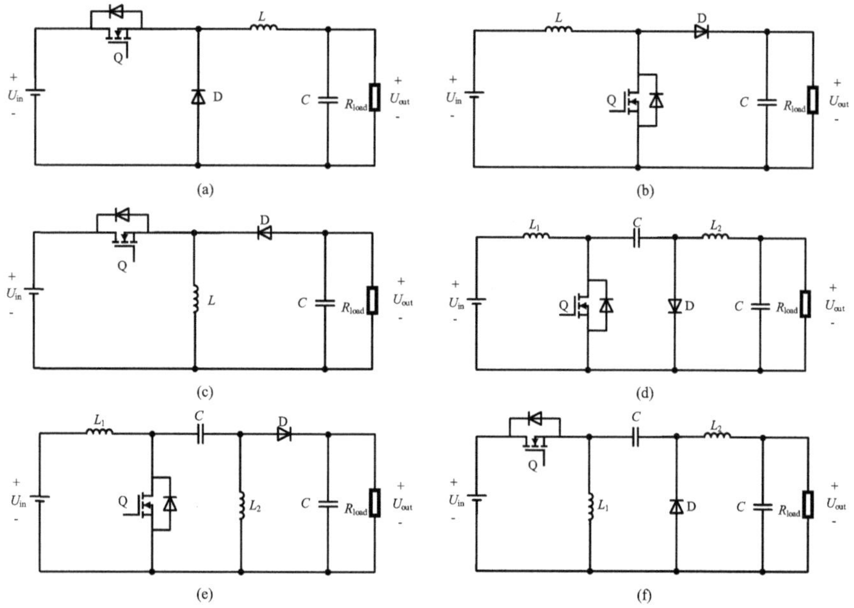

Fig. 1.3 Topologies of the conventional DC–DC converters: **a** Buck, **b** Boost, **c** Buck-Boost, **d** Cuk, **e** Sepic, and **f** Zeta

1.2.1 Conventional Non-isolated DC–DC Converters

A. Buck converter

The switching states of the Buck converter are presented in Fig. 1.4. When the power switch Q is turned ON, the diode D is OFF, and the inductor L is charged. The load R_{load} is powered by the input source and the inductor. The current in the inductor, denoted as i_{L}, increases, and its variation Δi_{L} during this state can be derived as follows, where d_{Buck} represents the duty cycle of the power switch Q.

$$\Delta i_{\text{L}} = \frac{U_{\text{in}} - U_{\text{out}}}{L} d_{\text{Buck}} T_{\text{s}} \tag{1.4}$$

When the power switch Q is turned OFF, the load R is powered by the inductor L and the output capacitor C. The inductor L discharges, causing the inductor current i_{L} to decrease. Depending on the condition of i_{L}, the operation of the Buck converter can be classified into Continuous Conduction Mode (CCM) and Discontinuous Conduction Mode (DCM), with the critical condition in between known as Boundary Conduction Mode (BCM). The variation in inductor current during the turn-off state under CCM can be obtained as follows.

$$\Delta i_{\text{L}} = \frac{U_{\text{out}}}{L}(1 - d_{\text{Buck}}) T_{\text{s}} \tag{1.5}$$

Under DCM, the current variation is expressed as (1.6), where d_{disch} represents the discharging duty cycle of inductor current i_{L}, and $d_{\text{Boost}} + d_{\text{disch}} < 1$.

$$\Delta i_{\text{L}} = \frac{U_{\text{out}}}{L} d_{\text{disch}} T_{\text{s}} \tag{1.6}$$

The CCM voltage gain of Buck converter M_{Buck} can be derived by combining (1.4) and (1.5).

$$M_{\text{Buck}} = \frac{U_{\text{out}}}{U_{\text{in}}} = d_{\text{Buck}} \tag{1.7}$$

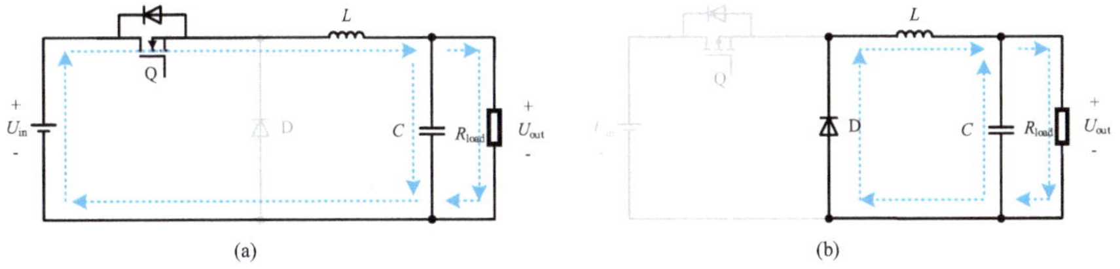

Fig. 1.4 Switching states of the Buck converter during **a** power switch turned ON and **b** power switch turned OFF

According to (1.4) and (1.6), the DCM voltage gain of Buck converter can be obtained as follows.

$$M_{\text{Buck}} = \frac{U_{\text{out}}}{U_{\text{in}}} = \frac{d_{\text{Buck}}}{d_{\text{Buck}} + d_{\text{disch}}} = \frac{2}{1 + \sqrt{1 + \frac{8L}{d_{\text{Buck}}^2 R_{\text{load}} T_s}}} \tag{1.8}$$

It is important to note that the expressions for voltage gain, both in CCM and DCM, are applicable for DCM.

B. Boost converter

In the Boost converter, the power switch Q regulates the charging and discharging of inductor L. When Q is turned ON, the inductor L charges from the input source, causing an increase in the inductor current i_{L}. The output load R is powered by the energy stored in the output capacitor C.

Referring to the operating states depicted in Fig. 1.5, the variation in inductor current Δi_{L} during the turn-on state can be calculated as follows. Here, U_{in} represents the input voltage, d_{Boost} is the duty cycle of the power switch Q, and T_{s} is the switching cycle.

$$\Delta i_{\text{L}} = \frac{U_{\text{in}}}{L} d_{\text{Boost}} T_{\text{s}} \tag{1.9}$$

When the power switch Q is turned OFF, the inductor L discharges, resulting in a decrease in the inductor current i_{L}. The output load R_{load} is then powered by both the input source and the inductor through the diode D. Similar to the Buck converter, the operation of the converter can be categorized into CCM, DCM, and BCM, based on the inductor current conditions during the turn-off state. The variation in inductor current Δi_{L} during the turn-off state of the Boost converter under CCM can be derived as follows.

$$\Delta i_{\text{L}} = \frac{U_{\text{in}} - U_{\text{out}}}{L} (1 - d_{\text{Boost}}) T_{\text{s}} \tag{1.10}$$

When the total variation of inductor current during a switching cycle is equal to zero, the converter achieves steady-state operation. Consequently, the voltage gain

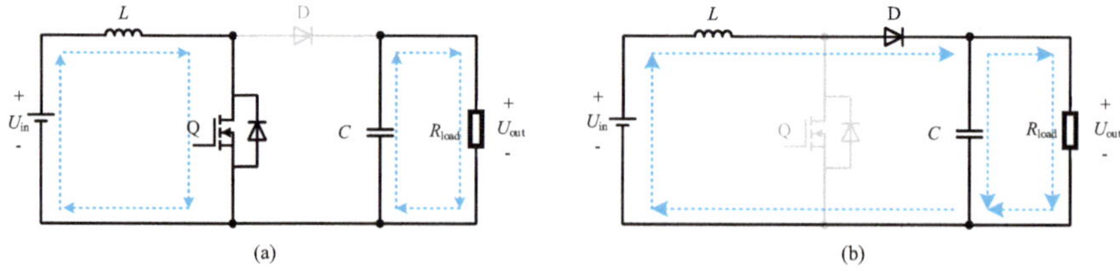

Fig. 1.5 Switching states of the Boost converter during **a** power switch turned ON and **b** power switch turned OFF

of the Boost converter M_{Boost} can be derived by combining (1.9) and (1.10) as shown below.

$$M_{\text{Boost}} = \frac{U_{\text{out}}}{U_{\text{in}}} = \frac{1}{1 - d_{\text{Boost}}} \tag{1.11}$$

Given the duty cycle range of $0 < d_{\text{Boost}} < 1$, the voltage step-up can be realized by the Boost converter, as indicated by (1.11). In the case of the Boost converter operating in DCM, the voltage gain can be expressed as (1.12), where d_{disch} is the duty cycle of the inductor discharging state that $d_{\text{Boost}} + d_{\text{disch}} < 1$.

$$M_{\text{Boost}} = \frac{U_{\text{out}}}{U_{\text{in}}} = \frac{d_{\text{Boost}} + d_{\text{disch}}}{d_{\text{disch}}} \tag{1.12}$$

The voltage gain in DCM can be further determined using (1.13). As evident, the voltage gain of the Boost converter in DCM is related to the value of inductance.

$$M_{\text{Boost}} = \frac{1 + \sqrt{1 + \frac{2d_{\text{Boost}}^2 R_{\text{load}} T_s}{L}}}{2} \tag{1.13}$$

1.2.2 Derivation and Classification of Non-isolated DC–DC Converters

While conventional Boost and Buck converters can achieve voltage step-up and step-down from the input to the output, practical applications with requirements for wide voltage gain are limited. This limitation arises from the high voltage stresses and extreme duty cycles associated with these converters. Additionally, the hard-switching operation tends to result in low efficiency and significant electromagnetic interference, especially at high power levels.

Therefore, to address these challenges, various non-isolated DC–DC converters have been proposed and studied. By combining basic topologies with additional energy storage components and power switches, these converters aim to achieve improved performance in terms of conversion efficiency, voltage gain range, voltage and current stresses, and power density. The derivation of novel topologies involves forming a voltage-boosting cell with inductors/capacitors and power switches and connecting capacitors in series to lower the voltage stresses.

A. Switched-inductor

The switched-inductor structure comprises inductors and diodes, with the inductors alternating between charging and discharging [16, 17]. Figure 1.6 illustrates both passive and active switched-inductor units. In Fig. 1.6a, the passive switched-inductor cell consists of two inductors sharing the same inductance and three diodes. The

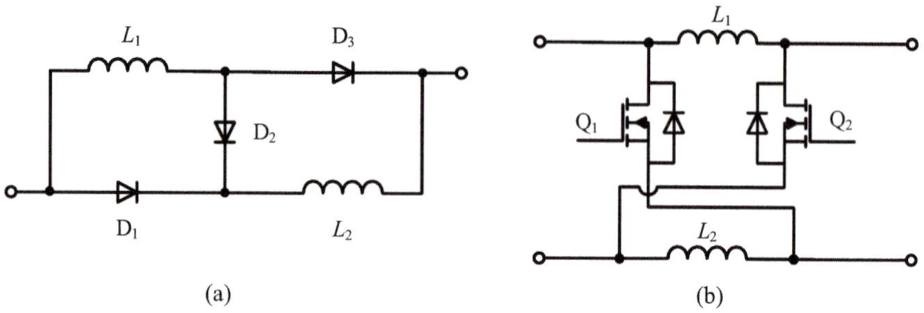

Fig. 1.6 Topologies of the **a** passive and **b** active switched-inductor cells

inductors charge in parallel when D_1 and D_2 are conducted, and D_3 is turned off. They discharge in series when D_1 and D_2 are reverse biased, and D_3 is conducted.

Figure 1.6b showcases the active switched-inductor cell, which includes two inductors and two power switches. The circuit's operation depends on the continuous switching of the power switches and the energy storage properties of the inductors. Efficient regulation of the output voltage is achieved by controlling the timing and duration of the switch states.

By replacing the inductor in the conventional Boost converter, the switched-inductor Boost converter topology can be derived, as illustrated in Fig. 1.7. Compared to the conventional Boost converter, this modification results in an increased voltage gain. Additionally, it allows for the sharing of average inductor current, facilitating a reduction in the size of each individual inductor. Moreover, with equal inductances, the inductors can be integrated into a single magnetic core, thereby enhancing power density. Nevertheless, the switched-inductor Boost converter faces a limitation wherein the voltage stresses across the diodes match the output voltage, and the use of multiple diodes exacerbates the issues related to reverse recovery.

B. Switched-capacitor

The switched-capacitor is a common method used for voltage boosting. The structure of a switched-capacitor cell is illustrated in Fig. 1.8, which consists of power

Fig. 1.7 Topology of the switched-inductor Boost converter

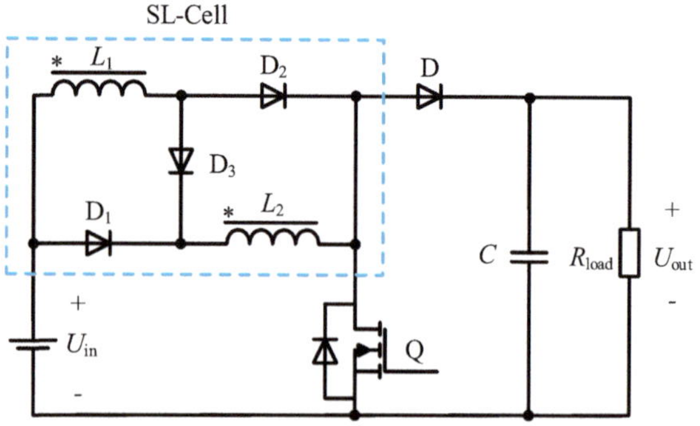

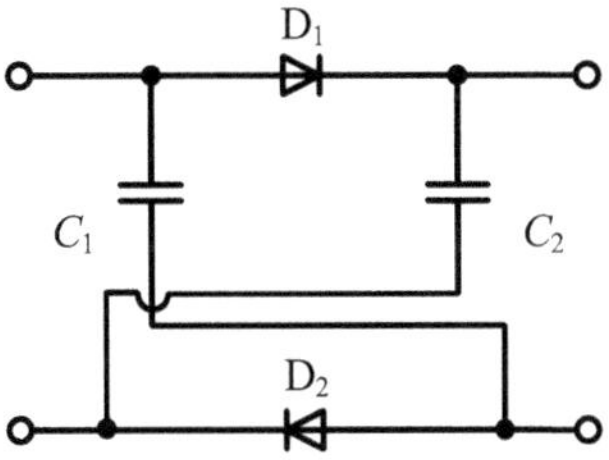

Fig. 1.8 Switched-capacitor structure

switches and capacitors. By transitioning between series and parallel connections of the capacitors, high voltage gain can be achieved through the combination with the switched-capacitor [18].

Figure 1.9 depicts the topology of the switched-capacitor Boost converter. The charging and discharging states of the switched-capacitor are governed by the main power switch. The switched-capacitor Boost converter presents several advantages, including reduced voltage stresses across power switches, a doubled voltage gain, and the self-balancing capability of capacitors. However, a notable drawback of the switched-capacitor lies in the pulse current generated by paralleled capacitors. This issue can be mitigated by introducing an auxiliary inductor connected in series with the capacitor.

C. Z-source

By replacing diodes with inductors in switched-capacitor cell, the Z-source structure can be obtained. As displayed in Fig. 1.10, the Z-source network comprises a diode, a power switch, two capacitors, and two inductors [19]. When the power switch Q is turned OFF, the inductors L_1 and L_2 are charged by the capacitors C_1 and C_2, respectively. Conversely, when Q is turned ON, the capacitors are charged by the input source, while the inductors are discharged to the output.

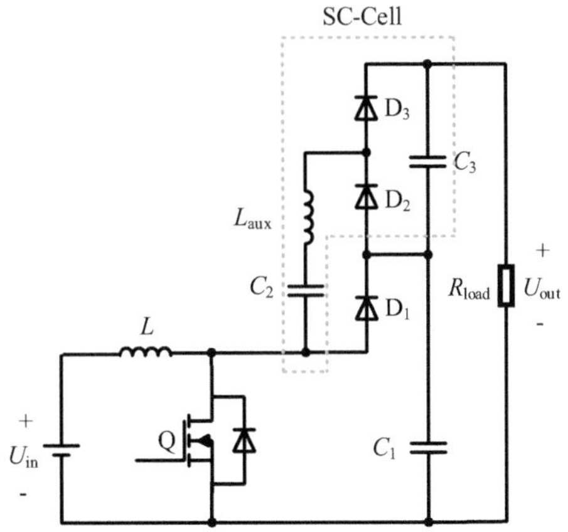

Fig. 1.9 Switched-capacitor Boost converter with auxiliary inductor

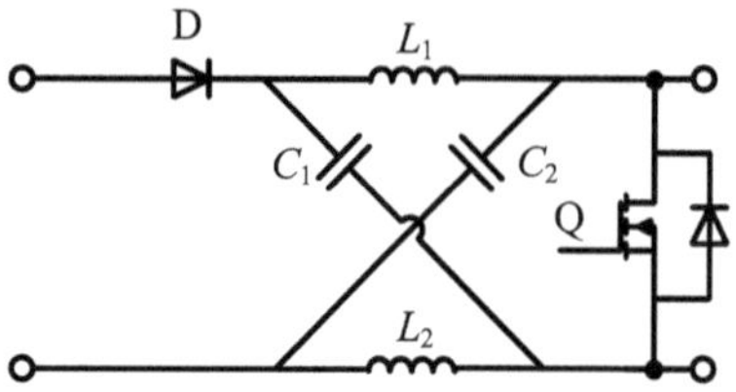

Fig. 1.10 Z-source structure

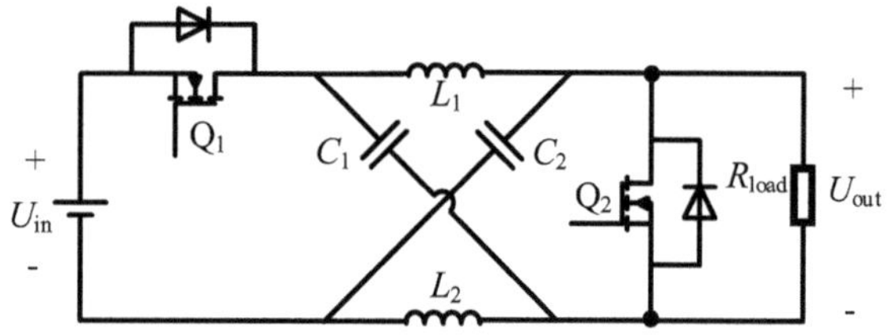

Fig. 1.11 Topology of the Z-source converter

The topology of the Z-source converter is depicted in Fig. 1.11, offering a wide voltage gain range for matching input and output voltages. However, it lacks a common ground between the input and output sides, potentially leading to extra issues with electromagnetic interference.

D. Quasi-Z-source

The quasi-Z-source is derived from the Z-source structure through impedance transformation, effectively overcoming the common ground problem. In Fig. 1.12, the structure of the quasi-Z-source cell is presented, encompassing the same components as the Z-source [20]. The operating principles of both the quasi-Z-source and the Z-source are also similar.

The topology of the basic quasi-Z-source DC–DC converter is shown in Fig. 1.13, exhibiting features such as continuous input current, low voltage stresses on capacitors, and a common ground between the input and output sides. Furthermore, a high voltage gain can be attained when the duty cycle approaches 0.5.

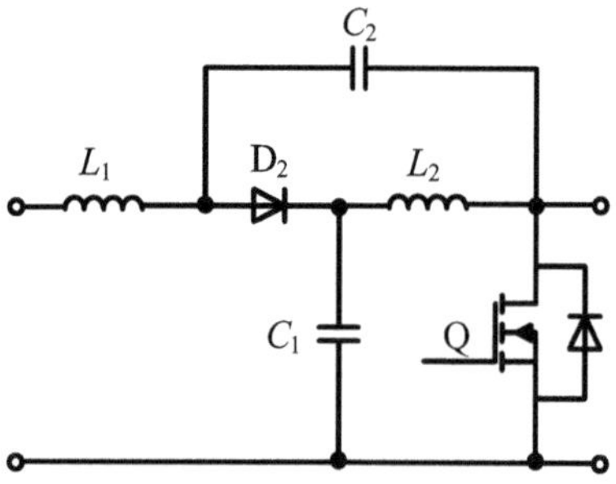

Fig. 1.12 Quasi-Z-source structure

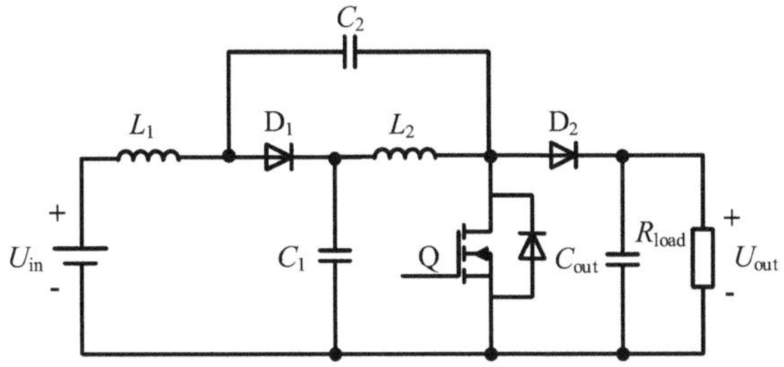

Fig. 1.13 Topology of the quasi-Z-source DC–DC converter

E. H-bridge

The multilevel converters have become a favored technique for high voltage applications, offering reductions in magnetic components and contributing to cost, volume, and voltage stress reductions. Figure 1.14 displays three basic submodules: the H-bridge, asymmetrical half-bridge, and symmetrical half-bridge [21]. In the H-bridge configuration, upper and lower switches operate complementarily with dead times to prevent shoot-through. The H-bridge's output results in either positive or negative voltage when the diagonal switches are activated. Otherwise, the output voltage is clamped at zero.

The topology of the isolated phase-shift full-bridge DC–DC converter is illustrated in Fig. 1.15, capable of achieving zero voltage switching [22]. As a result, it has been widely employed in practical applications due to its wide voltage gain range. The soft-switching of the converter depends on the energy stored in the leakage inductor of the transformer, which cannot be achieved at light loads. To tackle this limitation, some researchers have adopted auxiliary components to develop novel topologies with improved performance.

F. Three-level

Three-level DC–DC converters effectively halve the voltage stresses on power switches compared to the output voltage. The structure of a three-level unit, illustrated

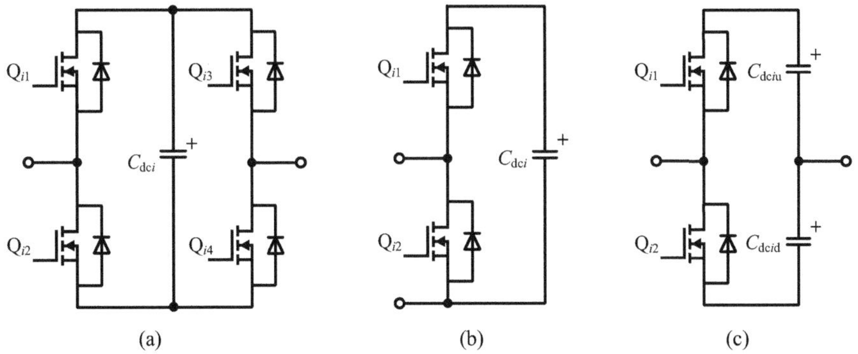

Fig. 1.14 **a** H-bridge, **b** asymmetrical half-bridge, and **c** symmetrical half-bridge submodules

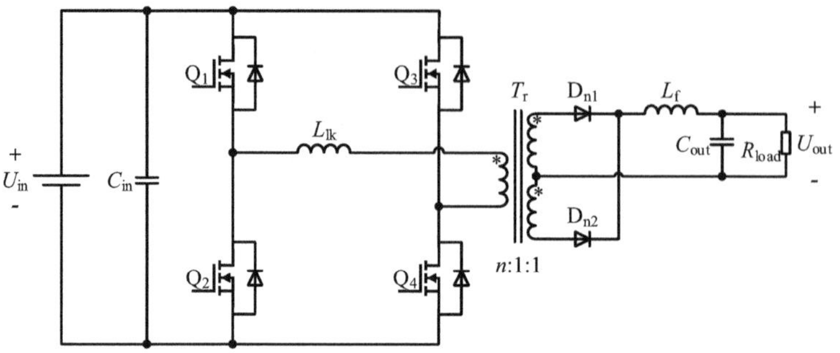

Fig. 1.15 Topology of the isolated phase-shift full-bridge DC–DC converter

in Fig. 1.16, comprises two power switches, two diodes, and two capacitors. Derived from the conventional Buck/Boost converter and featuring a three-level structure, the bidirectional three-level DC–DC converter's topology is depicted in Fig. 1.17. The converter employs a phase-shifted switching scheme, resulting in an effective inductor current ripple frequency twice the switching frequency [23].

The reduced voltage stresses across components make three-level converters suitable for high voltage applications. Additionally, since switching losses are closely

Fig. 1.16 Three-level structure

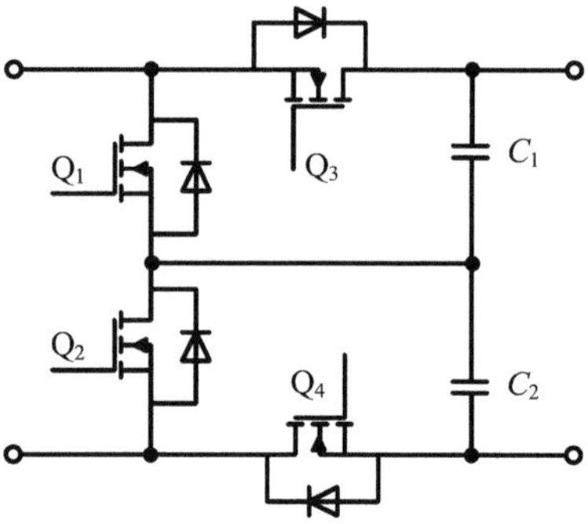

Fig. 1.17 Topology of the bidirectional three-level DC–DC converter

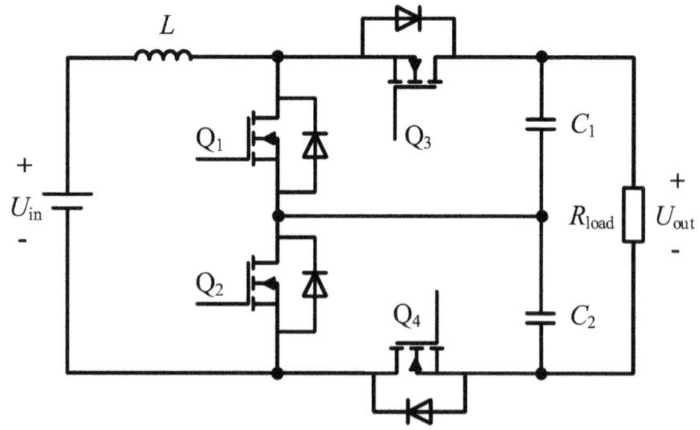

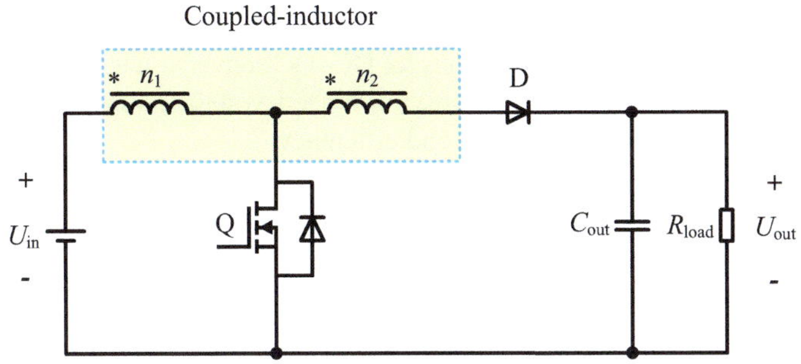

Fig. 1.18 Topology of high step-up converter with coupled-inductor

tied to voltage stresses, and low voltage switches exhibit lower parasitic parameters, the converter's switching loss can be significantly minimized. However, maintaining balance in voltage capacitors requires additional control, and the use of a large number of power switches imposes restrictions on further improvements in cost and power density.

G. Coupled-inductor

By integrating inductor windings into one magnetic core, the coupled-inductor can be employed to achieve high voltage gain. The combination of the basic Boost converter and coupled-inductor is shown in Fig. 1.18. The coupled-inductor structure is equivalent to a leakage inductance L_{lk}, a magnetizing inductance L_m, and an ideal transformer with a turns ratio of $1/N$. The magnetizing inductance L_m operates as the Boost inductor, while the secondary winding can be regarded as a controlled voltage source. The voltage gain of the converter is the sum of the basic Boost converter and the flyback converter [24].

Therefore, coupled-inductor converters can achieve high voltage gain with a simple structure by selecting a reasonable turns ratio. Moreover, the voltage stresses across power switches can be reduced, improving overall efficiency. However, the main issue with coupled-inductor is the leakage energy, which could result in high voltage spikes across switches due to the resonant phenomenon. Besides, the voltage stress across the diode in series with the second winding is larger than the output voltage.

1.2.3 High Efficiency Power Conversion with Wide Voltage Gain Range

The renewable energy system is comprised of distributed power sources, including PV panels and fuel cells, along with energy storage devices such as batteries and super-capacitors. DC–DC converters are employed to interface the power units with the common DC link, facilitating power conversion and voltage matching. The overall performance of the system is heavily reliant on the operation of these converters. Consequently, the desired characteristics for DC–DC conversion in renewable energy systems encompass low current ripple, minimized voltage stresses and cost, and maximized power density, reliability, and efficiency.

For the voltage gain of DC–DC converters in renewable energy applications, a wide voltage gain range is crucial for achieving high efficiency. Converters with both higher and lower step-up voltage gains allow for lower input current and better utilization of device capacity. Importantly, lower input current results in reduced conduction losses. At the same current level, a converter with a wide voltage gain range can achieve a higher level of power conversion. Therefore, achieving a wide voltage gain range also contributes to high operational efficiency.

As illustrated in Fig. 1.19, a wide voltage gain is not the sole required criterion for DC–DC converters in renewable energy applications. It is more convenient for converters with reduced voltage stresses and current ripples to select components, leading to lower switching losses. Furthermore, the implementation of synchronous rectification and soft-switching capabilities can further enhance conversion efficiency.

To achieve a wide voltage gain in DC–DC converters with high efficiency, various voltage-boosting techniques introduced in Sect. 1.2.2 have been widely adopted and combined. Additionally, advanced approaches such as cascaded and interleaved structures can be employed to create multistage topologies that further expand the voltage gain range.

A. Cascaded structure

Cascading connects the output port of one converter with the input port of the next. This can be achieved in two ways: one utilizes the same circuit topologies, combining them into a new structure with distinct characteristics through various connection modes; the other employs a hybrid cascade structure, connecting different circuits together [25].

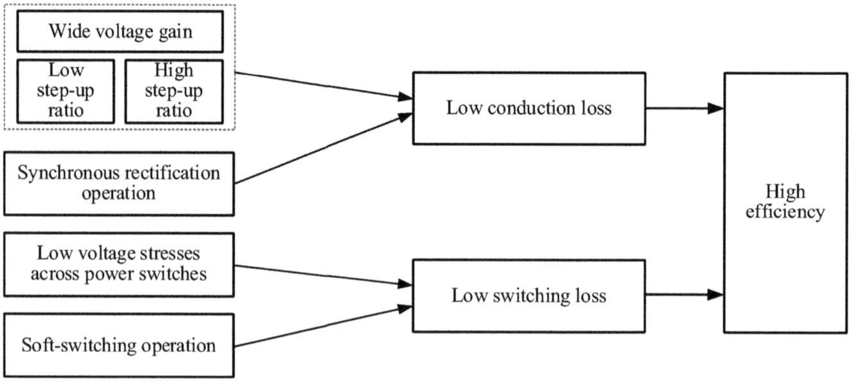

Fig. 1.19 Approaches to achieve high efficiency for DC–DC converters

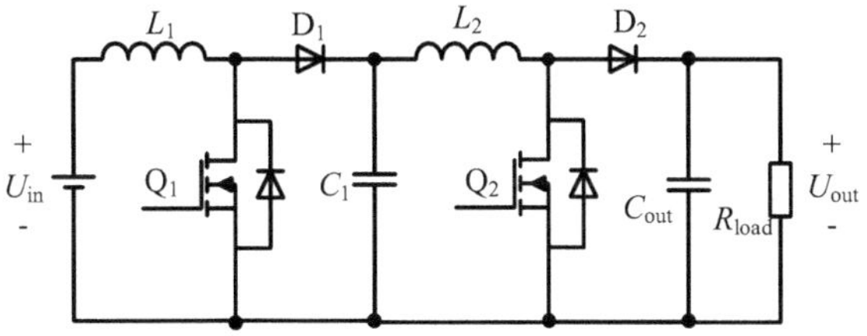

Fig. 1.20 Topology of the cascaded Boost converter

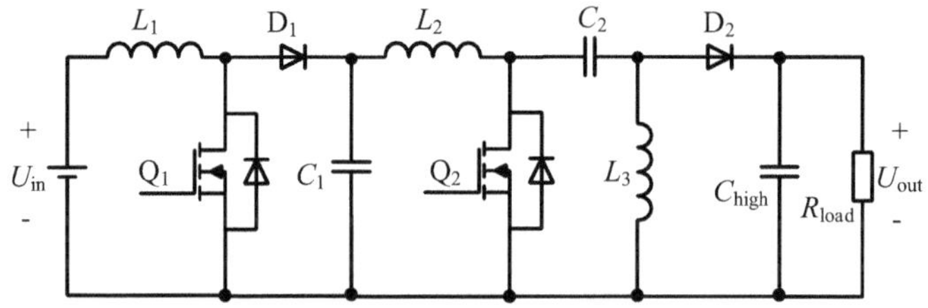

Fig. 1.21 Topology of the non-symmetric cascaded converter

The topology of the cascaded Boost converter is depicted in Fig. 1.20, illustrating the connection of two basic Boost converters to form a symmetric configuration. The resulting quadratic Boost converter achieves a higher voltage gain but comes at the expense of decreased efficiency. Figure 1.21 displays a non-symmetric cascaded DC–DC converter, comprising a basic Boost converter and a basic Sepic converter.

Within cascaded converters, each stage can operate at different switching frequencies. Specifically, the voltage stresses across the power switches in the first stage are lower, enabling operation at higher frequencies. Figure 1.22 illustrates three cascaded DC–DC converters utilizing a single power switch. The topology in Fig. 1.22a is a multistage combination of Boost converters, and the topology in Fig. 1.22b is derived from Sepic and Buck-Boost converters. The topology in Fig. 1.22c is integrated by Boost and Sepic converters. These converters can achieve a high voltage gain with non-extreme duty cycles by increasing stages. However, the cascaded structure involves an increased number of components, leading to higher costs and power losses. As a result, cascaded converters often fail to achieve high conversion efficiency.

B. Interleaved structure

Interleaving is an important technique used to expand DC–DC topologies. The conventional interleaved Boost converter is depicted in Fig. 1.23. The gate signals of the power switches, as well as the charging/discharging of inductors, are phase-shifted. Consequently, compared to the basic Boost converter operating under the same conditions and with the same inductances, the input current can be significantly reduced. The interleaved Boost converter exhibits the same voltage gain as the

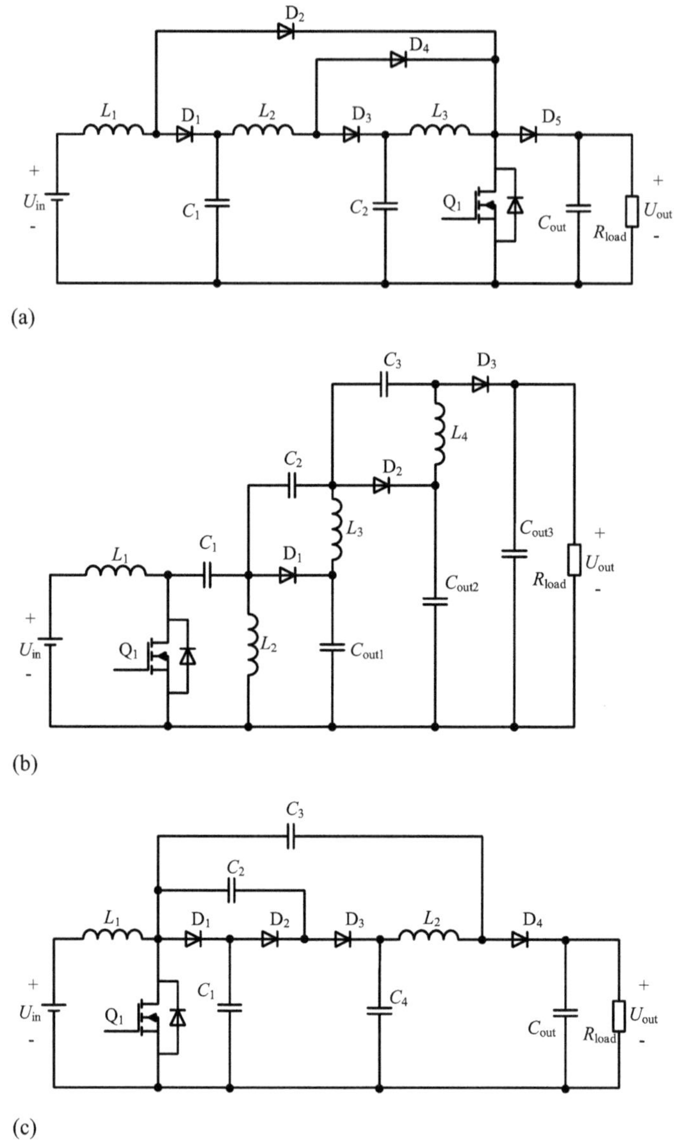

Fig. 1.22 Cascaded topologies of **a** multistage Boost converter, **b** Sepic-Buck-Boost converter, and **c** Boost-Sepic converter

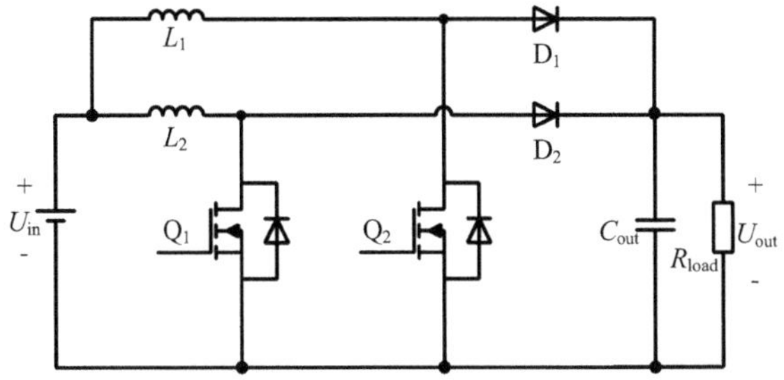

Fig. 1.23 Topology of the interleaved Boost converter

conventional Boost converter, while the input current is shared. Thus, the interleaved structure is well-suited for applications with high current levels.

Figure 1.24 illustrates two interleaved Buck-Boost converters with distinct configurations. In Fig. 1.24a, the Buck-Boost modules are connected in a parallel-input, series-output manner, enabling output voltage boosting. However, a notable drawback of this topology is the absence of a common ground between its input and output sides. On the other hand, the converter in Fig. 1.24b exhibits a common ground structure while maintaining the same voltage gain as the conventional Buck-Boost converter [26].

For the presented interleaved converters, the gate signals of each phase are half-cycle shifted, minimizing input current ripple when the duty cycle of the power switches is 0.5. Despite the advantages of interleaved structures, such as higher power levels and lower current ripple, the main problem is the increased number of required components. On the other hand, expanding the basic converter by simply parallel interleaving the phases cannot increase the voltage gain. To achieve a higher output voltage, further modifications are required.

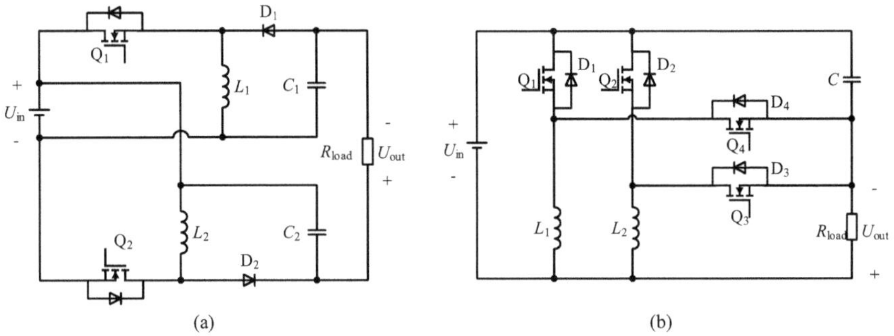

Fig. 1.24 Topologies of two interleaved Buck-Boost converters

1.3 Applications of Non-isolated DC–DC Converters for Renewable Energies

Nowadays, environmental pollution and the depletion of fossil fuels have led to an increasing demand for clean energies. As a result, renewable energies are emerging as potential solutions to contribute to energy generation. Conventional methods of energy generation have resulted in negative impacts on the global environment. In contrast, renewable energies are seen as both cost-effective and environmentally friendly. Additionally, the fossil fuel power generation sector is grappling with shortages caused by massive consumption. In response to this challenge, power generation techniques utilizing renewable energies have become a focused research field, gaining increasing recognition and attention.

Renewable energies are considered alternative energy sources due to their environmentally friendly characteristics and vast availability, including solar energy, wind energy, hydrogen energy, etc. However, the drawback of renewable energies is the difficulty in generating the required quantity of electric power within a short duration to meet the load demand. The random behavior of renewable energies makes achieving continuous and highly efficient power generation challenging without additional techniques.

The hybrid DC grid is regarded as one of the most popular approaches for achieving reliable power generation. It involves integrating renewable energies with backup devices connected to the DC bus via DC–DC converters. Backup devices, such as batteries, super-capacitors, etc., are employed to supply deficit power and address transient load demands. This arrangement ensures regulated power output from the sources, maintaining a constant voltage at the DC bus.

1.3.1 Non-isolated DC–DC Converters for Electric Vehicles

To meet dynamic performance and endurance distance requirements, the power systems of electric vehicles are expected to exhibit both high power density and high energy density. Additionally, fast response capability and a long operational life are essential. Figure 1.25 illustrates two power system architectures for electric vehicles: one with a combination of battery and super-capacitor, and the other with a fuel cell.

In Fig. 1.25a, the battery and super-capacitor are combined to constitute a hybrid energy storage system, each connected to the DC bus through a bidirectional DC–DC converter. Batteries are known for their high energy capacity, enabling continuous power supply but with limited ability to respond quickly. The DC–DC converter interfacing the battery and the DC bus is expected to exhibit low charging current ripple for the battery and high conversion efficiency. Furthermore, a high voltage gain of the converter can enable the DC bus to be matched with a lower battery voltage, thereby reducing the number of batteries that need to be connected in series.

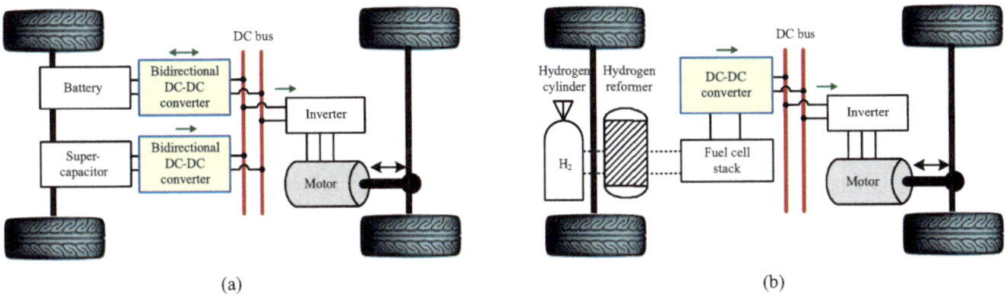

Fig. 1.25 Power system architecture of electric vehicle with **a** battery and super-capacitor, and **b** fuel cell

In contrast, super-capacitors are high power devices with larger capacity than normal capacitors, allowing them to release a large amount of power within a short time. In addition to these advantages, super-capacitors offer benefits such as low series resistance, high efficiency, low heating losses, and fast response. Consequently, electric vehicles can utilize super-capacitors as backup devices to provide additional power during acceleration and store regenerative power during braking. However, the terminal voltage of a super-capacitor varies during charging/discharging, dependent on the state of charge (SOC). For instance, a 75% discharging of the super-capacitor capacity results in a 50% drop in the terminal voltage. Therefore, it is necessary to employ a DC–DC converter with a wide voltage gain range to interface the super-capacitors with the DC bus.

The fuel cell is widely adopted in electric vehicles due to its pollution-free nature and relatively high efficiency, ranging from 35 to 45%. Additionally, fuel cells boast a large storage capacity, enabling power applications at high power levels. The operation of a fuel cell is similar to that of a battery. However, the output voltage of a fuel cell shows a significant degree of regulation with increasing current, as illustrated in Fig. 1.26.

Fig. 1.26 Output characteristic of fuel cell

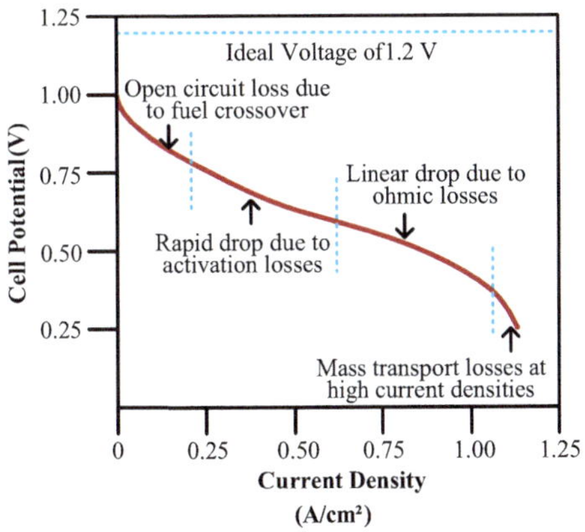

As the current of the fuel cell increases, the cell voltage drops significantly due to activation losses, ohmic losses, and concentration losses. Activation losses result from slow chemical reactions occurring on the electrode surfaces. Ohmic losses are caused by the resistance of the membrane, electrode material, and the interconnections of the fuel cell. The voltage drop is proportional to the current density within the linear ohmic losses region, where the fuel cell voltage ranges within 0.25–0.75 of the open circuit voltage.

Figure 1.25b presents the power system architecture of the electric vehicle with fuel cell. The fuel cell is connected with the onboard DC bus via a DC–DC converter, which acts as a regulator for the fuel cell output. Then DC–DC converter transfers the energy from the fuel cell toward the DC bus. In addition, since the fuel cell voltage varies according to the current density, the capability of dynamic voltage matching within a wide voltage gain range is necessary for the DC–DC converter. Besides, the operation performance and energy utilization efficiency are highly dependent on the efficiency of the DC–DC converter. Therefore, the design of DC–DC converter with high efficiency and light volume, and wide voltage gain range is paramount to achieve good performance for electric vehicles.

1.3.2 Non-isolated DC–DC Converters for PV Systems

Solar energy has emerged as a highly appealing renewable resource due to its quiet operation, cleanliness, flexibility, and minimal maintenance requirements. The PV cell harnesses solar energy through the photovoltaic effect, converting it into electrical energy. Multiple PV cells are interconnected in series and parallel to form PV panels, enabling the generation of the required output current and voltage.

The output characteristics of a PV cell depend on the irradiation level and temperature, as depicted in Fig. 1.27. Consequently, environmental conditions can adversely impact the efficiency of PV cells, presenting a challenge to widespread adoption. Thus, achieving maximum power point tracking (MPPT) holds paramount importance in extracting the utmost available energy from the PV cell.

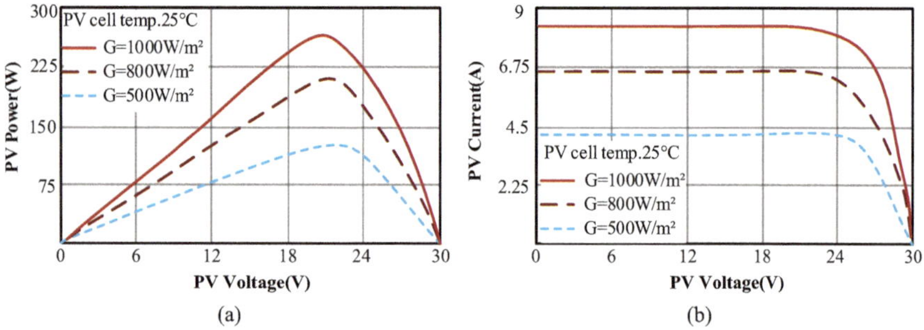

Fig. 1.27 **a** *P-U* and **b** *I-U* characteristics of PV cell under different environmental conditions

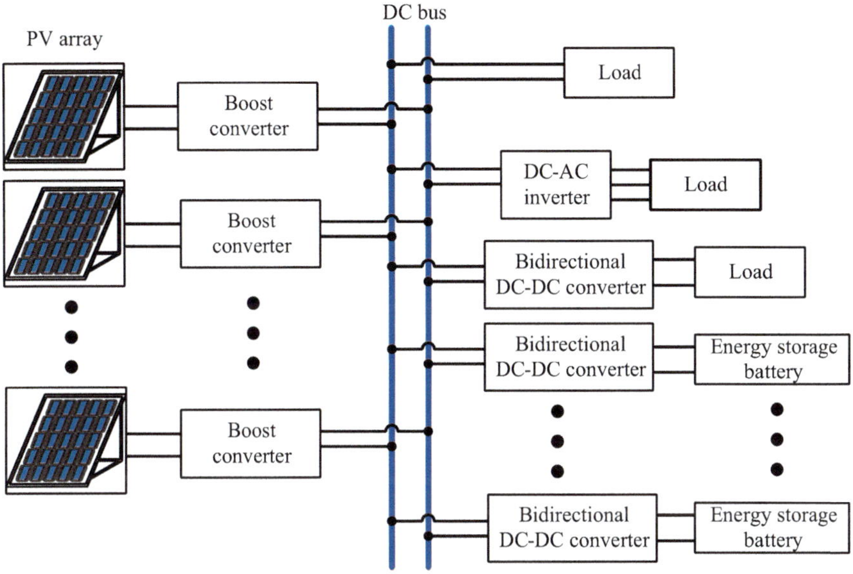

Fig. 1.28 Architecture of distributed PV system

The architecture of distributed PV system is illustrated in Fig. 1.28. The PV panel is connected to the DC bus through a DC–DC converter, serving as the MPPT controller to efficiently track and convert solar energy. As the operation conditions change frequently to track the maximum power point, the DC–DC converter is expected to have a simple control scheme and wide voltage gain. Furthermore, the conversion efficiency of the DC–DC converter is a key factor in the overall efficiency of the PV system. To harvest a large amount of solar energy and enhance power tracking performance, the PV panels are distributed and connected to the DC bus via separate DC–DC converters. Therefore, a high power density of the DC–DC converter can enable long endurance of the system.

Considering the randomness and volatility of solar energy, a battery is capable of supplying stable power during low irradiance conditions. When the maximum output power from the PV system exceeds the load demand, the excess power is absorbed by the battery bank to prevent overvoltage. Conversely, if the PV output power is insufficient to meet the load requirements, the battery bank discharges to protect the DC bus from voltage sag. This allows solar energy to be stored and effectively utilized, enabling the system to achieve continuous operation over nighttime.

To regulate the battery power, a bidirectional DC–DC converter is connected between the battery bank and the DC bus. Furthermore, achieving a wide voltage gain, high efficiency, and lightweight design of the bidirectional DC–DC converter can reduce the required capacity of the battery bank and the total weight of the system.

1.4 Conclusion

The operation performance of power systems utilizing renewable energies is highly dependent on the DC–DC converters, which serve as power interfaces. These converters are connected between the renewable energy source and the DC bus, enabling dynamic voltage matching, bidirectional power flow, and maximum power point tracking. Consequently, the design and selection of DC–DC converters for renewable energy applications should be based on factors such as voltage gain range, conversion efficiency, power density, etc.

Specifically, the voltage gain range affects the maximum utilization degree of renewable energy capacity, while conversion efficiency is key to the overall energy harvesting efficiency. Non-isolated DC–DC converters show great application potential for renewable energies due to their simple circuit structure and small volume. In this book, various non-isolated DC–DC converters for renewable energy applications are presented and analyzed. Additionally, the design methodology of the non-isolated DC–DC converters is explained, and their operation performances are validated through experimental results.

Chapter 2
H-Bridge DC–DC Converters

2.1 Introduction

In applications where galvanic isolation is required, the dual active full-bridge DC–DC converter has been used to achieve high voltage gain by adjusting the turns ratio of the high-frequency transformer [27]. However, at high power levels, this approach presents several challenges [28]. The large turns ratio of the transformer leads to reduced coupling, core losses due to non-sinusoidal excitation, and dielectric losses in the insulation. Additionally, the distributed capacitance of the winding turns adversely affects efficiency. When the input and output voltages do not match the transformer's turns ratio, switching losses escalate dramatically [29].

For applications that do not require isolation, DC–DC converters with coupled inductors offer higher voltage gain and reduced diode reverse recovery losses [30, 31]. However, designing coupled inductors and high-frequency transformers at high power levels proves challenging. Conventional bidirectional DC–DC converters exhibit extreme voltage stresses across power switches and diodes when extreme duty cycles are used, leading to EMI issues, reverse recovery losses, and lower conversion efficiency [32]. While some converters achieve higher voltage gain [33], they face limitations such as higher maximum voltage stress or the need for auxiliary circuits. Cascaded configurations of two power stages, such as Cuk and Sepic/Zeta converters, suffer from lower conversion efficiencies [34, 35]. A soft-switching converter with complex coupled inductors and associated circuits has been proposed [36], but it requires additional components and lacks cost-effectiveness. Interleaved converters with switched capacitors and coupled-inductor in series offer high conversion ratios, but require more active power switches, capacitors, and coupled inductors [37]. Interleaved switched-capacitor converters have been proposed to reduce input current ripple [38, 39], but they face challenges such as lack of soft-switching or significant current ripple in the low voltage side. A high voltage conversion ratio was achieved using a switched-capacitor cell [40], but the converter was unsuitable for energy storage systems due to the presence of high-frequency PWM voltage. Apart from

Y. Zhang and S. Gao, *High Efficiency Non-isolated DC-DC Converters with Wide Voltage Gain Range for Renewable Energies*, CPSS Power Electronics Series,
https://doi.org/10.1007/978-981-97-0648-8_2

the transformer-based, coupled-inductor, switched-capacitor, and Z-source DC–DC converters, there is another group of converters utilizing the Z-source structure to improve the step-up and step-down ratios [41]. However, these Z-source converters lack a common ground between the input and output sides, which is crucial for renewable power generation systems and hybrid energy source electric vehicles.

A high ratio bidirectional H-Bridge DC–DC converter is introduced in this chapter. The converter operates with synchronous rectification, maximizing the efficiency by utilizing the power switches without additional hardwares. A bidirectional DC–DC converter specifically designed for energy storage systems in renewable energy and electric vehicles is also presented. It incorporates one LC energy storage component and four active power switches with anti-parallel diodes, eliminating the need for transformers, coupled inductors, switched capacitors, or Z-source impedance net. The converter achieves high step-up and step-down ratios by employing non-extreme duty cycles for power switches. Additionally, the low and high voltage sides share a common ground, ensuring continuous input/output currents on the low voltage side.

2.2 H-Bridge Bidirectional DC–DC Converter

The bidirectional power cell illustrated in Fig. 2.1 comprises power switches Q_1 and Q_2. By integrating Q_3 and Q_4, a paralleled connection structure is established. In this configuration, the drain nodes of Q_1 and Q_3 serve as the positive port on the high voltage side, while the source nodes of Q_2 and Q_4 act as the negative port on the high voltage side. The input voltage is stabilized through the use of the high-side capacitor C_{high}. Additionally, a filter network is connected between the midpoints of the left and right bridges, denoted as node "a" and node "b". This filter network, consisting of inductor L and capacitor C_{low}, further refines the system. Consequently, the H-bridge bidirectional DC–DC converter is created, allowing for both high voltage gain and bidirectional power flow [42].

The topology of the bidirectional H-bridge DC–DC converter is illustrated in Fig. 2.2. By manipulating the main power switches Q_1 and Q_4, as well as the synchronous rectifiers Q_2 and Q_3, energy transfers from the high voltage side to

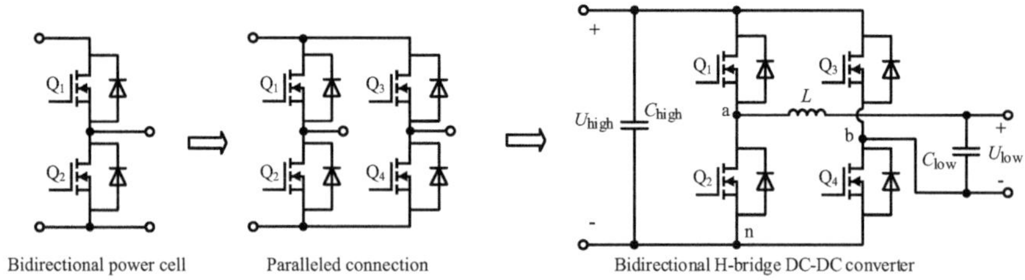

Fig. 2.1 Derivation of H-bridge bidirectional DC–DC converter

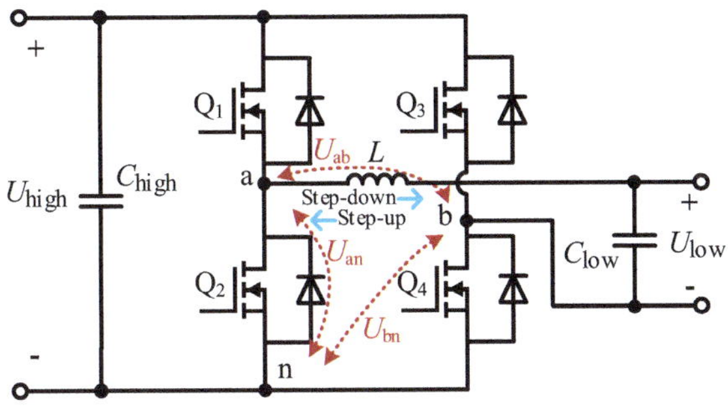

Fig. 2.2 Topology of H-bridge bidirectional DC–DC converter

the low voltage side, enabling step-down mode operation. Conversely, during step-up mode operation, Q_2 and Q_3 function as the main power switches, while Q_1 and Q_4 act as synchronous rectifiers, facilitating energy flow from the low voltage side to the high voltage side.

The direction of inductor current i_L in step-down mode and step-up mode is indicated in Fig. 2.2. Additionally, the voltages between points "a" and "n" (U_{an}) and between points "b" and "n" (U_{bn}) are represented. The voltage between the midpoint nodes "a" and "b" is denoted as U_{ab}. The operating principle of the converter is analyzed as follows.

2.2.1 Operating Principle of the H-Bridge Bidirectional DC–DC Converter

A. Step-down operating principle

During the step-down operation of the H-bridge bidirectional DC–DC converter, the output voltage U_{low} is filtered by L and C_{low} from the pulse voltage U_{ab}, and U_{ab} can be described as follows.

$$U_{ab} = U_{an} - U_{bn} \tag{2.1}$$

To achieve a high step-down voltage gain, the pulse voltage U_{ab} should be narrow. As stated in (2.1), even when the pulse voltages U_{an} and U_{bn} are wide, a narrow pulse voltage U_{ab} can be obtained by considering the voltage difference between U_{an} and U_{bn}.

Figure 2.3 shows the key-waveforms of the H-bridge bidirectional DC–DC converter in step-down mode, where T_s is the switching cycle, S_1 and S_4 are the gate signals of Q_1 and Q_4, respectively. As illustrated in Fig. 2.3a, b, the gate signals S_1 and S_4 are determined by the modulation indexes m_b and m_a, respectively. Besides, the converter operates under the conditions of $0 < m_b < 0.5 < m_a < 1$, and $m_a + m_b > 1$. As shown in Fig. 2.3d, e, the output pulse voltage U_{an} and U_{bn} are controlled by

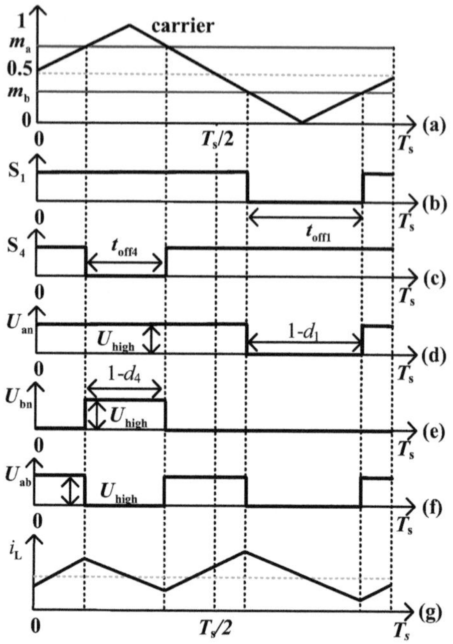

Fig. 2.3 Key-waveforms of the H-bridge bidirectional DC–DC converter in step-down mode

the gate signals S_1 and S_4. As a result, in step-down mode, m_a and m_b can be adjusted near 0.5 to obtain the narrow pulse voltage U_{ab}.

In the current continuous mode, according to the volt-second balance principle, (2.2) can be obtained, where t_{off1} and t_{off4} are the turn-off interval for Q_1 and Q_4, respectively.

$$\left(U_{high} - U_{low}\right) \times [(0.5T_s - t_{off4}) + (0.5T_s - t_{off1})] = U_{low} \times (t_{off1} + t_{off4}) \quad (2.2)$$

Hence, the step-down voltage gain M_{Buck} of the converter can be obtained as follows, where d_1 and d_4 are the duty cycles of Q_1 and Q_4, respectively.

$$M_{Buck} = \frac{U_{low}}{U_{high}} = d_1 + d_4 - 1 \quad (2.3)$$

$$\begin{cases} d_1 = 1 - m_b \\ d_4 = m_a \end{cases} \quad (2.4)$$

Combining (2.3) and (2.4), the relationship between M_{Buck} and m_a, m_b can be derived.

$$M_{Buck} = \frac{U_{low}}{U_{high}} = m_a - m_b, \ (0 < m_b < 0.5 < m_a < 1 \text{ and } m_a + m_b > 1) \quad (2.5)$$

B. Step-up operating principle

When the H-bridge bidirectional DC–DC converter is functioning in step-up mode, it produces the output voltage U_{high} using pulse voltage U_{ab} in combination with synchronous rectifiers Q_1 and Q_4, and the filter capacitor C_{high}. Similar to the step-down mode, a narrow pulse voltage U_{ab} is necessary to achieve a high step-up ratio. This waveform of U_{ab} is determined by the voltage difference between U_{an} and U_{bn}.

The key-waveforms of the H-bridge bidirectional DC–DC converter in step-up mode are shown in Fig. 2.4. The gate signals S_2 and S_3 for Q_2 and Q_3 are generated by the modulation indexes m_{c} and m_{d}, which are shown in Fig. 2.4a–c. Furthermore, m_{c} and m_{d} are adjusted within the range that $0 < m_d < 0.5 < m_{\text{c}} < 1$ and $m_{\text{c}} + m_{\text{d}} < 1$. As shown in Fig. 2.4d, the output PWM voltages U_{an} and U_{bn} are controlled via S_2 and S_3, respectively. Therefore, with m_{c} and m_{d} around 0.5, the narrow pulse voltage U_{ab} can be obtained.

In step-up mode, (2.6) can be obtained from the volt-second balance of inductor L, where t_{on2} and t_{on3} are the turn-on intervals of Q_2 and Q_3, respectively.

$$(U_{\text{high}} - U_{\text{low}}) \times [(0.5T_{\text{s}} - t_{\text{on2}}) + (0.5T_{\text{s}} - t_{\text{on3}})] = U_{\text{low}} \times (t_{\text{on2}} + t_{\text{on3}}) \quad (2.6)$$

Thus, the step-up voltage gain M_{Boost} of the H-bridge bidirectional DC–DC converter is obtained as below, where d_2 and d_3 are the duty cycles of Q_2 and Q_3, respectively.

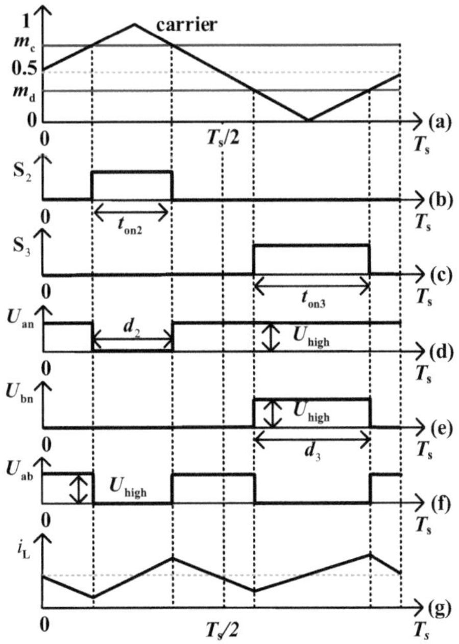

Fig. 2.4 Key-waveforms of the H-bridge bidirectional DC–DC converter in step-up mode

$$M_{\mathrm{Boost}} = \frac{U_{\mathrm{high}}}{U_{\mathrm{low}}} = \frac{T}{T_{\mathrm{s}} - (t_{\mathrm{on2}} + t_{\mathrm{on3}})} = \frac{1}{1 - (d_2 + d_3)} \tag{2.7}$$

$$\begin{cases} d_2 = 1 - m_{\mathrm{c}} \\ d_3 = m_{\mathrm{d}} \end{cases} \tag{2.8}$$

According to (2.7) and (2.8), the voltage step-up ratio M_{Boost} can be represented by m_{c} and m_{d} as follows.

$$M_{\mathrm{Boost}} = \frac{U_{\mathrm{high}}}{U_{\mathrm{low}}} = \frac{1}{m_{\mathrm{c}} - m_{\mathrm{d}}}, \ (0 < m_{\mathrm{d}} < 0.5 < m_{\mathrm{c}} < 1, \text{ and } m_{\mathrm{c}} + m_{\mathrm{d}} < 1) \tag{2.9}$$

2.2.2 *Control Strategy of the H-Bridge Bidirectional DC–DC Converter*

A. Synchronous rectification operation control

As presented in Fig. 2.2, the H-bridge bidirectional DC–DC converter consists of four MOSFETs with anti-parallel diodes. However, due to the relatively larger parasitic resistance of the anti-parallel diodes, the inductor current of the H-bridge bidirectional DC–DC converter can lead to considerable conducting loss during turn-off intervals. Therefore, the application of synchronous rectification is necessary to enhance the operating efficiency of the converter.

The synchronous rectification operation of the H-bridge bidirectional DC–DC converter in step-down and step-up modes is illustrated in Fig. 2.5a, b, respectively. In step-down mode, Q_1 and Q_4 operate as the main power switches, controlled by the gate signals S_1 and S_4, respectively. The inductor current i_{L} flows through the synchronous rectifiers Q_2 and Q_3 during the dead time t_{d}, where both Q_1 and Q_4 are turned OFF. Otherwise, i_{L} flows through the main power switches. Conversely, in step-up mode, Q_2 and Q_3 function as the main power switches, controlled by the gate signals S_2 and S_3. During the dead time t_{d}, when both Q_2 and Q_3 are turned OFF, the inductor current i_{L} flows through the synchronous rectifiers Q_1 and Q_4. And i_{L} flows through Q_2 and Q_3 when they are turned ON.

B. Bidirectional power flow operation control

When the H-bridge bidirectional DC–DC converter interfaces with the battery stacks and the DC bus, bidirectional power flow operation is required for battery charging and discharging. The control strategy of the bidirectional power flow in the converter is presented in Fig. 2.6. The converter operates as a current source with a current loop, where the operating condition is selected between the step-down and step-up modes via a power flow directional signal U_{c}. When $U_{\mathrm{c}} = 0$, the converter operates

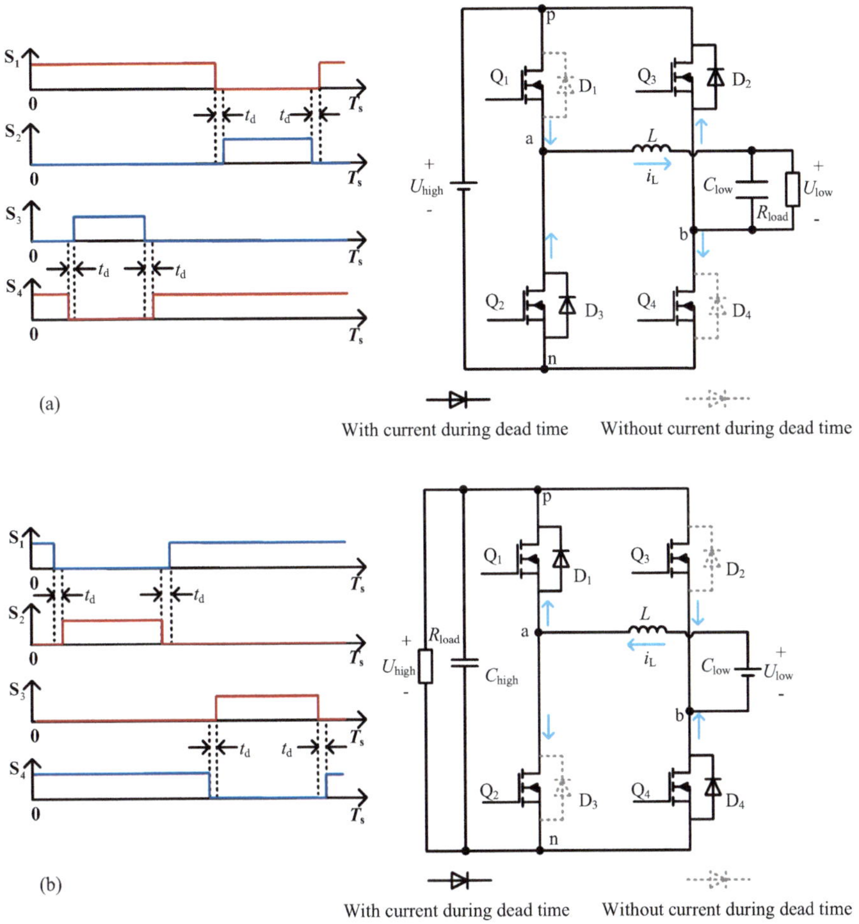

Fig. 2.5 Synchronous rectification of the H-bridge bidirectional DC–DC converter in **a** step-down mode and **b** step-up mode

in step-down mode, where the inductor current i_L is controlled to track the reference current $I_{\text{ref-Buck}}$. In this case, the PWM scheme in Figs. 2.3 and 2.5a is selected to generate the gate signals. Thus, the power flows from the DC bus on the HVS to the battery stacks on the LVS. On the other hand, when U_c is set to 1, i_L is controlled by the Boost current controller, and the reference current is $I_{\text{ref-Boost}}$ with the opposite direction to that in step-down mode. The corresponding PWM scheme in Figs. 2.4 and 2.5b is applied for the gate signal generation. During the mode transition between step-down and step-up modes, the inductor current will decrease to zero and rise reversely afterward. In this way, bidirectional power flow control is achieved.

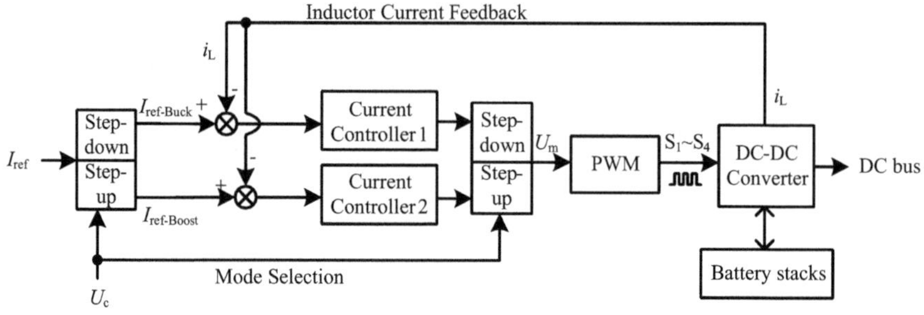

Fig. 2.6 Control strategy of bidirectional power flow of the H-bridge bidirectional DC-DC converter

2.2.3 *Small-Signal Modeling of the H-Bridge Bidirectional DC–DC Converter*

To simplify the small-signal modeling procedure of the H-bridge converter, the inductor and the capacitors, as well as the power switches, are considered as ideal components. In addition, the capacitances of C_{high} and C_{low} are large enough to neglect the voltage ripples, meaning that the high and low voltage sides are regarded as constant within a switching state interval.

In step-up mode, the H-bridge bidirectional DC–DC converter has three switching states.

State 1) $S_1S_2S_3S_4 = 0101$

The duration of State 1 is d_2T_s. And the state space average model is obtained as below, where R_{Boost} is the equivalent load resistance in the step-up mode.

$$\begin{bmatrix} \dfrac{\mathrm{d}i_L(t)}{\mathrm{d}t} \\ \dfrac{\mathrm{d}u_{high}(t)}{\mathrm{d}t} \end{bmatrix}_{S_2S_3=10} = \begin{bmatrix} 0 & 0 \\ 0 & -\frac{1}{C_{high}R_{Boost}} \end{bmatrix} \begin{bmatrix} i_L(t) \\ u_{high}(t) \end{bmatrix} + \begin{bmatrix} \dfrac{1}{L} \\ 0 \end{bmatrix} u_{low}(t) \tag{2.10}$$

State 2) $S_1S_2S_3S_4 = 1010$

The duration of State 2 is d_3T_s. And the state space average model is obtained, as shown below.

$$\begin{bmatrix} \dfrac{\mathrm{d}i_L(t)}{\mathrm{d}t} \\ \dfrac{\mathrm{d}u_{high}(t)}{\mathrm{d}t} \end{bmatrix}_{S_2S_3=01} = \begin{bmatrix} 0 & 0 \\ 0 & -\frac{1}{C_{high}R_{Boost}} \end{bmatrix} \begin{bmatrix} i_L(t) \\ u_{high}(t) \end{bmatrix} + \begin{bmatrix} \dfrac{1}{L} \\ 0 \end{bmatrix} u_{low}(t) \tag{2.11}$$

State 3) $S_1S_2S_3S_4 = 1010$

The duration of State 3 is $(1 - d_2 - d_3)T_s$. And the state space average model is obtained as follows.

$$\begin{bmatrix} \dfrac{di_L(t)}{dt} \\ \dfrac{du_{high}(t)}{dt} \end{bmatrix}_{S_2S_3=00} = \begin{bmatrix} 0 & -\frac{1}{L} \\ \frac{1}{C_{high}} & -\frac{1}{C_{high}R_{Boost}} \end{bmatrix} \begin{bmatrix} i_L(t) \\ u_{high}(t) \end{bmatrix} + \begin{bmatrix} \dfrac{1}{L} \\ 0 \end{bmatrix} u_{low}(t) \tag{2.12}$$

Hence, the average model of the H-bridge bidirectional DC–DC converter can be derived by combining (2.10)–(2.12) as below.

$$\begin{bmatrix} \dfrac{di_L(t)}{dt} \\ \dfrac{du_{high}(t)}{dt} \end{bmatrix} = \begin{bmatrix} 0 & -\frac{1-d_2(t)-d_3(t)}{L} \\ \frac{1-d_2(t)-d_3(t)}{C_{high}} & -\frac{1}{C_{high}R_{Boost}} \end{bmatrix} \begin{bmatrix} i_L(t) \\ u_{high}(t) \end{bmatrix} + \begin{bmatrix} \frac{1}{L} \\ 0 \end{bmatrix} u_{low}(t) \tag{2.13}$$

By introducing the small-signal disturbances $\hat{i}_L(t)$, $\hat{u}_{high}(t)$, $\hat{d}(t)$, and $\hat{u}_{low}(t)$ into the variables, (2.14) is obtained, where I_L, U_{high}, D, and U_{low} are the steady-state variables.

$$\begin{cases} i_L(t) = I_L + \hat{i}_L(t) \\ u_{high}(t) = U_{high} + \hat{u}_{high}(t) \\ d(t) = D + \hat{d}(t) \\ u_{low}(t) = U_{low} + \hat{u}_{low}(t) \end{cases} \tag{2.14}$$

In addition, the voltages on the LVS and HVS can be expressed as follows.

$$\begin{cases} U_{high} = (1 - d_2(t) - d_3(t))I_L \times R_{Boost} \\ U_{low} = U_{high} \times (1 - d_2(t) - d_3(t)) \end{cases} \tag{2.15}$$

Hence, the small-signal model of the H-bridge bidirectional DC–DC converter in the step-up mode can be derived as (2.16).

$$\begin{aligned} \begin{bmatrix} \frac{d\hat{i}_L(t)}{dt} \\ \frac{d\hat{u}_{high}(t)}{dt} \end{bmatrix} &= \begin{bmatrix} 0 & -\frac{1-D_2-D_3}{L} \\ \frac{1-D_2-D_3}{C_{high}} & -\frac{1}{C_{high}R_{Boost}} \end{bmatrix} \begin{bmatrix} \hat{i}_L(t) \\ \hat{u}_{high}(t) \end{bmatrix} + \begin{bmatrix} \frac{1}{L} \\ 0 \end{bmatrix} \hat{u}_{low}(t) \\ &+ \begin{bmatrix} 0 & \frac{1}{L} \\ -\frac{1}{C_{high}} & 0 \end{bmatrix} \begin{bmatrix} I_L \\ U_{high} \end{bmatrix} \hat{d}_2(t) + \begin{bmatrix} 0 & \frac{1}{L} \\ -\frac{1}{C_{high}} & 0 \end{bmatrix} \begin{bmatrix} I_L \\ U_{high} \end{bmatrix} \hat{d}_3(t) \end{aligned} \tag{2.16}$$

Then, the dynamic mathematical model of the converter in step-up mode can be obtained, which is presented below, where $\hat{u}_{high}(s)$ is the output voltage disturbance.

$$\begin{cases} G_{u_{\text{high}}d_2}(s) = \left.\dfrac{\hat{u}_{\text{high}}(s)}{\hat{d}_2(s)}\right|_{\hat{u}_{\text{low}}=0,\hat{d}_3(s)=0} = -\dfrac{I_L Ls + U_{\text{high}}(D_2+D_3-1)}{LC_{\text{high}}s^2 + \frac{L}{R_{\text{Boost}}}s + (D_2+D_3-1)^2} \\ G_{u_{\text{high}}d_3}(s) = \left.\dfrac{\hat{u}_{\text{high}}(s)}{\hat{d}_3(s)}\right|_{\hat{u}_{\text{low}}=0,\hat{d}_2(s)=0} = -\dfrac{I_L Ls + U_{\text{high}}(D_2+D_3-1)}{LC_{\text{high}}s^2 + \frac{L}{R_{\text{Boost}}}s + (D_2+D_3-1)^2} \end{cases} \tag{2.17}$$

When the H-bridge bidirectional DC–DC converter operates in step-down mode, the main power switches are Q_1 and Q_4, each having three switching states represented as $S_1S_2S_3S_4 = [0101, 1010, 1001]$. The average model of the converter is symmetric to that of step-up mode and is obtained as follows, where R_{Buck} is the equivalent load resistance in the step-down mode.

$$\begin{bmatrix} \dfrac{di_L(t)}{dt} \\ \dfrac{du_{\text{low}}(t)}{dt} \end{bmatrix} = \begin{bmatrix} 0 & -\frac{1}{L} \\ \frac{1}{C_{\text{low}}} & -\frac{1}{C_{\text{low}}R_{\text{Buck}}} \end{bmatrix} \begin{bmatrix} i_L(t) \\ u_{\text{low}}(t) \end{bmatrix} + \begin{bmatrix} \frac{d_1(t)+d_4(t)-1}{L} \\ 0 \end{bmatrix} u_{\text{high}}(t) \tag{2.18}$$

Thus, the small-signal mode can be derived as below.

$$\begin{aligned} \begin{bmatrix} \frac{d\hat{i}_L(t)}{dt} \\ \frac{d\hat{u}_{\text{low}}(t)}{dt} \end{bmatrix} &= \begin{bmatrix} 0 & -\frac{1}{L} \\ \frac{1}{C_{\text{low}}} & -\frac{1}{C_{\text{low}}R_{\text{Buck}}} \end{bmatrix} \begin{bmatrix} \hat{i}_L(t) \\ \hat{u}_{\text{low}}(t) \end{bmatrix} + \begin{bmatrix} \frac{D_1+D_4-1}{L} \\ 0 \end{bmatrix} \hat{u}_{\text{high}}(t) \\ &+ \begin{bmatrix} \frac{D_1+D_4-1}{L} \\ 0 \end{bmatrix} U_{\text{high}}\hat{d}_1(t) + \begin{bmatrix} \frac{D_1+D_4-1}{L} \\ 0 \end{bmatrix} U_{\text{high}}\hat{d}_4(t) \end{aligned} \tag{2.19}$$

On the basis of (2.19), the dynamic mathematic model of the converter in step-down mode can be obtained as follows.

$$\begin{cases} G_{u_{\text{low}}d_1}(s) = \left.\dfrac{\hat{u}_{\text{low}}(s)}{\hat{d}_1(s)}\right|_{\hat{u}_{\text{high}}(s)=0,\hat{d}_4(s)=0} = \dfrac{U_{\text{high}}(D_1+D_4-1)}{LC_{\text{low}}s^2 + \frac{Ls}{R_{\text{Buck}}} + 1} \\ G_{u_{\text{low}}d_4}(s) = \left.\dfrac{\hat{u}_{\text{low}}(s)}{\hat{d}_4(s)}\right|_{\hat{u}_{\text{high}}(s)=0,\hat{d}_1(s)=0} = \dfrac{U_{\text{high}}(D_1+D_4-1)}{LC_{\text{low}}s^2 + \frac{Ls}{R_{\text{Buck}}} + 1} \end{cases} \tag{2.20}$$

Based on the small-signal analysis above, the control scheme for the H-bridge bidirectional DC–DC converter, consisting of an outer voltage loop and an inner current loop, can be established using (2.17) and (2.20), as shown in Fig. 2.7. The transfer function $G_{u\text{HL}di}(s)$ is represented by (2.17) in step-up mode, and (2.20) in step-down mode. $G_{\text{m}}(s)$ is the PWM transfer function, $H(s)$ is the feedback transfer function, and $G_{\text{cv}}(s)$, and $G_{\text{ci}}(s)$ are the transfer functions of the voltage and current controllers, respectively. The voltage and current loops are implemented using two PI controllers, which can be appropriately designed to achieve good static and dynamic performance of the converter.

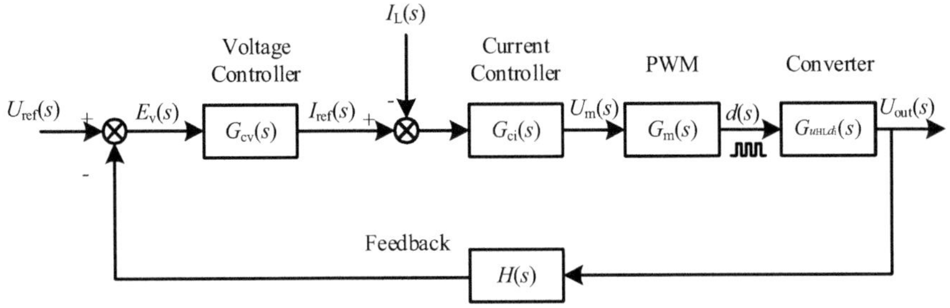

Fig. 2.7 Control scheme with voltage and current loops

2.3 Asymmetric H-Bridge Bidirectional DC–DC Converter

The topology of the asymmetric H-bridge bidirectional DC–DC converter can be derived from the H-bridge bidirectional DC–DC converter in Sect. 2.2 through the following steps [43].

Firstly, as illustrated in Fig. 2.8a, disconnect the negative terminal on the low voltage side from the midpoint of the right half-bridge, namely cut node "d" from node "b". Then, connect node "b" to the negative terminal on the high voltage side, denoted as node "g".

Secondly, as shown in Fig. 2.8b, disconnect the left end of inductor L from the midpoint of the left half-bridge, namely cut node "e" from node "a". Next, connect node "e" to the midpoint of the right half-bridge, identified as node "b".

Finally, as shown in Fig. 2.8c, disconnect the drain of Q_3, referred to as node "c", from the drain of Q_1. Connect node "c" to node "a".

The topology of the asymmetric H-bridge bidirectional DC–DC converter with a common ground is presented in Fig. 2.9, where Q_1–Q_4 are power switches. D_1–D_4 and S_1–S_4 are the anti-parallel diodes and gate signals of Q_1–Q_4, respectively. U_{high} is the HVS voltage, and U_{low} is the LVS voltage. L is the energy storage inductor, while C_{high} and C_{low} are the filter capacitors on the HVS and LVS, respectively. Furthermore, the output characteristic of the asymmetric H-bridge bidirectional DC–DC converter is similar to that of the H-bridge bidirectional DC–DC converter. The narrow PWM voltage U_{bg} is obtained from the difference between U_{ag} and U_{ab} as shown below.

$$U_{bg} = U_{ag} - U_{ab} \tag{2.21}$$

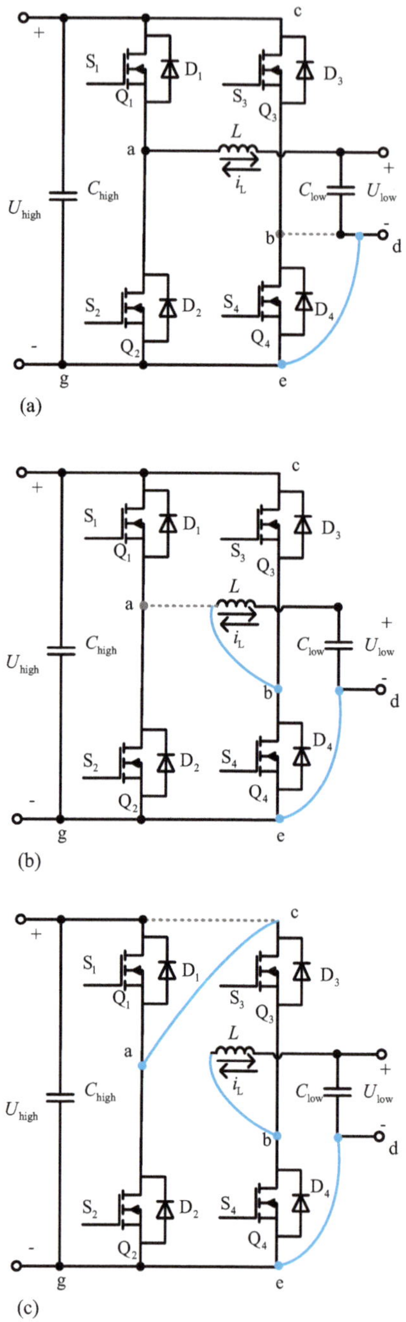

Fig. 2.8 Derivation of the asymmetric H-bridge bidirectional DC–DC converter **a** Step 1 **b** Step 2 **c** Step 3

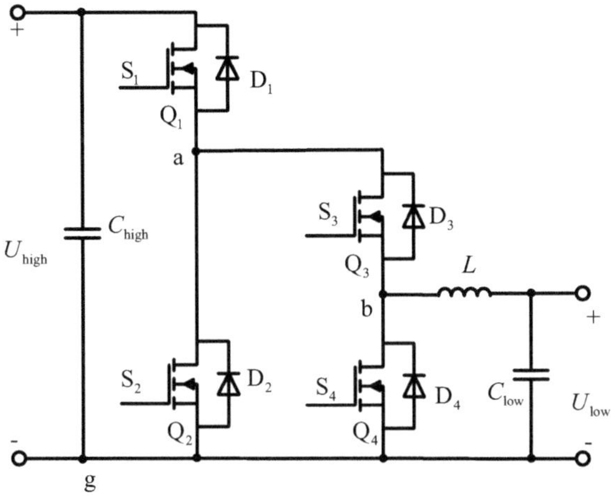

Fig. 2.9 Topology of the asymmetric H-bridge bidirectional DC–DC converter

2.3.1 *Operating Principle of the Asymmetric H-Bridge Bidirectional DC–DC Converter*

The asymmetric H-bridge DC–DC converter is capable of bidirectional power flow. The operating principle of the converter in step-down and step-up modes is analyzed as follows.

A. Step-down operating principle

When the asymmetric H-bridge bidirectional DC–DC converter operates in step-down mode, the power flows from U_{high} to U_{low}. In this mode, Q_1 and Q_3 are the main power switches, and there are three switching states, as shown below.

State 1) $S_1S_2S_3S_4 = 1010$

When Q_1 and Q_3 are turned ON while Q_2 and Q_4 are turned OFF, the inductor current i_L passes through Q_1 and Q_3. The inductor L is charged by C_{high}, and U_{bg} is equal to U_{high}.

State 2) $S_1S_2S_3S_4 = 1000$

When Q_1 is turned ON and Q_2–Q_3 are turned OFF, i_L flows through D_4. During this state, the inductor L is discharged and $U_{\text{bg}} = 0$.

State 3) $S_1S_2S_3S_4 = 0010$

When Q_3 is turned ON while Q_1, Q_2, and Q_4 are turned OFF, the inductor current i_L is divided into D_2 and D_4. L continues to be discharged and $U_{\text{bg}} = 0$.

B. Step-up operating principle

In step-up mode, the power flows from the LVS to the HVS in the asymmetric bidirectional DC–DC converter. In this mode, Q_2 and Q_4 operates as the main power switches and there are three switching states that symmetrical to that in step-down mode.

State 1) $S_1S_2S_3S_4 = 0000$

When all power switches are turned OFF, the inductor current i_L passes through the diodes D_1 and D_3. The inductor L is discharged and $U_{bg} = U_h$.

State 2) $S_1S_2S_3S_4 = 0100$

When Q_2 is turned ON and Q_1, Q_3 and Q_4 are turned OFF, i_L flows through D_3 and Q_2. L is charged by C_{low}, and $U_{bg} = 0$.

State 3) $S_1S_2S_3S_4 = 0001$

When Q_4 is turned ON and Q_1–Q_3 are turned OFF, the inductor L is charged by C_{low} via Q_4, and $U_{bg} = 0$.

C. Wide-voltage-gain conversion

The generation of the gate signals S_1 and S_3 for power switches Q_1 and Q_3 in step-down mode are depicted in Fig. 2.10a, which can be expressed as follows, where m_a and m_b are the modulation indexes for the left and right half-bridges of the asymmetric H-bridge respectively.

$$\begin{cases} m_a > U_{carrier}, & S_1 = 1 \\ m_b < U_{carrier}, & S_3 = 1 \end{cases} \tag{2.22}$$

The modulation indexes are limited within the range of $0 < m_b < 0.5 < m_a < 1 < m_a + m_b$. The magnitude of the carrier is 1. Thus, the duty cycles of Q_1 and Q_3 are equal to the values of m_a and m_b. In the asymmetric H-bridge BDC, the narrower the pulse width of U_{bg}, the shorter the inductor charging duration, namely the higher the step-down voltage gain. Besides, the voltage frequency of U_{bg}, as well as the inductor current frequency are both twice the switching frequency.

According to the volt-second balance law, (2.23) can be derived, where t_{off1} and t_{off3} are the turn-off interval of Q_1 and Q_3 within a switching cycle, respectively.

$$(U_{high} - U_{low}) \times (T_s - t_{off1} - t_{off3}) = U_{low} \times (t_{off1} + t_{off3}) \tag{2.23}$$

The step-down voltage gain M_{Buck} of the converter can in turn be obtained as follows.

$$\begin{cases} M_{Buck} = \frac{U_{low}}{U_{high}} = m_a - m_b \\ 0 < m_b < 0.5 < m_a < 1 < m_a + m_b \end{cases} \tag{2.24}$$

Fig. 2.10 Key-waveforms of the asymmetric H-bridge bidirectional DC–DC converter in **a** step-down and **b** step-up mode

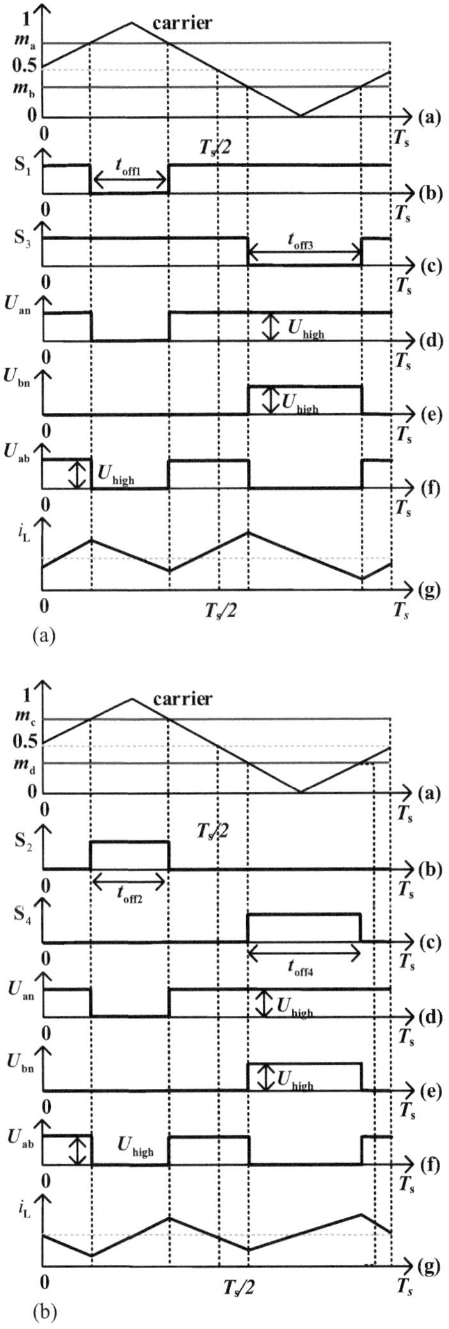

On the basis of (2.24), m_a and m_b can be expressed by M_{Buck} as below.

$$\begin{cases} m_a = 0.5 + 0.51 M_{Buck} \\ m_b = 0.5 - 0.49 M_{Buck} \end{cases} \tag{2.25}$$

According to (2.25), the high step-down voltage ratio of asymmetric H-bridge BDC can be achieved with non-extreme duty cycles for power switches.

Similarly, the gate signals S_2 and S_4 for power switches Q_2 and Q_4 are generated by modulation indexes m_a and m_b in step-up mode, as shown in Fig. 2.10b.

$$\begin{cases} m_a < U_{carrier}, \ S_2 = 1 \\ m_b > U_{carrier}, \ S_4 = 1 \end{cases} \tag{2.26}$$

The modulation indexes are also limited within the range of $0 < m_b < 0.5 < m_a < 1 < m_a + m_b$. In step-up mode, the duty cycles of Q_2 and Q_4 are equal to the values of m_a and m_b. Consequently, the narrower the pulse width of U_{bg}, the shorter the inductor discharging duration, leading to a higher voltage gain. What's more, both the voltage frequency of U_{bg} and the inductor current frequency are twice the switching frequency, mirroring the conditions in step-down mode. As a result, with the same amount of inductance, the inductor current ripple of the asymmetric H-bridge BDC is only half of that in the Buck/Boost converter.

According to the volt-second balance law, the step-up voltage gain M_{Boost} can be determined as follows.

$$\begin{cases} M_{Boost} = \frac{U_{high}}{U_{low}} = \frac{1}{m_a - m_b} \\ 0 < m_b < 0.5 < m_a < m_a + m_b < 1 \end{cases} \tag{2.27}$$

Subsequently, the modulation indexes can be expressed in the following manner.

$$\begin{cases} m_a = 0.5 + \frac{0.49}{M_{Boost}} \\ m_b = 0.5 - \frac{0.51}{M_{Boost}} \end{cases} \tag{2.28}$$

In summary, the relationship between U_{high} and U_{low} in the converter can be unified as follows.

$$\begin{cases} U_{low} = U_{high} \times (m_a - m_b), \\ \text{step-down: } m_a + m_b > 1, \text{ and} \begin{cases} m_a = 0.5 + 0.51 M_{Buck} \\ m_b = 0.5 - 0.49 M_{Buck} \end{cases} \\ \text{step-up: } m_a + m_b < 1, \text{ and} \begin{cases} m_a = 0.5 + 0.49 / M_{Boost} \\ m_b = 0.5 - 0.51 / M_{Boost} \end{cases} \end{cases} \tag{2.29}$$

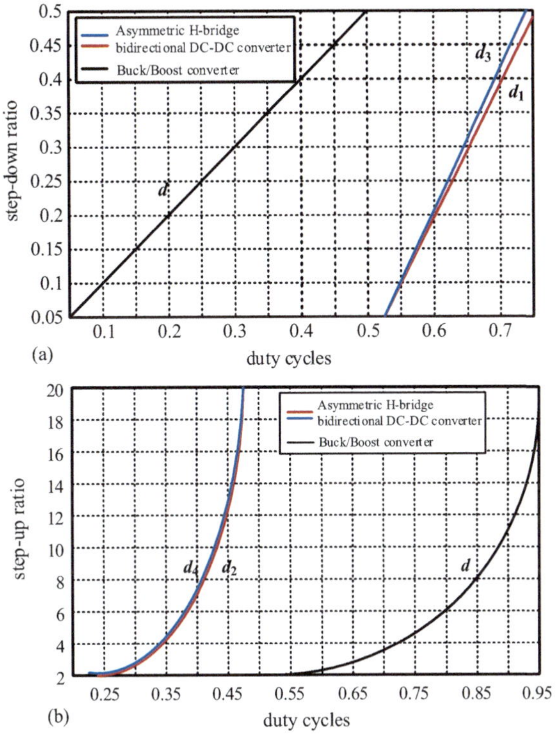

Fig. 2.11 Voltage gain comparison between the asymmetric H-bridge and Buck/Boost converter in **a** step-down and **b** step-up modes

By combining (2.25) and (2.28), the asymmetric H-bridge bidirectional DC–DC converter achieves a wide voltage gain range without extreme duty cycles. Figure 2.11 illustrates the voltage gain comparison between the conventional Buck/Boost converter and the asymmetric H-bridge bidirectional DC–DC converter.

Upon observation, it is evident that the former encounters issues with extreme duty cycles, when operating at high voltage gains. Conversely, the latter requires duty cycles closer to 0.5 at the same voltage gain. In step-down mode, for a 0.05 voltage ratio, the duty cycles for the Buck/Boost converter and the asymmetric H-bridge converter are approximately 0.05 and 0.525, respectively. Similarly, for a 20 voltage gain in step-up mode, the duty cycles for the Buck/Boost converter and the asymmetric H-bridge converter are approximately 0.95 and 0.475, respectively.

Therefore, the asymmetric H-bridge converter boasts a wider voltage gain compared to the conventional Buck/Boost converter. This characteristic proves advantageous in enhancing the operational stability and efficiency of the system.

2.3.2 *Control Strategy of the Asymmetric H-Bridge Bidirectional DC–DC Converter*

A. Synchronous rectification operation control

As analyzed previously, the asymmetric H-bridge bidirectional DC–DC converter exhibits three operation states in each operating mode. In step-down mode, the main power switches are Q_1 and Q_3, whereas in step-up mode, the main power switches are Q_2 and Q_4. The operation states of the asymmetric H-bridge bidirectional DC–DC converter are summarized as follows.

$$\begin{cases} [S_1 S_2 S_3 S_4]_{\text{step - down - DR}} = [1010,\ 1000,\ 0010] \\ [S_1 S_2 S_3 S_4]_{\text{step - up - DR}} = [0000,\ 0100,\ 0001] \end{cases} \tag{2.30}$$

To enable synchronous rectification operation, the power switches are controlled in a complementary manner, as outlined below.

$$\begin{cases} S_1 = \overline{S_2} \\ S_3 = \overline{S_4} \end{cases} \tag{2.31}$$

Consequently, the switching states during synchronous rectification are as follows.

$$\begin{cases} [S_1 S_2 S_3 S_4]_{\text{step - down - SR}} = [1010,\ 1001,\ 0110] \\ [S_1 S_2 S_3 S_4]_{\text{step-up-SR}} = [1010,\ 0110,\ 1001] \end{cases} \tag{2.32}$$

This approach leads to a significant reduction in conducting losses, achieving both ZVS ON and ZVS OFF for Q_2 and Q_4 in step-down mode, as well as Q_1 and Q_3 in step-up mode, respectively. It is important to note that a dead time t_d is necessary between S_1 and S_2, as well as S_3 and S_4.

In step-down mode, the synchronous rectifiers for the asymmetric H-bridge bidirectional DC–DC converter are Q_2 and Q_4, as illustrated in Fig. 2.12a. These rectifiers achieve ZVS ON and ZVS OFF. Conversely, in step-up mode, the synchronous rectifiers are Q_1 and Q_3, as depicted in Fig. 2.12b, where ZVS ON and ZVS OFF are also achieved.

B. Bidirectional operation control

The control strategy for enabling bidirectional power flow in the asymmetric H-bridge bidirectional DC–DC converter is presented in Fig. 2.13. Both step-down and step-up modes are governed by a double-loop structure featuring an outer voltage loop and an inner current loop. The current loop references for inductor current i_L in step-down and step-up modes are denoted as $I_{\text{ref-Buck}}$ and $I_{\text{ref-Boost}}$, respectively. Within the outer voltage loop, U_{low} and U_{high} track the voltage references $U_{\text{ref-Bcuk}}$ and $U_{\text{ref-Boost}}$ in step-down and step-up modes, respectively. The converter's mode transition is implemented by a binary indicator U_c. The controller and PWM generation scheme undergo modification based on the value of U_c. During the mode

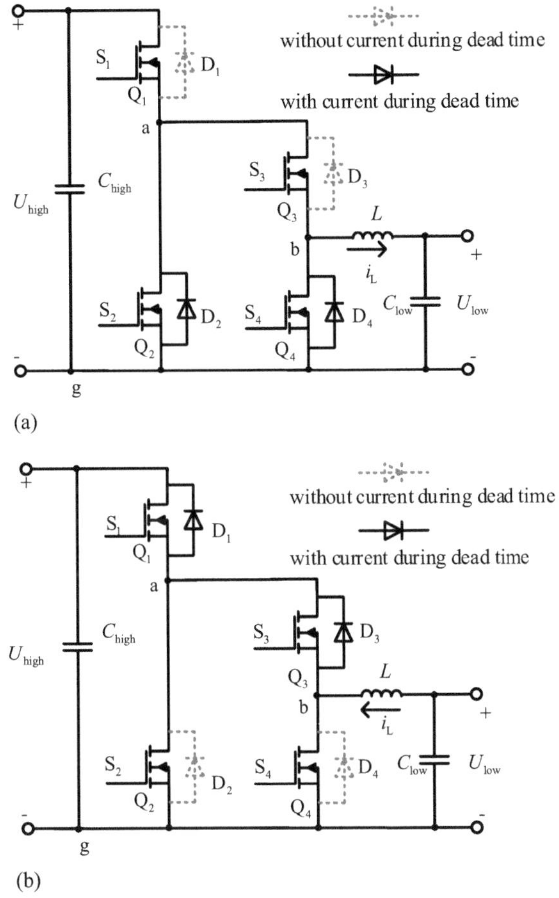

Fig. 2.12 Synchronous rectification of the asymmetric H-bridge bidirectional DC–DC converter in **a** step-down mode and **b** step-up mode

transition process, the inductor current i_L drops to zero and then gradually increases in the reverse direction. As a result, the bidirectional operation of the asymmetric H-bridge converter can be achieved.

2.3.3 Comparison of the Asymmetric H-Bridge Bidirectional DC–DC Converter with Counterparts

The comparison results of the asymmetric H-bridge bidirectional DC–DC converter with other bidirectional solutions are listed in Table 2.1. The asymmetric H-bridge converter adopts the same number of power switches as the converters in [126] and [40] but achieves a higher voltage gain range. However, this comes at the cost of requiring two inductors and facing high voltage stresses across power switches,

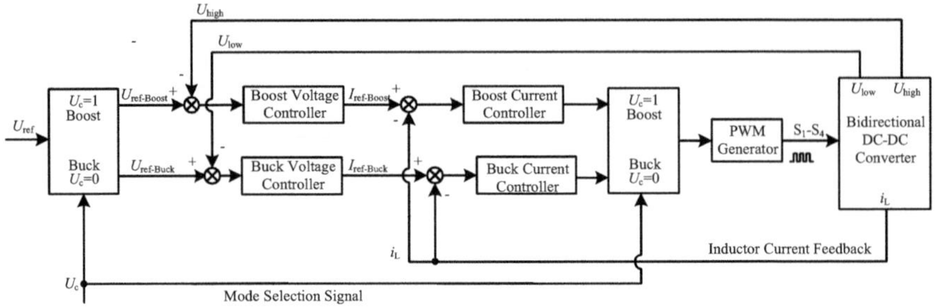

Fig. 2.13 Control strategy of bidirectional power flow of the asymmetric H-bridge bidirectional DC-DC converter

limiting operational efficiency. What's more, both [40] and [42] exhibit pulse voltage differences between the negative terminals of input and output, leading to extra EMI issues. In contrast, the conventional Buck/Boost converter in [44] employs only two power switches, achieving a maximum conversion efficiency of 94.4%. However, its voltage gain may be constrained in practical applications due to parasitic resistances and extreme duty cycles.

In conclusion, the asymmetric H-bridge converter demonstrates a relatively higher conversion efficiency, comparable power switch stresses to the Buck/Boost converter, and the added advantage of reducing EMI issues through its common ground feature. Furthermore, it allows for a high step-up/step-down voltage gain. Additionally, the reduction in inductor current ripple results in smaller required inductances for the asymmetric H-bridge converter.

2.4 Efficiency and Application Analysis

To validate the feasibility of the H-bridge and asymmetric H-bridge bidirectional DC–DC converters, comprehensive experiments have been implemented to fully evaluate the efficiency and practical application potential of the topologies.

2.4.1 Efficiency Analysis

In the experimental setup of the H-bridge bidirectional DC–DC converter, during step-down mode, the high voltage side of the DC–DC converter is connected to a DC voltage source at $U_{\text{high}} = 150$ V, representing battery stacks on the DC bus. Simultaneously, the low voltage side of the DC–DC converter is connected to a load resistor R_{Load}. In step-up mode, the low voltage side and high voltage side of the DC–DC converter are respectively connected to a 48 V DC voltage source and a load resistor.

Table 2.1 Comparative analysis of the asymmetric H-bridge bidirectional DC–DC converter and other bidirectional solutions

Bidirectional solutions	Step-up voltage gain	Amount of semiconductors	Amount of inductors	Maximum voltage stress across semiconductors	Maximum current stress across semiconductors	Common ground	Efficiency (%)
Converter in [126]	$\frac{1}{(1-d)^2}$	4	2	$U_{\mathrm{high}} + U_{\mathrm{high}}(1-d)$	$\left(\frac{1}{1-d} + \frac{d}{(1-d)^2}\right) I_{\mathrm{h}}$	Yes	95–97.1
Converter in [40]	$\frac{1+d}{1-d}$	3	2	$\frac{1}{1+d} U_{\mathrm{high}}$	$\frac{1+d}{1-d} I_{\mathrm{h}}$	No	86–98
Converter in [42]	$\frac{1}{1-(d_2+d_3)}$	4	1	U_{high}	$\frac{1}{1-(d_2+d_3)} I_{\mathrm{h}}$	No	71–93.6
Buck/Boost converter in [44]	$\frac{1}{1-d}$	2	1	U_{high}	$I_{\mathrm{h}}/(1-d)$	Yes	83.5–94.4
Asymmetric H-bridge bidirectional DC–DC converter	$\frac{1}{1-(d_2+d_4)}$	4	1	U_{high}	$\frac{1}{1-(d_2+d_4)} I_{\mathrm{h}}$	Yes	83–93.5

The efficiency curves of the H-bridge bidirectional DC–DC converter are illustrated in Fig. 2.14, including four operational conditions: step-up mode with synchronous rectification (i.e., Buck-SR) and diode rectification (i.e., Buck-DR), and step-down mode with synchronous rectification (i.e., Boost-SR) and diode rectification (i.e., Boost-DR).

In step-down mode, under DR condition, the inductor current flows through one MOSFET and one anti-parallel diode. Conversely, in SR condition, the inductor current flows through two MOSFETs. Due to the lower on-state resistance of the MOSFET compared to the anti-parallel diode and the achievement of ZVS operation in SR, the efficiency of Buck-SR condition surpasses that of Buck-DR condition.

Similarly, in step-up mode, the inductor is discharged via two diodes in series and charged via one MOSFET and one anti-parallel diode. Consequently, akin to step-down mode, the operational efficiency of the converter in Boost-SR condition is higher than that of Boost-DR condition. The efficiency increase between DR and SR is approximately 4.8% and 3.4% in step-down and step-up modes, respectively. As depicted in the figure, the operational efficiency of the converter varies with the voltage gain. Specifically, when the low-side voltage is at 50 V, it reaches the maximum efficiency at 94.1% in step-down mode and 93.6% in step-up mode.

For the asymmetric H-bridge bidirectional DC–DC converter, a 300 W prototype was developed for experimental demonstration. In step-down mode, a DC source is connected to the high voltage side of the converter to simulate battery stacks, while a load resistor is connected to the low voltage side. In step-up mode, the high voltage side and low voltage side of the converter are connected to a load resistor and a DC voltage source, respectively. The reference voltage for the high voltage side is set at 200 V.

In the experimental setup of the asymmetric H-bridge bidirectional DC–DC converter, the prototype was built with a rated power of 300 W. The high voltage side is set at 200 V, considered as the DC bus in practical applications. The low voltage side varies from 24 to 48 V. Moreover, a switching frequency of 10 kHz is chosen.

The operational efficiency of the prototype in both step-up and step-down modes at different power levels is measured and plotted in Fig. 2.15. In both step-up and step-down modes, efficiency increases as the low-side voltage increases, namely, as

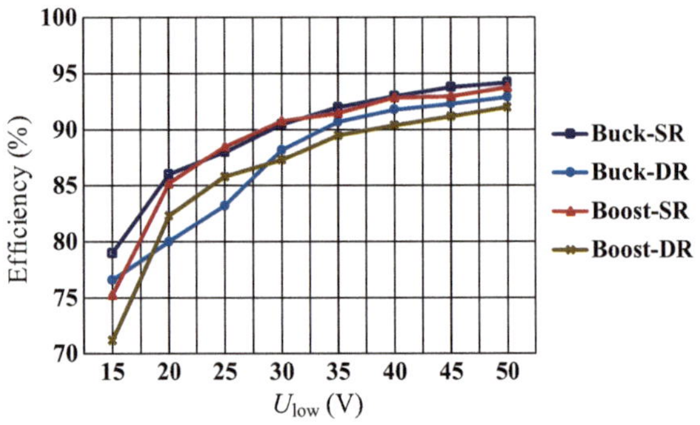

Fig. 2.14 Efficiency curves of the H-bridge bidirectional DC–DC converter

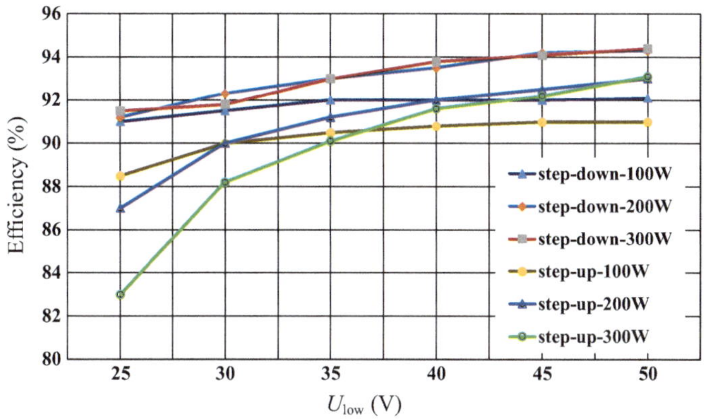

Fig. 2.15 Efficiency curves of the asymmetric H-bridge bidirectional DC–DC converter

the voltage gain decreases. The overall efficiency of the prototype in step-down mode is relatively higher than that in step-up mode.

In addition, the prototype demonstrates a higher maximum conversion efficiency at the rated power level of 300 W compared to light load conditions. As depicted in the figure, the minimum measured efficiency in step-down mode is 90.7% with an output voltage of 24 V and an output power of 300 W, while the maximum efficiency is 94.7% with an output voltage of 48 V and an output power of 300 W. In step-up mode, with a fixed output voltage of 200 V, the minimum efficiency is 83% when the input voltage is 24 V and the output power is 300 W. The maximum efficiency is 93.5% with an input voltage of 48 V and an output power of 200 W.

2.4.2 Application Analysis

Based on the efficiency analysis of both the H-bridge and asymmetric H-bridge bidirectional DC–DC converters mentioned above, it is evident that both topologies can effectively operate in both step-down and step-up modes. Consequently, these converters prove to be suitable for applications in energy storage systems, specifically for source charging and discharging purposes. Given that the voltage of batteries and super-capacitors tends to be lower than the DC bus voltage, these converters can serve as a power interface between the energy source and the DC bus. In step-down mode, the energy source can be charged through the converter, while in step-up mode, the converter facilitates the transfer of power flow from the energy source to the DC link.

Nevertheless, the efficiency curves of the H-bridge bidirectional DC–DC converter highlight a notable increase in operation losses with lower voltage on the low side. Specifically, the conversion ratio falls below 80% when the voltage gain exceeds 7.5. In contrast, the asymmetric H-bridge topology exhibits superior conversion performance at higher voltage gain. Based on experimental results, the operational efficiency of the asymmetric H-bridge bidirectional DC–DC converter consistently stays above 80%, when the voltage gain exceeds 8.

Therefore, as both the H-bridge and asymmetric H-bridge bidirectional DC–DC converters discussed in this chapter can be utilized in energy storage systems, the asymmetric H-bridge topology is well-suited for high voltage gain scenarios, like power interfaces for super-capacitors and fuel cells. Additionally, owing to its common ground structure and better electromagnetic performance, the asymmetric H-bridge topology holds promise for higher-power applications.

2.5 Conclusion

This chapter introduces and analyzes two non-isolated DC–DC converters: the H-bridge bidirectional DC–DC converter and the asymmetric H-bridge DC–DC converter. Both converters can achieve a high step-up/step-down voltage gain with non-extreme duty cycles, approaching 0.5, and exhibit a doubled equivalent switching frequency. The application of a synchronous rectification control scheme reduces the conducting losses and reverse recovery losses of diodes, significantly improving the operational efficiency of the converters.

Compared to the H-bridge bidirectional converter, the asymmetric H-bridge converter comprises the same number of power components and features a common ground structure. This design helps to alleviate EMI issues. Consequently, both the H-bridge and asymmetric H-bridge converters are deemed suitable for renewable energy applications.

Chapter 3
Three-Level DC–DC Converters

3.1 Introduction

Electric vehicles with hybrid energy sources incorporate super-capacitors connected in parallel to battery stacks via a bidirectional DC–DC converter. This configuration is instrumental in ensuring a consistent and stable energy supply, thereby extending the driving range of electric vehicles. Super-capacitors discharge during acceleration and recharge during braking, effectively managing instantaneous pulse powers. As a result, the bidirectional DC–DC converter assumes a pivotal role in maintaining the optimal performance of these vehicles.

Nevertheless, the voltage across series-connected super-capacitors, typically remaining low, experiences significant variations during charging and discharging. Consequently, the necessity arises for a bidirectional DC–DC converter capable of operating within a wide voltage conversion range. While DC–DC converters equipped with transformers or coupled inductors can address this challenge [45, 46], they come with larger volumes and lower efficiencies. Alternatively, conventional bidirectional three-level DC–DC converters can alleviate switching losses through lower blocking voltages of power switches, enabling the use of low-rated on-state resistance switches in high voltage and power converters [47]. However, these power switches encounter extreme duty cycles when operating at high voltage gains.

The conventional choice for voltage step-up in DC–DC converters is the Boost two-level converter, primarily due to its simple structure [48, 49]. However, this converter faces limitations, including a restricted voltage-gain range and high voltage stress on its power semiconductors. To counter the issue of voltage level mismatch, it becomes necessary to increase the rated voltage of the fuel cell stack, thereby complicating the assembly of the stack. Additionally, the use of higher-rated blocking voltage power semiconductors is required, resulting in increased conduction losses.

In an effort to alleviate the high voltage stress on power semiconductors, Boost three-level DC–DC converters have been proposed, effectively halving the voltage stress [50, 51]. Nevertheless, two critical problems persist in the interface between

Y. Zhang and S. Gao, *High Efficiency Non-isolated DC-DC Converters with Wide Voltage Gain Range for Renewable Energies*, CPSS Power Electronics Series,
https://doi.org/10.1007/978-981-97-0648-8_3

the fuel cell stack and the DC-link bus: limited voltage gain, similar to the Boost two-level converter, and the intricate control required to maintain flying-capacitor voltage balance in the Boost three-level converter, particularly during transient states [52]. The latter issue may lead to power semiconductor failure. Consequently, it is imperative to address these challenges when deploying Boost three-level DC–DC converters with flying capacitors in fuel cell vehicles.

In the realm of non-isolated step-up DC–DC converters with high voltage gain, voltage multiplier circuits have been incorporated to extend the voltage gain [53]. Switched-inductor structures and switched-capacitor DC–DC converters also offer avenues to achieve high voltage gain [54, 55]. Nevertheless, these high voltage-gain step-up converters often exhibit complexity, posing challenges for cost and size reduction.

The quadratic Boost DC–DC converter is capable of achieving high voltage gain [56], but the power semiconductors on the output side (high voltage side) experience elevated voltage stresses and generate high dv/dt during switching. Although a high voltage gain interleaved Boost DC–DC converter, composed of two stages in parallel and one series multiplier stage, can convert 24–200 V [48], there remain two diodes in the multiplier stage subjected to full output voltage stress.

A family of diode-coupled-winding Boost DC–DC converters with high voltage gain outperforms their active-clamp counterparts due to the recycling of leakage energy, achieving a maximum efficiency of ~ 91.7%. Expanding on the concepts in [57, 58], a high voltage-gain interleaved Boost DC–DC converter, magnetically coupled to a voltage-double circuit, was proposed in [59]. Additionally, another high voltage-gain Boost DC–DC converter, based on the three-state commutation cell with two additional transformers (six windings), can achieve even higher efficiency [60].

The utilization of Z-source networks in conventional step-up DC–DC converters has been explored to achieve higher voltage gain. However, these converters face the drawback of having input and output sides that lack a common ground. This deficiency can lead to concerns related to maintenance safety and EMI issues. An alternative strategy involves replacing the output diode with an inductor in Z-source DC–DC converters [61]. Regrettably, this modification unexpectedly reduces the voltage gain.

Another circuit structure, known as the diode rectification quasi-Z-source circuit, has been proposed for integrating a low voltage DC source and an inverter [62, 63]. It can also be applied in step-up DC–DC converters, offering the advantages of lower capacitor voltages and a common ground [64]. However, its voltage gain remains the same as that of conventional Z-source DC–DC converters, and the power switch still experiences high voltage stress equal to the output voltage.

To achieve high voltage gain, coupled inductor-based Z-source DC–DC converters have been developed by adjusting the turn ratio of the coupled inductor [65]. However, these converters may experience significant spike voltages across the power switches due to the leakage inductance of the coupled inductor. The analysis in [66] indicates that the voltage stress on the power semiconductors is reduced within the range of half of the output voltage to nearly the output voltage when the duty cycle (voltage gain) is increased. However, as the duty cycle is increased, the current stress on the power switch becomes several times higher than the output current.

3.2 Transformerless Buck Three-Level DC–DC Converter

To overcome the problem associated with high transient voltage stress, three-level converter topologies serve as alternative choices for DC–DC power conversion. These converters mostly employ a phase-shift control method and inevitably provide a circulating current path through the transformer and primary power stage during the freewheeling period, resulting in an increase in conducting losses in the converter.

In order to eliminate or minimize the circulating current, various auxiliary circuits with active switches and passive components are utilized on the secondary side of the converter to improve the overall efficiency of the converter. However, the auxiliary circuit also introduces additional power losses and costs, especially at high switching frequencies.

Figure 3.1 illustrates the topology of a half-bridge Buck three-level DC–DC converter, featuring two power stages divided by a transformer [67]. The low voltage state is connected to the center-tapped windings of the transformer, accompanied by an output filter comprising L_{out} and C_{out}. The blocking diode D_b is series connected to the load side to prevent reverse power flow. On the input side, capacitors C_1 and C_2 establish a voltage midpoint between zero and the input DC voltage. Two switch pairs, composed of four power switches Q_1–Q_4, are employed. The flying capacitor C_f provides charging and discharging paths during freewheeling time to balance the input capacitors.

To address challenges arising from transformers in three-level DC–DC converters, this section introduces a transformerless Buck three-level DC–DC converter. This

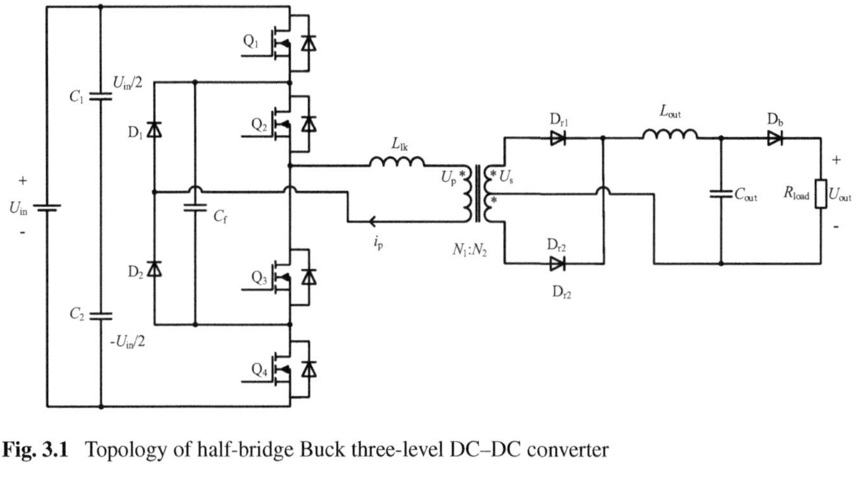

Fig. 3.1 Topology of half-bridge Buck three-level DC–DC converter

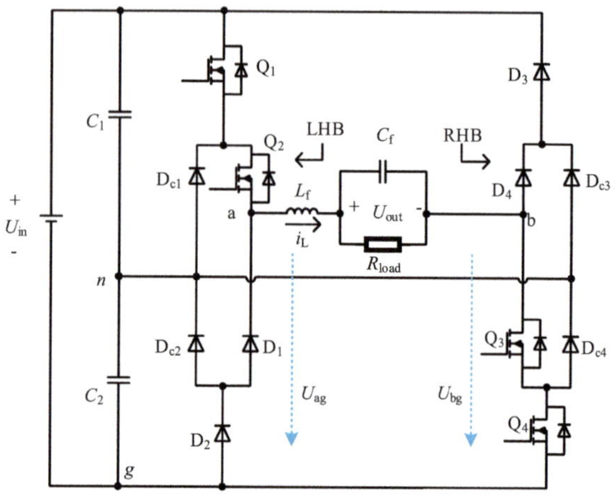

Fig. 3.2 Topology of the transformerless Buck three-level DC–DC converter

converter exhibits the capability of capacitor voltage self-balance and can achieve a high step-down voltage gain. Moreover, by eliminating the need for a transformer, the converter can efficiently operate across a wide voltage gain range, thereby preventing the occurrence of extreme duty cycles.

The topology of the transformerless Buck three-level DC–DC converter is illustrated in Fig. 3.2, where U_{in} represents the input source divided by input filtering capacitors C_1 and C_2. The converter consists of a left half-bridge and a right half-bridge, offering three different voltage levels. The left half-bridge incorporates power switches Q_1 and Q_2, along with diodes D_1 and D_2. Symmetrically, the right half-bridge features power switches Q_3 and Q_4, along with diodes D_3 and D_4. Blocking voltages are clamped at $U_{in}/2$ by adopting the clamped diodes D_{c1}–D_{c4}. The ground of the converter is denoted as "g", while the midpoints of the left and right half-bridges are labeled as points "a" and "b", respectively. The output voltage U_{out} is derived from the difference voltage between points "a" and "b" via the output filter, where L_f and C_f represent the filter inductor and filter capacitor, respectively.

3.2.1 Operating Principle of the Buck Three-Level DC–DC Converter

The power switches of the transformerless Buck three-level DC–DC converter are controlled by gate signals S_1–S_4, with the turn-on and turn-off switching states denoted as "1" and "0", respectively. The left half-bridge is characterized by three switching states, which are analyzed as follows.

State 1) $S_1S_2 = 01$

When the power switch Q_1 is turned OFF and Q_2 is turned ON, the clamp diode D_{c1} is conducted. The voltage stress across power switch Q_1 is $U_{in}/2$, which is equal to the capacitor voltage of C_1. The voltage difference between “a” and “g”, namely U_{ag} is $U_{\mathrm{in}}/2$, which is equal to the capacitor voltage of C_2.

State 2) $S_1S_2 = 00$

When the power switches Q_1 and Q_2 are both turned OFF, the diodes D_1 and D_2 are freewheeling. The voltage stresses across Q_1 and Q_2 are both clamped as $U_{\mathrm{in}}/2$ by the capacitor voltages of C_1 and C_2, respectively.

State 3) $S_1S_2 = 11$

When the power switches Q_1 and Q_2 are both turned ON, the voltage U_{ag} is equal to U_{in}.

Similarly, there are also three switching states for the right half-bridge, which are explained as below.

State 1) $S_3S_4 = 11$

When the power switches Q_3 and Q_4 are both turned ON, the voltage difference between “b” and “g”, namely U_{bg} is zero.

State 2) $S_3S_4 = 10$

When the power switch Q_3 is turned ON and Q_4 is turned OFF, the clamp diode D_{c4} is conducted. The voltage U_{bg} is equal to $U_{\mathrm{in}}/2$, which is also the capacitor voltage of C_2 and the voltage stress across Q_4.

State 3) $S_3S_4 = 00$

When the power switches Q_3 and Q_4 are both turned OFF, the diodes D_3 and D_4 are freewheeling. Meanwhile, the voltage stress across Q_3 is equal to the capacitor voltage of C_1 and the voltage stress across Q_4 is equal to the capacitor voltage of C_2.

Thereby, the pulse voltage U_{ab} of the Buck three-level DC–DC converter can be derived from the voltage difference between “a” and “b”, as shown in (3.1).

$$U_{ab} = U_{ag} - U_{bg} \tag{3.1}$$

By using (3.1) and the output filter, the desired DC output voltage of the Buck three-level DC–DC converter can be obtained by controlling the switching states of the left and right half-bridges. It is assumed that the converter operates only in continuous current mode. As a result, the operation of the converter consists of six switching states, as shown in Fig. 3.3.

State 1) $S_1S_2S_3S_4 = 0110$

During this state, power switches Q_2 and Q_3 are turned ON and series connected with the clamped diodes D_{c1} and D_{c4}. The inductor L_f is discharged and the pulse voltage U_{ab} is zero.

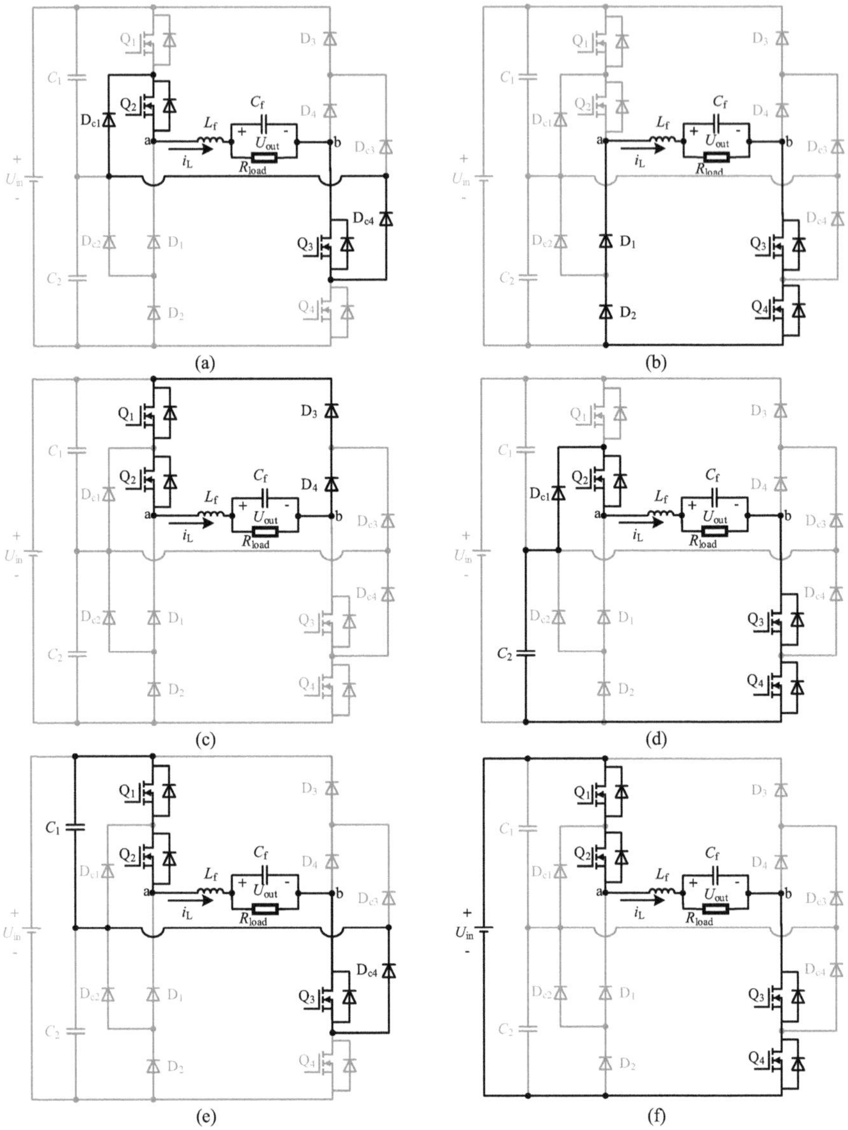

Fig. 3.3 Switching states of the transformerless Buck three-level DC–DC converter during **a** state 1, **b** state 2, **c** state 3, **d** state 4, **e** state 5, and **f** state 6

State 2) $S_1S_2S_3S_4 = 0011$

The current loop in this state consists of the power switches Q_3 and Q_4, and the diodes D_1 and D_2. The inductor L_f is discharged, and the pulse voltage U_{ab} is zero.

State 3) $S_1S_2S_3S_4 = 1100$

This switching state is similar to state 2, where the current loop consists of power switches Q_1 and Q_2, and the diodes D_3 and D_4. Meanwhile, the inductor L_f is discharged and the pulse voltage U_{ab} is also zero.

State 4) $S_1S_2S_3S_4 = 0111$

During this state, the inductor current i_L flows through the power switches Q_2, Q_3, and Q_4, as well as the clamped diode D_{c1}. The filtering bottom capacitor C_2 is discharged. The inductor L_f is charged and the pulse voltage U_{ab} is obtained as $U_{in}/2$.

State 5) $S_1S_2S_3S_4 = 1110$

When the inductor current i_L flows through the power switches Q_1, Q_2, and Q_3, as well as the clamped diode D_{c4}, the pulse voltage U_{ab} is also obtained as $U_{in}/2$. The inductor L_f is charged while the upper filtering capacitor C_1 is discharged.

State 6) $S_1S_2S_3S_4 = 1111$

When all power switches in the Buck three-level DC–DC converter are turned ON, the inductor current i_L flows through the upper arm of the left half-bridge and the lower arm of the right half-bridge. The inductor L_f is charged, and the pulses voltage U_{ab} is derived as U_{in}.

According to the analysis above, the states 1–3 are all redundant states where U_{ab} is equal to zero, due to the continuous inductor current. Besides, the states 4 and 5 are other redundant switching states where U_{ab} is equal to $U_{in}/2$.

3.2.2 *Control Strategy of the Buck Three-Level DC–DC Converter*

A. Modulation strategy with capacitor voltages self-balancing

The modulation strategy for the transformerless Buck three-level DC–DC converter is employed to regulate the switching states of the power switches, with the goal of achieving the desired high voltage gain. Figure 3.4 illustrates the modulation strategy based on the operating principle of the converter. Two modulation indices, m_a and m_b, are utilized for the left and right half-bridges, respectively. Additionally, carriers 1 and 2 are phase-shifted by π. The gate signals are derived from the following equation, where $U_{carrer1}$ and $U_{carrer2}$ represent the values of carrier1 and carrier2, respectively.

$$\begin{cases} m_b > U_{carrier1}, & S_1 = 0 \\ m_a > U_{carrier2}, & S_2 = 1 \\ m_a > U_{carrier1}, & S_3 = 1 \\ m_b > U_{carrier2}, & S_4 = 0 \end{cases} \tag{3.2}$$

As a result, the operation of the power switches can be regulated, which includes five distinct switching states denoted as $S_1S_2S_3S_4 = [0110, 1110, 1100, 1110, 0110]$. U_{ag} and U_{bg} are generated as three-level voltages through the operation of the half-bridges. The pulse voltage U_{ab} is subsequently derived from these three-level voltages.

During the switching states that $S_1S_2S_3S_4 = 1110$ and $S_1S_2S_3S_4 = 0111$, the pulse voltage U_{ab} can be obtained as $U_{in}/2$ from the difference between U_{ag} and U_{bg}, and the inductor current i_L increases. Whereas, the pulse voltage U_{ab} remains zero during the other switching states due to the equal values of U_{ag} and U_{bg}. Meanwhile, the

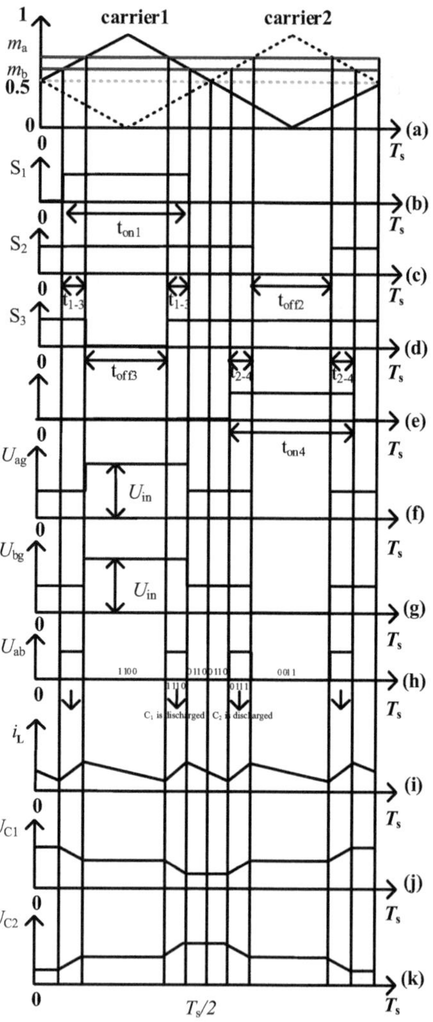

Fig. 3.4 Modulation strategy with capacitor voltage self-balance

inductor current i_L decreases. The capacitors C_1 and C_2 discharge to supply power to the load in the first and second halves of the switching cycle, respectively.

As depicted in Fig. 3.4, each transition in the switching period involves only one changed gate signal. Consequently, the switching period of the power switches aligns with the carrier period, allowing for minimal switching losses. In each switching period, capacitors C_1 and C_2 undergo charging and discharging with a half-periodic phase-shift. Moreover, the modulation indexes m_a and m_b remain constant, and the average inductor current I_L can be considered constant.

Simultaneously, due to the symmetrical charging and discharging intervals of C_1 and C_2, and the constancy of the charging/discharging current (i.e., the average inductor current), the capacitor voltages achieve self-balancing. Additionally, the equivalent frequency of the pulse voltage U_{ab} is twice the switching frequency.

In steady-state operation, the energy stored in inductor L_f remains balanced during each switching period. Based on the earlier analysis of switching states, inductor L_f undergoes charging when U_{ab} is $U_{in}/2$ and discharging when U_{ab} is zero. The charged energy W_{ch} and discharged energy W_{disch} of L_f can be represented as follows, where U_{out} is the output DC voltage across capacitor C_f, t_{on1} is the turn-on duration of Q_1, and t_{off3} is the turn-off duration of Q_3.

$$\begin{cases} W_{ch} = \left(\frac{U_{in}}{2} - U_{out}\right) \times I_L \times 2 \times (t_{on1} - t_{off3}) \\ W_{disch} = U_{out} \times I_L \times [T - 2 \times (t_{on1} - t_{off3})] \\ W_{ch} = W_{disch} \end{cases} \tag{3.3}$$

On the basis of (3.3), the duty cycles for the power switches can be derived as follows, where d_1–d_3 represent the duty cycles for power switches Q_1–Q_4, respectively.

$$\begin{cases} d_1 = d_4 = \frac{t_{on1}}{T} = 1 - m_b \\ d_3 = d_2 = \frac{T - t_{off3}}{T} = m_a \end{cases} \tag{3.4}$$

The voltage gain, denoted as M, of the transformerless Buck three-level DC–DC converter can be determined.

$$M = \frac{U_{out}}{U_{in}} = (m_a - m_b) = (d_1 + d_2 - 1) \tag{3.5}$$

B. Transformerless operation without extreme duty cycles

The voltage gain of conventional transformerless Buck three-level DC–DC converters is constrained by extreme duty cycles. Specifically, achieving a voltage gain > 10 requires duty cycles lower than 0.1, while the ideal range for duty cycle is around 0.5. As demonstrated in (3.5), the transformerless Buck three-level DC–DC converter discussed in this section overcomes this limitation, enabling high voltage gain with non-extreme duty cycles. For instance, by setting modulation indexes as $m_a = 0.6$ and $m_b = 0.5$, a step-down voltage gain of 0.1 can be attained with duty cycles at 0.6 and 0.5.

As previously analyzed, the output voltage of the transformerless Buck three-level DC–DC converter is determined by the difference between the midpoint voltages of the half-bridges. The pulse width of U_{ab} is contingent on the overlap of duty cycles. The overlap interval between S_1 and S_3 is denoted as t_{1_3}, while the overlap interval between S_2 and S_4 is denoted as t_{2_4}, as illustrated in (3.6).

$$\begin{cases} t_{1\_3} = \dfrac{t_{on1} - t_{off3}}{2} \\ t_{2\_4} = \dfrac{t_{on4} - t_{off2}}{2} \end{cases} \tag{3.6}$$

Given the symmetry of the switching states within each period, the overlapped intervals t_{1_3} and t_{2_4} are equal. The pulse width of U_{ab} depends on the difference between modulation indexes. Consequently, there exist infinite pairs of solutions for m_a and m_b that yield any desired voltage gain. Duty cycles can be regulated around 0.5 to achieve a high step-down voltage gain. The modulation indexes can be designed by fixing m_b (e.g., $m_b = 0.55$), allowing for the adjustment of m_a using the current PI controller. The value of m_b should be determined with reference to the ideal duty cycle and the inductor current ripple.

3.2.3 Filter Design of the Buck Three-Level DC–DC Converter

The transformerless Buck three-level DC–DC converter generates DC output voltage through a filter circuit from a pulse voltage. Therefore, the effective design of the filtering components, including inductor L_f and capacitor C_f, plays a pivotal role in shaping the converter's operational performance.

Determining the minimum parameters for L_f and C_f involves considering the maximum inductor current ripple ΔI_{L_max} and the output voltage ripple ΔU_{out_max}, respectively. The inductor current ripple is contingent on factors such as charging/discharging duration and inductor voltage. Calculating the duration of inductor discharging t_{disch} entails examining the duty cycle of Q_2 and the switching period T_s.

$$t_{disch} = (1 - d_2) \times T_s \tag{3.7}$$

Meanwhile, the voltage across inductor u_L can be represented as below.

$$u_L = -(d_1 + d_2 - 1) \times U_{in} \tag{3.8}$$

The inductor current ripple ΔI_L can be obtained from the following formula.

$$\Delta I_{\mathrm{L}} = -\frac{u_{\mathrm{L}} t_{\mathrm{f}}}{L_{\mathrm{f}}} \tag{3.9}$$

Hence, the inductor current can be derived from (3.7), (3.8), and (3.9), which is written as follows.

$$\Delta I_{\mathrm{L}} = \frac{(d_1 + d_2 - 1)(1 - d_2) T_{\mathrm{s}} U_{\mathrm{in}}}{L_{\mathrm{f}}} \tag{3.10}$$

When d_1 is set as 0.5, d_2 is calculated as 0.75. According to (3.10), the inductor current ripple can be calculated.

$$\Delta I_{L\_\mathrm{max}} = \frac{U_{\mathrm{in}} T_{\mathrm{s}}}{16 L_{\mathrm{f}}} \tag{3.11}$$

Thus, the minimum inductor parameter $L_{\mathrm{f\_min}}$ can be derived from (3.11) as below.

$$L_{\mathrm{f\_min}} = \frac{U_{\mathrm{in}} T_{\mathrm{s}}}{16 \Delta I_{L\_\mathrm{max}}} \tag{3.12}$$

The output voltage ripple ΔU_{out} can be obtained from the inductor current ripple as follows.

$$\Delta U_{\mathrm{out}} = \frac{1}{C_{\mathrm{f}}} \times \frac{\Delta I_{\mathrm{L}}}{2} \times \frac{T_{\mathrm{s}}}{8} = \frac{T_s \Delta I_{\mathrm{L}}}{16 C_{\mathrm{f}}} \tag{3.13}$$

Therefore, the minimum capacitor parameter $C_{\mathrm{f\_min}}$ can be determined by the following formula, with $\Delta U_{\mathrm{out\_max}}$ representing the maximum output voltage ripple.

$$C_{\mathrm{f\_min}} == \frac{T_{\mathrm{s}} \Delta I_{\mathrm{L}}}{16 \Delta U_{\mathrm{out\_max}}} \tag{3.14}$$

3.3 Hybrid Boost Three-Level DC–DC Converter

The topology of a conventional single-phase diode-clamped three-level inverter consists of four power switches and two clamp diodes, as illustrated in Fig. 3.5a [68]. Building upon this architecture, the three-level DC–DC converters shown in Fig. 3.5b and c are deduced. Furthermore, the converters showcased in Fig. 3.6 also stem from the conventional single-phase diode-clamped three-level inverter. However, these two converters exhibit an issue of imbalanced capacitor voltages.

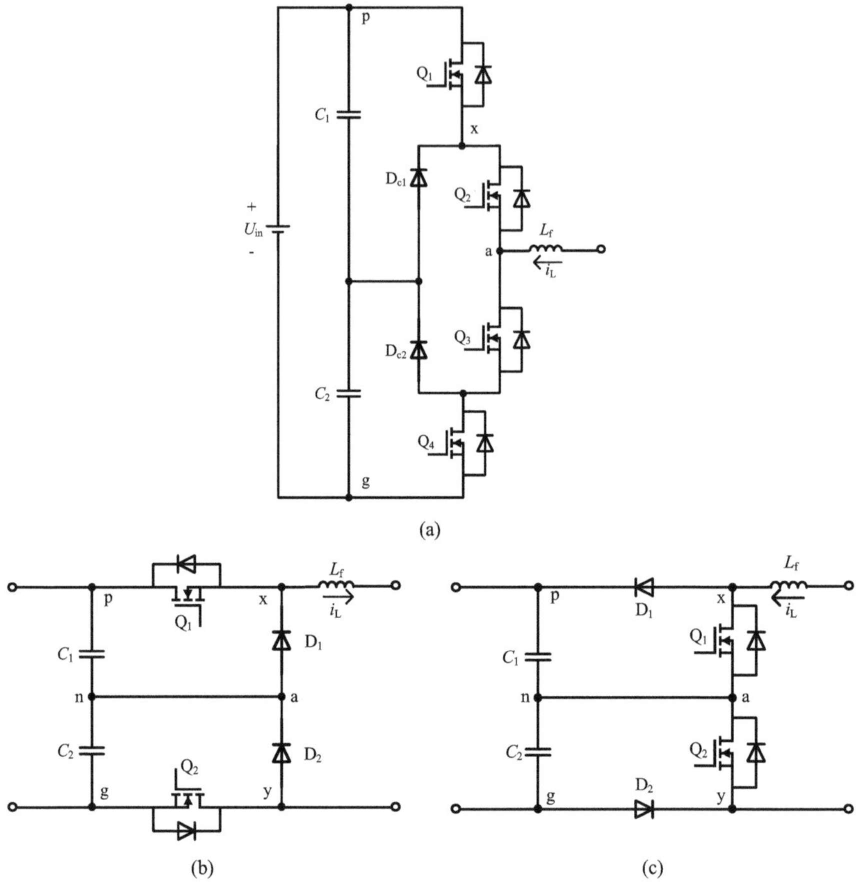

Fig. 3.5 Topologies of **a** conventional single-phase diode-clamped three-level inverter, **b** Buck three-level DC–DC converter, and **c** Boost three-level DC–DC converter

To overcome this challenge, a novel hybrid Boost three-level DC–DC converter is derived based on the three-level DC–DC converters. Tailored for PV applications requiring high output voltage and power levels, this converter adopts the concept of obtaining a narrower pulse voltage from the discrepancy between wider pulse voltages. Consequently, the hybrid Boost three-level DC–DC converter achieves high voltage gain without extreme duty cycles.

The topologies of the Boost three-level DC–DC converters are presented in Fig. 3.6, where U_{in1}, U_{in2}, $L_{f1,}$ and L_{f2} denote the DC input sources and filtering inductors of converter I and converter II, respectively. The derivation of the hybrid Boost three-level DC–DC converter is detailed in Fig. 3.7. To enhance the input power level of the hybrid converter, the input terminals of the Boost three-level DC–DC converters in are connected in series. Simultaneously, the output power level is

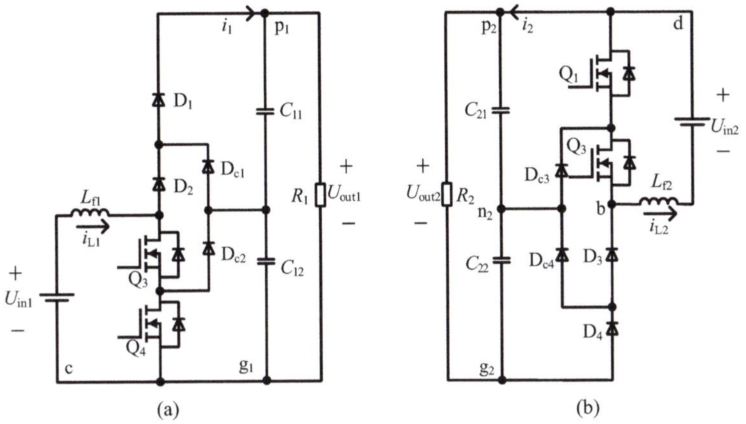

Fig. 3.6 Topologies of Boost three-level DC–DC converters. **a** Converter I. **b** Converter II

augmented by connecting the output terminals in parallel. Consequently, the hybrid Boost three-level DC–DC converter can be obtained through the following steps:

Cut I: Disconnect the negative input terminal (node "c") of Converter I from its negative output terminal (node "g_1").

Cut II: Similarly, sever the connection between the positive input terminal (node "d") of Converter II and its positive output terminal (node "p_2").

Connection I: Establish a connection between node "c" and node "d", ensuring that the inputs of Converter I and II are in series.

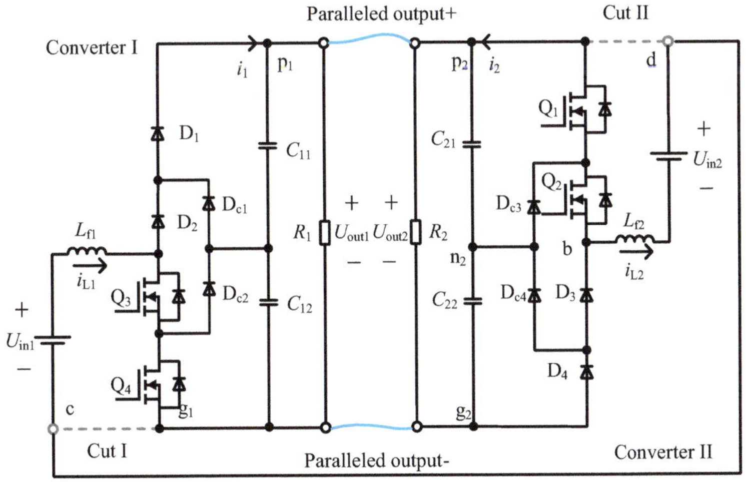

Fig. 3.7 Derivation of the hybrid Boost DC–DC converter

Connection II: Connect the negative output terminals (nodes "g_1" and "g_2") as well as the positive output terminals (nodes "p_1" and "p_2"), effectively linking the outputs of converters I and II in parallel.

3.3.1 Operating Principle of the Hybrid Boost Three-Level DC–DC Converter

The hybrid Boost three-level DC–DC converter is depicted in Fig. 3.8, featuring four power switches Q_1–Q_4, four diodes D_1–D_4, and four clamp diodes D_{c1}–D_{c4}. It also includes a filtering inductor L_f and filtering capacitors C_{f1}, C_{f2}. The hybrid converter is derived by combining converter I and converter II, thus the filtering capacitors C_{f1} and C_{f2} can be considered as the equivalent capacitors of C_{11} and C_{12} in converter I, and C_{21} and C_{22} in converter II. Similarly, the load resistor R_{load} of the hybrid converter can be regarded as the equivalent resistor of R_1 and R_2. Additionally, the neutral points of converters I and II, namely nodes "n_1" and "n_2", are connected together. Consequently, the neutral point "n" in the hybrid converter results in voltage stresses across power switches, acting as the capacitor voltages.

The hybrid Boost three-level DC–DC converter consists of two half-bridges. The pulse voltage U_{ab} is obtained from the difference between the midpoint voltages of the half-bridges, namely U_{ag} and U_{bg}.

$$U_{ab} = U_{ag} - U_{bg} \tag{3.15}$$

The DC output voltage U_{out} of the hybrid converter is then obtained from the pulse voltage U_{ab} through filtering capacitors. It is assumed that the voltages across filtering capacitors C_{f1} and C_{f2} are equal.

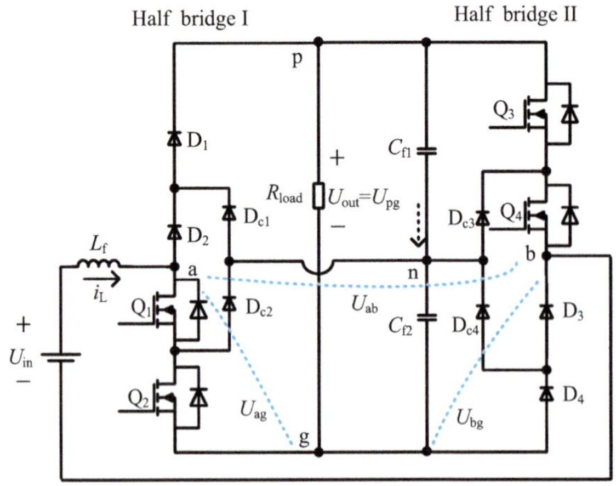

Fig. 3.8 Topology of the hybrid Boost three-level DC–DC converter

The hybrid Boost three-level DC–DC converter operates in six possible switching states, denoted by $S_1S_2S_3S_4 = [0000, 0001, 1000, 1100, 1001, 0011]$, where S_1–S_4 represent the gate signals of power switches Q_1–Q_4, respectively. The operating principle during each switching state is analyzed as follows.

State 1) $S_1S_2S_3S_4 = 0000$

When the power switches are all turned OFF, both the filtering capacitors are charged by the DC input source U_{in} and the inductor L_f through the diodes D_1–D_4. The output voltage U_{out} is equal to the pulse voltage U_{ab}.

State 2) $S_1S_2S_3S_4 = 0001$

When only the power switch Q_4 is turned ON, the filtering capacitor C_{f1} is charged by the DC input source U_{in} and the inductor L_f through the diodes D_1, $D_{2,}$ and D_{c3}. Meanwhile, the filtering capacitor C_{f2} is discharged to power the load. The output voltage U_{ab} is twice the pulse voltage U_{ab}, namely $U_{ab} = U_{out}/2$.

State 3) $S_1S_2S_3S_4 = 1000$

When only the power switch Q_1 is turned ON, the filtering capacitor C_{f2} charges via the DC input source and the inductor L_f through diodes D_3, D_4, and D_{c2}. Simultaneously, the output load is powered by the filtering capacitor C_{f1}, resulting in the output voltage U_{out} being twice the pulse voltage U_{ab}. Consequently, this switching state is considered redundant under the condition $U_{ab} = U_{out}/2$.

State 4) $S_1S_2S_3S_4 = 1100$

When both power switches Q_1 and Q_2 are turned ON, inductor L_f is charged, and filtering capacitors C_{f1} and C_{f2} are discharged in series. Simultaneously, diodes D_3 and D_4 are conducted. The output voltage U_{out} equals the pulse voltage U_{ab} when it is zero.

State 5) $S_1S_2S_3S_4 = 1001$

When both power switches Q_1 and Q_4 are turned ON, inductor L_f is charged by the DC input source while filtering capacitors C_{f1} and C_{f2} discharge to power the load. This state is considered redundant when the pulse voltage U_{ab} is zero.

State 6) $S_1S_2S_3S_4 = 0011$

When both power switches Q_3 and Q_4 are turned ON, the operation conditions of C_{f1} and C_{f2}, and L_f are consistent with those in state 5. Thus, this state is also considered a redundant state for the condition where U_{ab} equals zero.

3.3.2 Control Strategy of the Hybrid Boost Three-Level DC–DC Converter

A. Modulation strategy

According to the operating principle of the hybrid Boost three-level DC–DC converter, the midpoint voltages of the half-bridges can be described by the gate signals, as shown below, where U_{Cf1} and U_{Cf2} represent the capacitor voltages of C_{f1} and C_{f2}, respectively.

$$\begin{cases} U_{\mathrm{ag}} = (1 - S_1 \cdot S_2) \cdot (U_{\mathrm{Cf1}} + U_{\mathrm{Cf2}}) - (S_1 - S_2) \cdot U_{\mathrm{Cf1}} \\ U_{\mathrm{bg}} = S_3 \cdot U_{\mathrm{Cf1}} + S_4 \cdot U_{\mathrm{Cf2}} \end{cases} \tag{3.16}$$

Substituting (3.16) into (3.15), the output voltage U_{ab} can be written as

$$U_{\mathrm{ab}} = [(1 - S_1) \cdot (1 + S_2) - S_3] \cdot U_{\mathrm{Cf1}} + (1 - S_1 \cdot S_2 - S_4) \cdot U_{\mathrm{Cf2}} \tag{3.17}$$

The control scheme for the hybrid converter is developed based on the transition of switching states and the balance of capacitor voltages for C_{f1} and C_{f2}. The modulation strategy is illustrated in Fig. 3.9, where m_a and m_b represent the modulation indexes with two carriers, and t_{on1}–t_{on4} denotes the turn-on intervals of Q_1–Q_4. Carrier 1 and Carrier 2 are phase-shifted to achieve coordinated operation for the half-bridges. As a result, the gate signals can be obtained as follows.

$$\begin{cases} m_b > U_{\mathrm{carrier1}}, & S_1 = 0 \\ m_a > U_{\mathrm{carrier2}}, & S_2 = 1 \\ m_a > U_{\mathrm{carrier1}}, & S_3 = 1 \\ m_b > U_{\mathrm{carrier2}}, & S_4 = 0 \end{cases} \tag{3.18}$$

As a result, the hybrid converter can achieve three-level voltages, namely U_{ag}, U_{bg}, and U_{ab}. In switching states where U_{ab}, the inductor L_{f} is charged, causing the inductor current i_{L} to increase linearly. Conversely, in other switching states, the inductor current decreases. During the switching states represented by $S_1S_2S_3S_4$ = [0000, 1100, 0011], the filter capacitors C_{f1} and C_{f2} are charged and discharged synchronously. This ensures balanced charging and discharging energy between C_{f1} and C_{f2}. However, when $S_1S_2S_3S_4$ = [1000, 0001], $\mathrm{C_{f1}}$ and $\mathrm{C_{f2}}$ are discharged separately. It's crucial to note that if the discharging intervals of C_{f1} and C_{f2} are not equal, the capacitor voltages may become imbalanced, potentially impacting the operational performance of the converter.

According to the modulation strategy, the discharging interval of capacitors can be expressed by the turn-on intervals of power converters as follows.

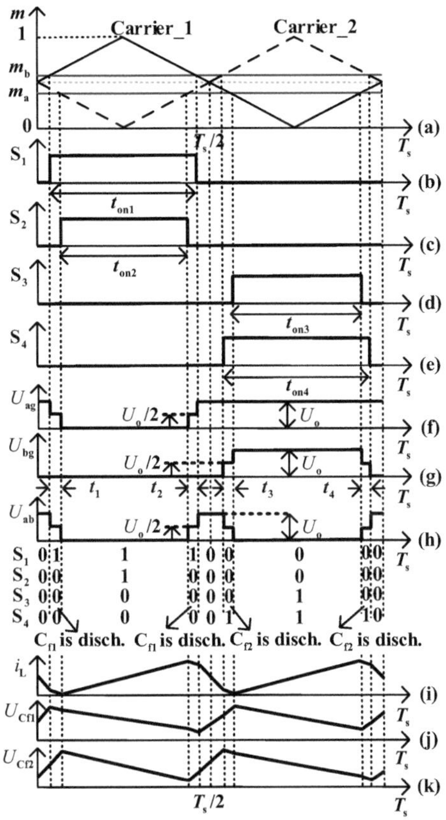

Fig. 3.9 Modulation strategy for the hybrid Boost three-level DC–DC converter

$$\begin{cases} t_1 = t_2 = \dfrac{t_{on1} - t_{on2}}{2} \\ t_3 = t_4 = \dfrac{t_{on4} - t_{on3}}{2} \end{cases} \tag{3.19}$$

Furthermore, given the carriers' symmetry at 0.5, the relationship of turn-on intervals can be described as below.

$$\begin{cases} t_{on1} = t_{on4} \\ t_{on2} = t_{on3} \end{cases} \tag{3.20}$$

Combing (3.19) and (3.20), the discharging time of C_{f1} and C_{f2} is equal, as presented below.

$$t_1 + t_2 = t_3 + t_4 \tag{3.21}$$

In consideration of the equivalent charging/discharging duration and the constant load current, the capacitor voltages of C_{f1} and C_{f2} in the hybrid Boost three-level DC–DC converter can be self-balanced.

B. High voltage gain operation

When the converter operates in a steady state, the energy charged from the input source and discharged to the load through the inductor is equal. It is assumed that the hybrid converter operates in CCM, and the average inductor current I_L remains constant during steady state. Thus, the charged energy W_{ch} and discharged energy W_{disch} of inductor L_f can be calculated as follows.

$$\begin{cases} W_{ch} = 2U_{in}I_L t_{on2} \\ W_{disch} = 2(U_{out} - U_{in})I_L\left(\frac{T_s}{2} - t_{on1}\right) + 2\left(\frac{U_{out}}{2} - U_{in}\right)I_L(t_1 + t_2) \\ W_{ch} = W_{disch} \end{cases} \tag{3.22}$$

Utilizing (3.19) and (3.22), the voltage gain M of the hybrid converter can be obtained as below, where d_1 and d_2 represent the duty cycles of power switches Q_1 and Q_2, respectively.

$$M = \frac{U_{out}}{U_{in}} = \frac{1}{1 - (d_1 + d_2)} \tag{3.23}$$

On the basis of the modulation strategy in Fig. 3.9, the relationship among duty cycles and modulation indexes can be described as follows, where d_3 and d_4 are the duty cycles of power switches Q_3 and Q_4, respectively.

$$\begin{cases} d_1 = d_4 = 1 - m_b \\ d_2 = d_3 = m_a \end{cases} \tag{3.24}$$

Hence, the voltage gain of the converter can be expressed by the modulation indexes as

$$M = \frac{U_{out}}{U_{in}} = \frac{1}{m_b - m_a} \tag{3.25}$$

Consequently, attaining a high voltage gain for the hybrid converter involves configuring modulation indexes with minimal difference, thereby averting the occurrence of extreme duty cycles. Referring to (3.24), it is crucial to adjust the modulation indexes within the range $m_a < 0.5 < m_b$ and $m_a + m_b < 1$. There exists an infinite number of solutions for the selection of m_a and m_b to achieve any desired voltage gain. To mitigate extreme duty cycles and concurrently reduce inductor current ripple, careful consideration must be given to the selection of modulation indexes.

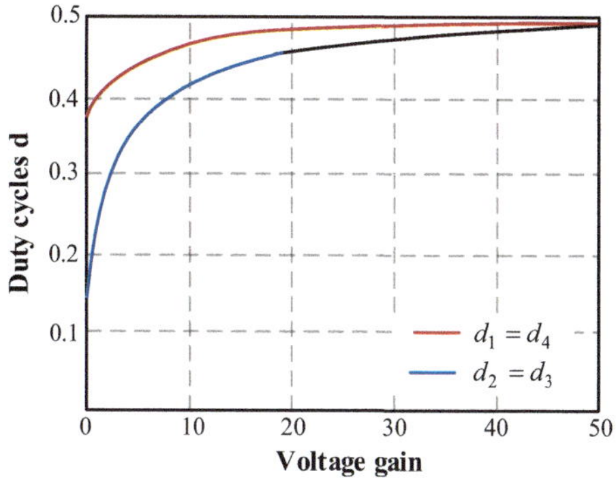

Fig. 3.10 Voltage gain range of the hybrid Boost three-level DC–DC converter

$$\begin{cases} m_{\mathrm{a}} = 0.5 - \dfrac{3}{4M} \\ m_{\mathrm{b}} = 0.5 + \dfrac{1}{4M} \end{cases} \tag{3.26}$$

As depicted in Fig. 3.10, the converter exhibits a wide voltage gain range while maintaining non-extreme duty cycles. Additionally, it is noteworthy that as the voltage gain increases, the duty cycles approach 0.5. Consequently, the hybrid Boost three-level DC–DC converter proves to be well-suited for applications requiring high voltage gain.

3.4 Bidirectional Three-Level DC–DC Converter

In the preceding section of this chapter, the Buck three-level DC–DC converter was presented and analyzed. This converter comprises two asymmetric half-bridges utilizing four power switches. The output pulse voltage is generated from the three-level voltages between the midpoints of the half-bridges and the negative input terminal. Subsequently, the DC output voltage is obtained through an LC filter circuit from the pulse voltage. The Buck three-level DC–DC converter exhibits low voltage stresses across power switches, which are maintained at half the input voltage. The step-down voltage gain is expressed as the following formula, where d_1 and d_2 are the duty cycles of power switches Q_1 and Q_2, respectively.

$$M_{\mathrm{Buck}} = d_1 + d_2 - 1 \tag{3.27}$$

As previously analyzed, the Buck three-level DC–DC converter can achieve a high step-down voltage gain without resorting to extreme duty cycles through proper modulation index adjustment.

Additionally, as discussed in the preceding section, the Boost three-level DC–DC converter exhibits symmetry with the Buck three-level DC–DC converter.

Comprising two half-bridges, this converter generates its output voltage from the three-level midpoint voltages. Similar to the Buck counterpart, the voltage stresses across the power switches in the Boost three-level DC–DC converter are consistently half of the input voltage. The mathematical expression for the step-up voltage gain of the converter is presented below, where d_3 and d_4 are the duty cycles of powers witches Q_3 and Q_4, respectively. Consequently, high voltage gain with non-extreme duty cycles can be achieved in the Boost three-level DC–DC converter.

$$M_{\text{Boost}} = \frac{1}{1 - d_3 + d_4} \tag{3.28}$$

Both the Buck and Boost three-level DC–DC converters are utilized in renewable energy systems, such as fuel cells and PV systems. However, energy storage devices require bidirectional power flow to accommodate both charging and discharging operations. Given that the Buck and Boost three-level DC–DC converters inherently possess symmetrical topologies and operational states, a bidirectional three-level DC–DC converter is derived by combining these configurations [69].

The bidirectional three-level DC–DC converter topology, depicted in Fig. 3.11, comprises eight power switches Q_1–Q_8 and four clamp diodes D_{c1}–D_{c4}. In step-down mode, where energy transfers from the high voltage side to the low voltage side, the main power switches are Q_1, Q_2, Q_7, and Q_8. Conversely, in step-up mode, with power flowing from the low voltage side to the high voltage side, the main power switches are Q_3–Q_6.

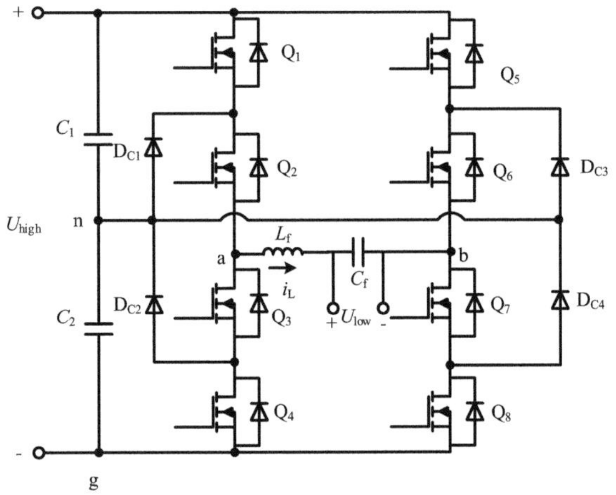

Fig. 3.11 Topology of the bidirectional three-level DC–DC converter

3.4.1 Operating Principle of the Bidirectional Three-Level DC–DC Converter

In a hybrid power system integrating renewable energies and energy storage devices, the bidirectional three-level DC–DC converter serves as the power regulator for devices like batteries and super-capacitors. This converter operates in either step-up or step-down mode to achieve charging or discharging control. Consequently, a bidirectional power control scheme is essential, employing a command signal for mode selection.

The modulation scheme for the bidirectional three-level DC–DC converter is depicted in Fig. 3.12, with m_{a} and m_{b} representing the modulation indexes, while S_1–S_8 denote the gate signals for power switches Q_1–Q_8. Additionally, i_{L} represents the inductor current, and U_{c1} and U_{c2} represent the capacitor voltages of C_1 and C_2, respectively. The two carriers, carrier 1 and carrier 2, are phase-shifted by π. As a result, the generation of the gate signals can be described as follows.

$$\begin{cases} m_b > U_{\text{carrier1}}, & S_1 = 0 \\ m_a > U_{\text{carrier2}}, & S_2 = 1 \\ m_a > U_{\text{carrier1}}, & S_3 = 1 \\ m_b > U_{\text{carrier2}}, & S_4 = 0 \end{cases} \tag{3.29}$$

The duty cycles for power switches in step-down mode can be derived as follows.

$$\begin{cases} d_1 = d_8 = 1 - m_{\text{b}} \\ d_2 = d_7 = m_{\text{a}} \end{cases} \tag{3.30}$$

Conversely, in step-up mode, the duty cycles are determined by the following equation, where d_1–d_8 are the duty cycles of power switches Q_1–Q_8, respectively. In addition, the range of modulation indexes should be $m_a < 0.5 < m_b$ and $m_a + m_b < 1$

$$\begin{cases} d_3 = d_6 = 1 - m_{\text{b}} \\ d_4 = d_5 = m_{\text{a}} \end{cases} \tag{3.31}$$

Therefore, the voltage gain M_{Buck} of the converter in step-down mode can be derived from (3.30) as follows.

$$M_{\text{Buck}} = \frac{1}{m_{\text{a}} - m_{\text{b}}} \tag{3.32}$$

Accordingly, the voltage gain M_{Boost} of the converter in step-up mode can be derived from (3.31) as below.

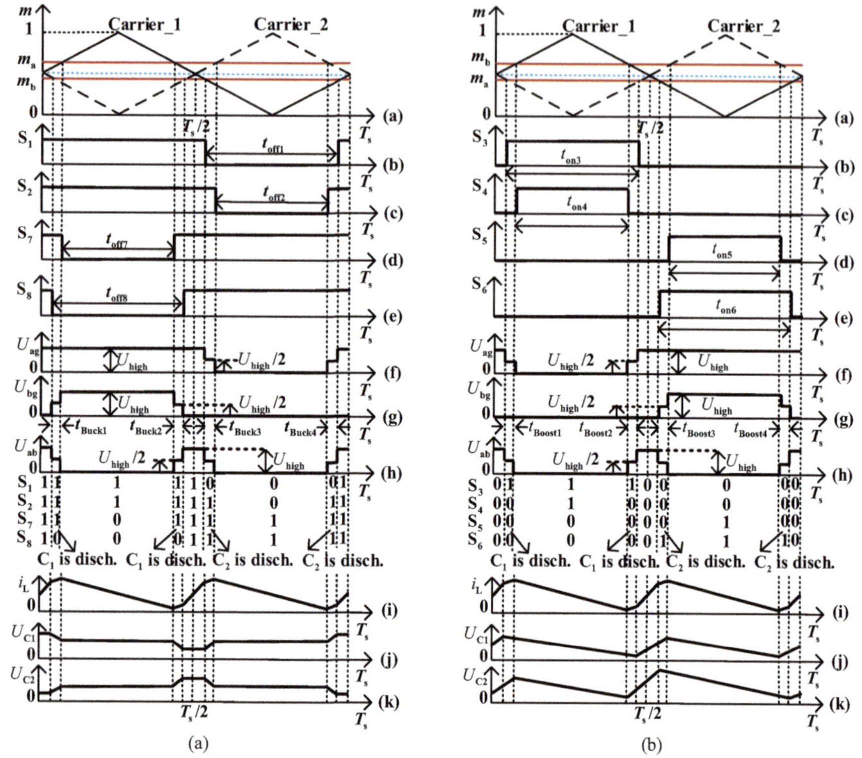

Fig. 3.12 Modulation scheme for the bidirectional three-level DC–DC converter in **a** step-down and **b** step-up mode

$$M_{\text{Boost}} = \frac{1}{m_{\text{b}} - m_{\text{a}}} \tag{3.33}$$

In terms of (3.30) to (3.33), the bidirectional three-level DC–DC converter exhibits a wide voltage gain range in both step-up and step-down modes.

3.4.2 Control Strategy of the Bidirectional Three-Level DC–DC Converter

A. Coordinate control between inductor current ripples and non-extreme duty cycles

The application of conventional DC–DC converters is often constrained by their limited usability at high voltage gain conditions, primarily due to the associated extreme duty cycles. These extreme duty cycles not only contribute to increased power losses but also pose a threat to system stability. In contrast, the bidirectional three-level DC–DC converter offers the advantage of achieving high voltage gain without necessitating extreme duty cycles.

However, as depicted in Fig. 3.12, it is evident that the variation in inductor current is closely tied to the duty cycles. Specifically, longer turn-off intervals in step-down mode or turn-on intervals in step-up mode result in larger inductor current ripples. Therefore, a careful consideration of the trade-off between inductor current ripples and duty cycles is essential. The selection of modulation indexes should be grounded in the desired voltage gain and duty cycle range.

The restriction function of modulation indexes is outlined as follows, where k_{Buck} is the restriction factor in step-down mode.

$$\begin{cases} m_a = 0.5 + \dfrac{0.5}{M_{\text{Buck}}} + \dfrac{k_{\text{Buck}}}{M_{\text{Buck}}} \\ m_b = 0.5 - \dfrac{0.5}{M_{\text{Buck}}} + \dfrac{k_{\text{Buck}}}{M_{\text{Buck}}} \end{cases} \tag{3.34}$$

Accordingly, the restriction function dictating duty cycles in step-down mode is written as (3.35). Then the restriction factor can be calculated by substituting the desired voltage gain and duty cycles.

$$\begin{cases} d_1 = d_8 = 0.5 + \dfrac{0.5 - k_{\text{Buck}}}{M_{\text{Buck}}} \\ d_2 = d_7 = 0.5 + \dfrac{0.5 + k_{\text{Buck}}}{M_{\text{Buck}}} \end{cases} \tag{3.35}$$

Similarly, the modulation indexes in step-up mode are written as the following equation, where k_{Boost} is the restriction factor in step-up mode.

$$\begin{cases} m_a = 0.5 - \dfrac{0.5}{M_{\text{Boost}}} - \dfrac{k_{\text{Boost}}}{M_{\text{Boost}}} \\ m_b = 0.5 + \dfrac{0.5}{M_{\text{Boost}}} - \dfrac{k_{\text{Boost}}}{M_{\text{Boost}}} \end{cases} \tag{3.36}$$

The restriction function for duty cycles in Boost mode in turn can be derived as follows.

$$\begin{cases} d_3 = d_6 = 0.5 + \dfrac{k_{\text{Boost}} - 0.5}{M_{\text{Boost}}} \\ d_4 = d_5 = 0.5 - \dfrac{0.5 + k_{\text{Boost}}}{M_{\text{Boost}}} \end{cases} \tag{3.37}$$

Hence, the restriction factor k_{Boost} can be obtained from the desired range of voltage gain and duty cycles in step-up mode.

Referencing (3.35) and (3.37), it is observed that larger restriction factors lead to shorter inductor discharging intervals in step-down mode and charging intervals in step-up mode. This, in turn, results in smaller inductor current ripple. However, it should be noted that as the restriction factors increase, the duty cycles move further from 0.5. Thus, the optimal approach is to minimize inductor current ripple while ensuring that duty cycles remain within a non-extreme range.

B. Duty cycle perturbation control for capacitor voltages balance

In an ideal scenario, the bidirectional three-level DC–DC converter operates with power switches precisely following the modulation scheme depicted in Fig. 3.12, and the capacitors C_1 and C_2 have identical parameters. Under these conditions, capacitor voltages can naturally achieve balance without the need for additional control. Nevertheless, practical applications introduce inherent disparities in the characteristics and parameters of components within the converter. Factors such as variations in capacitance due to aging and discrepancies in the rise and fall times of power switches, influenced by parasitic parameters, contribute to these differences.

The charging and discharging processes of capacitors C_1 and C_2 in the bidirectional three-level DC–DC converter are inherently complementary. In the step-down mode, C_1 undergoes charging while C_2 experiences discharging when the switch configuration is $S_1S_2S_7S_8 = 0111$. Conversely, during the same step-down mode, C_1 discharges, and C_2 charges when the switch configuration is $S_3S_4S_5S_6 = 1110$. Furthermore, owing to the symmetrical characteristics of the carrier signals, (3.38) can be deduced in the step-down mode.

$$\begin{cases} t_{\text{off1}} = t_{\text{off8}} \\ t_{\text{off2}} = t_{\text{off7}} \end{cases} \tag{3.38}$$

Hence, the charging and discharging durations of capacitors C_1 and C_2 are equal. Moreover, the inductor current remains consistent throughout the first and second half of the switching cycle. As a result, with identical capacitances for C_1 and C_2, the capacitor voltages can self-balance without the need for auxiliary control.

In step-up mode, when $S_1S_2S_7S_8 = 0001$, C_1 is charged while C_2 is discharged, and conversely, when $S_3S_4S_5S_6 = 1000$, C_1 is discharged, and C_2 is charged. Notably, the durations and currents for charging and discharging C_1 and C_2 are equal. Thus, with identical capacitances, the capacitor voltages of C_1 and C_2 can self-balance in step-up mode.

When considering differences in capacitances and power switch rise/fall times, the charging and discharging energy of C_1 and C_2 is unbalanced, necessitating auxiliary voltage balancing control. In step-down mode, for example, if the turn-on instants of Q_1 and Q_7 lag behind the corresponding gate signals, the turn-off intervals of t_{off1} and t_{off7} are extended, as depicted in Fig. 3.13. Consequently, the discharging duration of C_2 becomes shorter than that of C_1, resulting in C_2 releasing more energy than C_1 during each switching period. This imbalance causes the voltage of C_1 to be higher than that of C_2, indicating unbalanced capacitor voltages.

The condition of capacitor voltage unbalancing in step-up mode is presented in Fig. 3.14, where the turn-on instants of Q_4 and Q_5 lag behind the corresponding

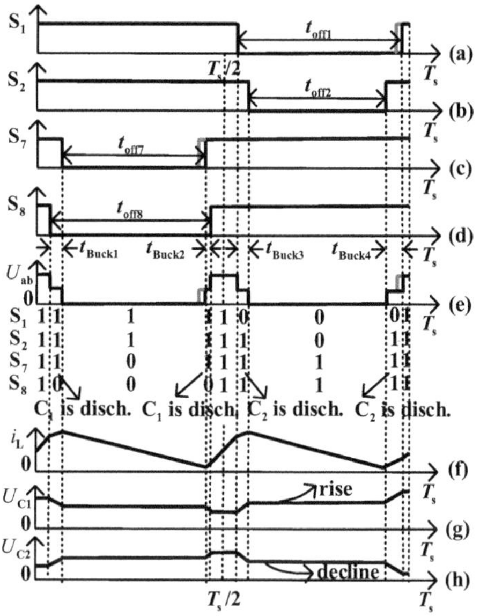

Fig. 3.13 Imbalance of capacitor voltages in step-down mode

gate signals, which resulted in shortened turn-on time of t_{on4} and t_{on5}. Therefore, the discharging time of C_1 is longer than that of C_2, so that C_1 releases more energy than C_2 during each switching period. Thereby, the voltage of C_2 will become higher than that of C_1 finally, that the capacitor voltages are balanced.

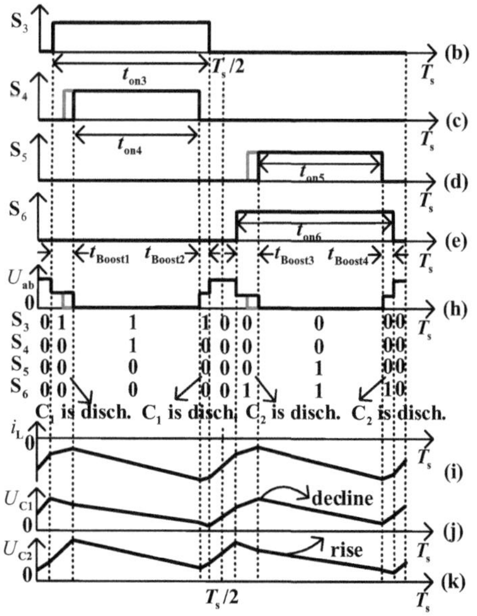

Fig. 3.14 Condition of capacitor voltage unbalancing in Boost mode

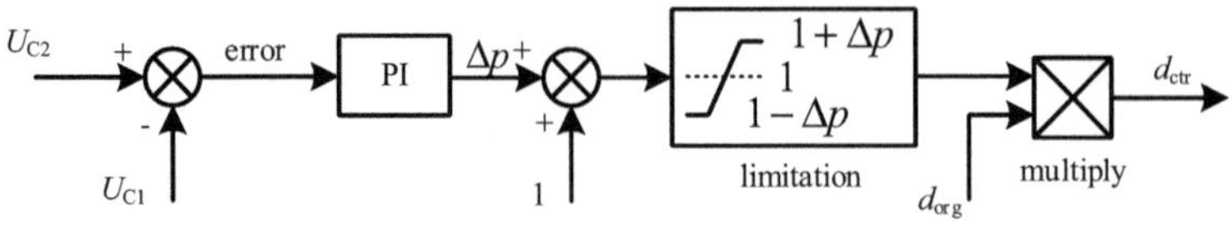

Fig. 3.15 Control diagram of the duty cycle perturbation control for capacitor voltage balancing

Based on the previous analysis, the non-ideal operation of power switches has the potential to result in unbalanced capacitor voltages, posing a significant threat to the system's reliability. To address this issue and rectify the capacitor voltages, an auxiliary control scheme is implemented. This scheme aims to prevent the unbalanced charging and discharging of capacitors by incorporating duty cycle perturbation.

To achieve capacitor voltage balancing in step-down mode, the duty cycle d_2 is chosen as the perturbed parameter. As illustrated in Fig. 3.13, the risk of over-discharge in C_2 is mitigated through negative perturbation of d_2, with the perturbation step calculated from the voltage error between C_1 and C_2. Consequently, the discharging duration of C_2 is restrained by increasing t_{off2}, effectively balancing the capacitor voltages. The control diagram in Fig. 3.15 depicts the process, where Δp represents the perturbation factor generated by the PI controller based on the capacitor voltage error. Moreover, Δp is constrained to prevent oscillation or extreme duty cycles. As a result, the original duty cycle d_{org} is linearly adjusted to d_{ctr}, thereby modifying the charging/discharging durations of the capacitors.

In the case that C_1 is over-discharged in step-up mode, d_4 is positively perturbed to suppress the voltage error, namely d_{ctr} is larger than d_{org}. With the extended t_{on4}, the discharging duration of C_1 in each switching period is shortened. As a result, the capacitor voltage of C_1 and C_2 can be balanced through the duty cycle perturbation strategy in step-up mode as well.

C. Control strategy of bidirectional power flow

The bidirectional three-level DC–DC converter employs distinct modulation schemes in Buck and Boost modes. This design allows for effective charging and discharging regulation of energy storage devices in hybrid power systems. For instance, in electric vehicles, super-capacitors absorb recycled energy and provide pulse power supply, significantly enhancing system efficiency and performance. To realize bidirectional power flow, a bidirectional control strategy is adopted, as illustrated in Fig. 3.16. Here, $SOC_{C\text{-ref}}$ represents the State of Charge (SOC) reference, while SOC_C is the measured SOC. Additionally, $I_{C\text{-ref}}$ denotes the current reference, and I_C and U_C are the sampled current and voltage on the low voltage side, respectively.

The duty of super-capacitors is to provide instantaneous power during acceleration and absorb recycled energy during braking. Consequently, when the vehicle operates at low velocity, the SOC reference of super-capacitors is set higher to store buffer energy for vehicle acceleration. Conversely, at high velocity, the SOC reference of super-capacitors is set lower to reserve capacity for absorbing energy during vehicle

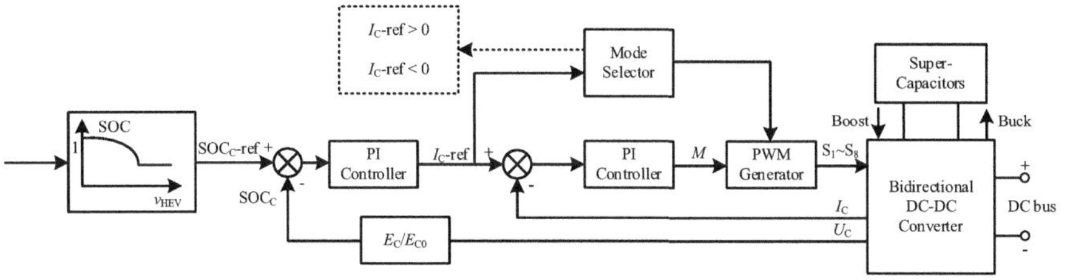

Fig. 3.16 Control strategy of bidirectional power flow

braking. The energy E_c stored in super-capacitors can be obtained as follows, where C and U_c represent the capacitance and voltage of super-capacitors, respectively.

$$E_C = \frac{1}{2} C U_C^2 \tag{3.39}$$

The SOC of the super-capacitors is determined by the ratio E_C/E_{C0}, where E_{C0} represents the energy of the super-capacitors at full charge. A PI controller is employed to generate the reference for the super-capacitors, serving as the SOC regulator. The error between $SOC_{C\text{-ref}}$ and SOC_C is utilized as the input for the outer PI controller, while the error between $I_{C\text{-ref}}$ and I_C serves as the input for the inner PI controller. The mode index for selecting the modulation scheme is subsequently derived from the output of the inner PI controller.

3.5 Quasi-Z-Source Boost Three-Level DC–DC Converter

To address the challenge of limited output voltage, the Z source network can be inte grated into DC–DC converters, enabling higher voltage gain. Figure 3.17 illustrates the topology of a commonly grounded Z-source DC–DC converter, derived from the fundamental Z-source DC–DC converter without the inclusion of extra components. Notably, both the input source and the load are situated on the same side of the Z-source network, as opposed to being on opposite sides. Additionally, the voltage stress on the power switches remains within the range from half of the output voltage to near the output voltage.

The flying-capacitor Buck three-level DC–DC converter depicted in Fig. 3.18a exhibits lower voltage stresses across the power switches compared to the conventional Buck converter. The left terminal of the inductor is denoted as node "x", and the voltage of node "x" with respect to the ground is designated as U_x. When both switches Q_1 and Q_2 are conducting, the voltage U_x equals the input voltage U_{in}. However, when either Q_1 and D_2, or Q_2 and D_1 are conducting, the voltage U_x equals $0.5U_{in}$. In the scenario where both diodes D_1 and D_2 are conducting, the voltage U_x is clamped at zero. Consequently, the converter operates at three voltage levels: zero, $0.5U_{in}$, and U_{in}. This three-level operation allows for a reduction in

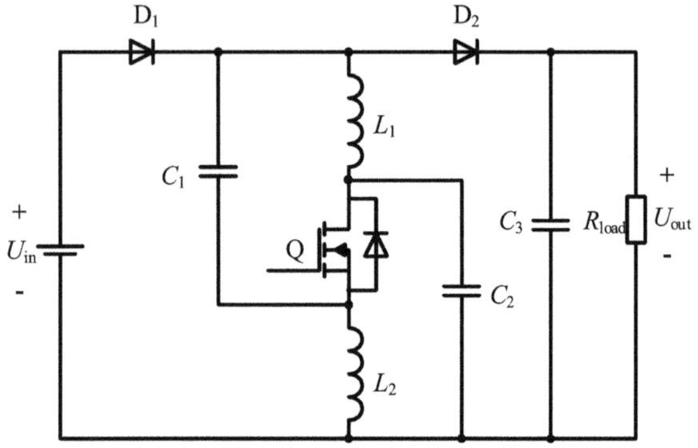

Fig. 3.17 Topology of common grounded Z-source DC–DC converter

the inductor volume for the same output current ripple compared to the conventional Buck converter. Additionally, the topology doubles the effective switching frequency, leading to a smaller magnetic component size. Furthermore, the design enables high efficiency in DC–DC power conversion. Conversely, the topology of the flying-capacitor Boost three-level DC–DC converter is illustrated in Fig. 3.18b.

Based on the commonly grounded Z-source DC–DC converter depicted in Fig. 3.17, the quasi-Z-source (QZS) cell can be derived. This cell comprises two inductors L_1 and L_2, and two capacitors C_1 and C_2, along with a diode D_1. By combining the QZS cell with the flying-capacitor Boost three-level DC–DC converter, the resulting Boost three-level DC–DC converter is illustrated in Fig. 3.19 [70]. This configuration significantly extends the step-up voltage gain range. In addition to the widened voltage gain range, the flying-capacitor C_{fly} reduces voltage stresses across power switches, and the common ground structure ensures higher operational reliability. Consequently, the QZS Boost three-level DC–DC converter is suitable for fuel cell and super-capacitor applications.

3.5.1 Operating Principle of the Quasi-Z-Source Boost Three-Level DC–DC Converter

A. Operating principle

The operation of QZS Boost three-level DC–DC converter contains four switching states that $S_1S_2 = [11, 10, 01, 00]$, where S_1 and S_2 are the gate signals of power switches Q_1 and Q_2, respectively. When the duty cycles d_1 and d_2 of Q_1 and Q_2 operate within the range of $0 < d_1 = d_2 < 0.5$, there are three states during each switching period that $S_1S_2 = [10, 01, 00]$. Whereas, when the range of duty cycles is $0.5 < d_1 = d_2 < 1$, the states during each switching period are $S_1S_2 = [10, 01, 11]$ instead. Since the inductors L_1 and L_2 are charged only during $S_1S_2 = 11$, the duty cycles are supposed to be limited within the range of $0.5 < d_1 = d_2 < 1$.

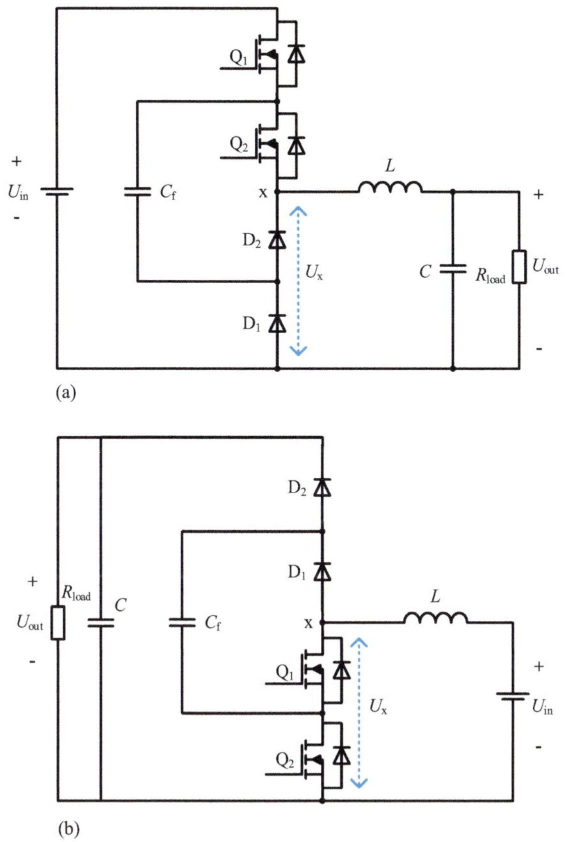

Fig. 3.18 Topology of **a** flying-capacitor Buck three-level DC–DC converter. **b** flying-capacitor Boost three-level DC–DC converter

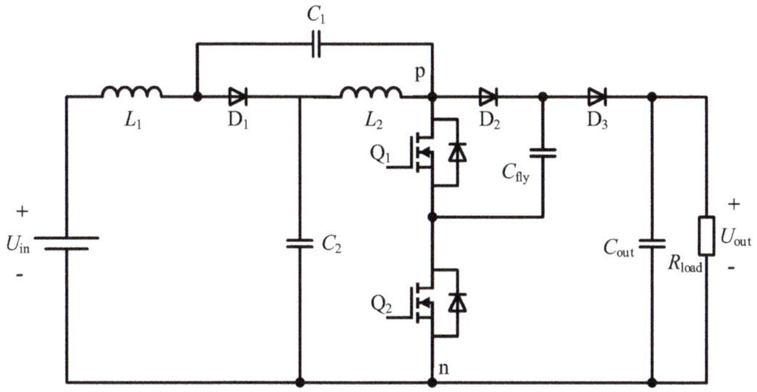

Fig. 3.19 Topology of quasi-Z-source Boost three-level DC–DC converter

The switching states of the QZS Boost three-level DC–DC converter are presented in Fig. 3.20, and the key waveforms are depicted in Fig. 3.21. The switching states of the converter are explained in detail as follows:

State 1) $S_1S_2 = 01$

When Q_1 is turned OFF and Q_2 is turned ON, both the inductors L_1 and L_2 are discharged during this state. Meanwhile, the capacitors C_1 and C_2 are charged, and the flying-capacitor C_{fly} is charged as well. The pulse voltage U_{pn} is equal to the capacitor voltage of C_{fly}.

State 2) $S_1S_2 = 10$

When Q_1 is turned ON and Q_2 is turned OFF, the discharging of inductors L_1 and L_2, as well as the charging of capacitors C_1 and C_2 are consistent with that in State 1. Whereas, the flying-capacitor is discharged to the load through diode D_3. During this state, the pulse voltage U_{pn} is obtained as below, where U_{out} is the output voltage of converter and U_{Cfly} is the capacitor voltage of C_{fly}.

$$U_{pn} = U_{\text{out}} - U_{\text{Cfly}} \tag{3.40}$$

State 3) $S_1S_2 = 11$

When the power switches Q_1 and Q_2 are both turned ON, the inductor L_1 is charged by the input source and C_1, while the L_2 is charged by C_2. The node “p” is connected to “n” during this switching state, and thus, the pulse voltage U_{ab} is equal to zero.

B. Wide voltage gain range operation

To simplify the analysis, the capacitor voltages and the average inductor currents in the QZS Boost three-level DC–DC converter are assumed as constant. According to the operating principle analyzed previously, the voltage of the flying-capacitor C_{fly} is half of the output voltage U_{out}, and the discharging of L_1 and L_2 are synchronous. Specifically, during the switching states that $S_1S_2 = [10, 01]$, the following equations are established by means of KVL, where $u_{\text{L1\_disch}}$ and $u_{\text{L2\_disch}}$ are the discharging voltages of L_1 and L_2, respectively.

$$\begin{cases} U_{\text{in}} + u_{L1\_\text{disch}} + u_{L2\_\text{disch}} = \dfrac{U_{\text{out}}}{2} \\ u_{L2\_\text{disch}} = U_{C1} \\ U_{\text{in}} + u_{L1\_\text{disch}} = U_{C2} \end{cases} \tag{3.41}$$

In the case of $S_1S_2 = [11]$, the inductors L_1 and L_2 are charged. The charging voltages are obtained as follows, where $u_{\text{L1\_ch}}$ and $u_{\text{L2\_ch}}$ are the charging voltages of L_1 and L_2, respectively.

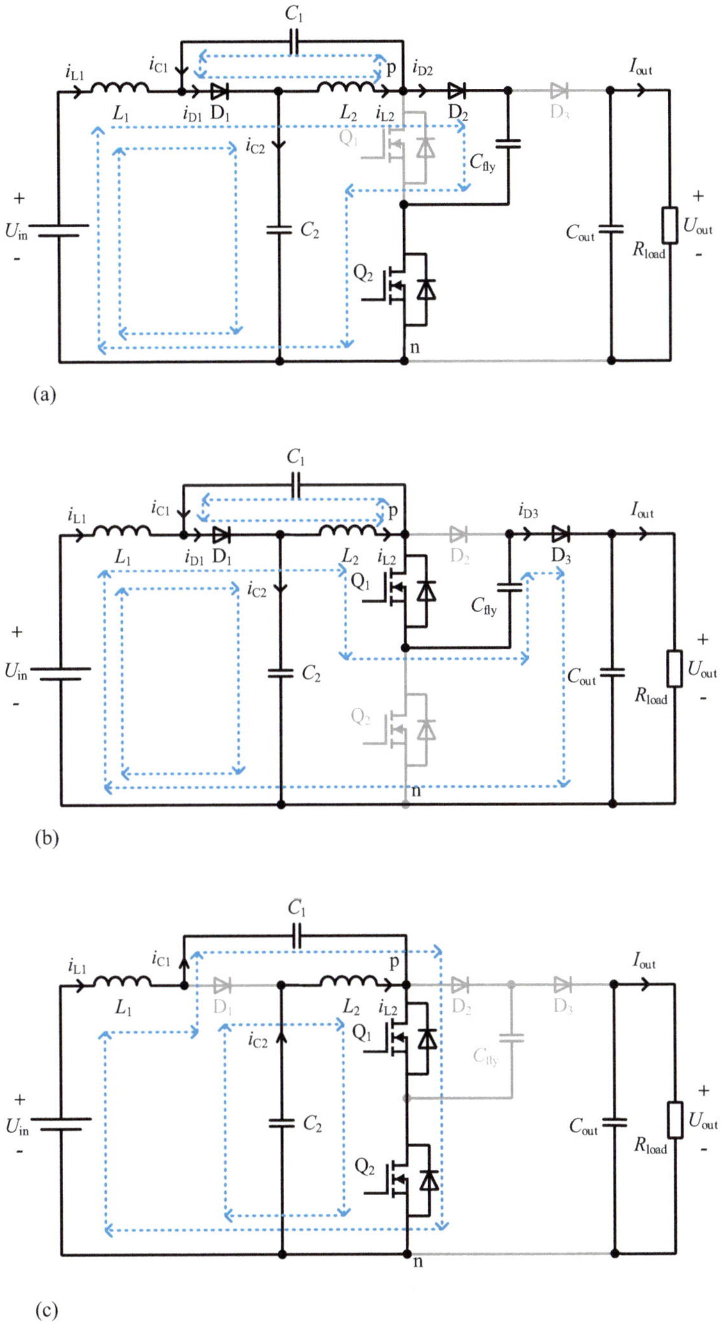

Fig. 3.20 Switching states of the quasi-Z-source Boost three-level DC–DC converter during **a** State 1, **b** State 2, and **c** State 3

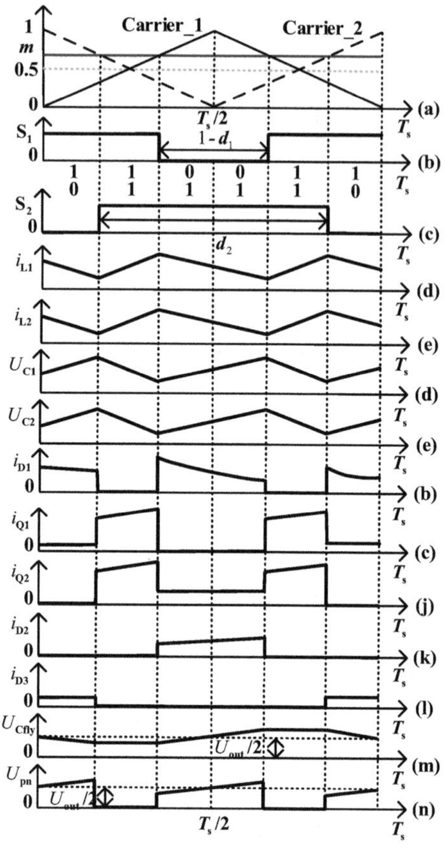

Fig. 3.21 Key waveforms of the quasi-*Z*-source Boost three-level DC–DC converter

$$\begin{cases} U_{\text{in}} + U_{C1} = u_{L1\_\text{ch}} \\ U_{C2} = u_{L2\_\text{ch}} \end{cases} \tag{3.42}$$

Considering that the discharging voltages of inductors L_1 and L_2 are equal, the discharging voltage of L_1 can be derived from (3.41) as below.

$$u_{L1\_\text{disch}} = \frac{\frac{U_{\text{out}}}{2} - U_{\text{in}}}{2} \tag{3.43}$$

In turn, the charging voltage of L_1 can be obtained, which is written as below.

$$u_{L1\_\text{ch}} = \frac{\frac{U_{\text{out}}}{2} + U_{\text{in}}}{2} \tag{3.44}$$

According to the modulation strategy shown in Fig. 3.21, the discharging interval $t_{\text{L1\_disch}}$ of inductor L_1 is derived as follows, where m is the modulation index, T_s

is the switching period, d_1 and d_2 are the duty cycles of Q_1 and Q_2, respectively. Moreover, the value of the equal duty cycles is denoted by d.

$$\begin{cases} t_{L1\_disch} = [(1-d_1)+(1-d_2)]T_s \\ d_1 = d_2 = m = d \end{cases} \tag{3.45}$$

By applying the voltage-second balance principle to inductor L_1, which ensures the conservation of energy during the charging and discharging processes, the following equation can be derived.

$$u_{L1\_disch} t_{L1\_disch} = u_{L1\_ch} t_{L1\_ch} \tag{3.46}$$

Combing (3.43)–(3.46), the step-up voltage gain M of the QZS Boost three-level DC–DC converter can be obtained as below, where the duty cycle d is limited by $0.5 < d < 0.75$.

$$M = \frac{U_{\text{out}}}{U_{\text{in}}} = \frac{2}{3-4d} \tag{3.47}$$

Furthermore, the capacitor voltages of C_1 and C_2 are obtained from (3.42), (3.43), and (3.47) as follows.

$$\begin{cases} U_{C1} = (d-0.5)U_{\text{out}} \\ U_{C2} = (1-d)U_{\text{out}} \end{cases} \tag{3.48}$$

The comparison of voltage gain ranges among the QZS Boost three-level DC–DC converter, the conventional Boost three-level DC–DC converter, the interleaved converter in [57], and the common ground converter in [66] is illustrated in Fig. 3.22. In contrast to previous works, the QZS Boost three-level DC–DC converter exhibits a wider step-up voltage gain within the duty cycle range of [0.5, 0.75). Both the converter in [57] and the conventional Boost three-level converter encounter challenges with extreme duty cycles when the voltage gain exceeds 10. Conversely, the compared converters face difficulty in achieving low voltage gain due to excessively low duty cycles. In contrast, the QZS Boost three-level DC–DC converter can achieve either high or low voltage gain with non-extreme duty cycles, showing significant potential for applications in renewable energies.

C. Self-balance of flying-capacitor voltage

During the switching states that $S_1S_2 = [01, 10]$, the inductor L_2 is discharged and the discharge voltage $u_{\text{L2\_disch}}$ is equal to the capacitor voltage of C_1, namely

$$u_{L2\_disch} = U_{C1} \tag{3.49}$$

Moreover, the flying-capacitor voltage U_{Cfly} in $S_1S_2 = 01$ is written as below.

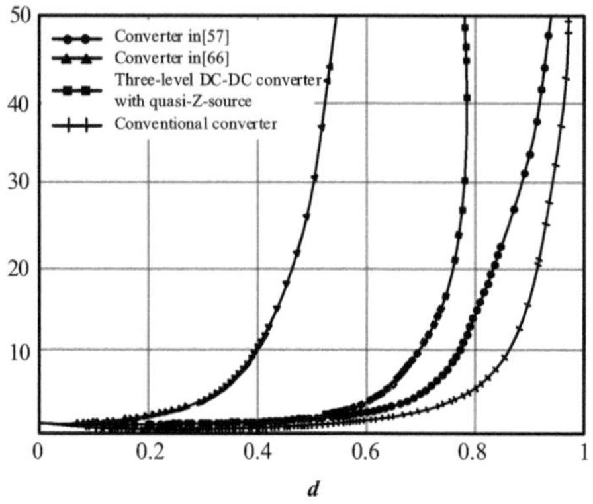

Fig. 3.22 Comparison of voltage gain ranges among the quasi-Z-source Boost three-level DC–DC converter with counterparts

$$U_{\text{Cfly}} = U_{C1} + U_{C2} \tag{3.50}$$

When $S_1S_2 = 10$, U_{Cfly} is written as

$$U_{\text{Cfly}} = U_{\text{out}} - (U_{C1} + U_{C2}) \tag{3.51}$$

During the switching states that $S_1S_2 = 11$, U_{Cfly} remains constant. Therefore, the flying-capacitor voltage can be derived from (3.48), (3.50), and (3.51) as

$$U_{\text{Cfly}} = \frac{U_{\text{out}}}{2} \tag{3.52}$$

According to the preceding analysis of the operating principles of the converter, the voltage of the flying capacitor is clamped by the sum of U_{C1} and U_{C2} from the QZS cell. Remarkably, the flying-capacitor voltage can self-balance at half the output voltage during either transient or steady states without additional voltage balancing control. Furthermore, the voltage stresses across all semiconductors are also equal to half the output voltage.

3.5.2 Steady-State Characteristics of the QZS Boost Three-Level DC–DC Converter

In terms of the flying-capacitor voltage, U_{Cfly} is half the output voltage U_{out} while equal to the sum of U_{C1} and U_{C2}, as described in (3.52). Based on the analysis on switching states of the converter, the voltage stresses across power switches and diodes can also be obtained. When Q_1 is turned OFF and Q_2 is turned on (i.e., $S_1S_2 = 01$), the diode D_3 is OFF. During this interval, the voltages across Q_1 and D_3 are equal to U_{Cfly} and (U_{out}–U_{Cfly}), respectively. Whereas, when Q_1 is turned ON and Q_2 is turned OFF, the diode D_2 is OFF. Meanwhile, the voltages across Q_2 and D_2

are clamped by $(U_{out} - U_{Cfly})$ and U_{Cfly}, respectively. However, when Q_1 and Q_2 are both turned ON, the diodes D_1–D_3 are all OFF. Thus, the voltages across D_1–D_3 are $(U_{C1} + U_{C2})$, $U_{Cfly,}$ and $(U_{out} - U_{Cfly})$, respectively.

As a result, the voltage stresses across power switches and diodes are listed as below, where U_{Q1}, $U_{Q2,}$ and U_{D1}–U_{D3} are the voltage stresses across Q_1, $Q_{2,}$ and D_1–D_3, respectively.

$$\begin{cases} U_{Q1} = U_{Cfly} = \dfrac{U_{out}}{2} \\ U_{Q2} = U_{out} - U_{Cfly} = \dfrac{U_{out}}{2} \\ U_{D1} = U_{C1} + U_{C2} = \dfrac{U_{out}}{2} \\ U_{D2} = U_{Cfly} = \dfrac{U_{out}}{2} \\ U_{D3} = U_{out} - U_{Cfly} = \dfrac{U_{out}}{2} \end{cases} \tag{3.53}$$

By applying ampere-second principle to the capacitors C_{fly} and C_{out}, the current stresses of power switches and diodes are obtained as follows, where I_{out} is the output load current, I_{Q1}, $I_{Q2,}$ and I_{D1}–I_{D3} are the current stresses of Q_1, $Q_{2,}$ and D_1–D_3, respectively.

$$\begin{cases} I_{Q1} = \dfrac{4I_{out}}{3-4d} \\ I_{Q2} = \dfrac{4I_{out}}{3-4d} \\ I_{D1} = \left(\dfrac{4}{3-4d} \quad \dfrac{1}{1-d}\right) I_{out} \\ I_{D2} = \dfrac{I_{out}}{1-d} \\ I_{D3} = \dfrac{I_{out}}{1-d} \end{cases} \tag{3.54}$$

Moreover, the current stress of D_{FC} is equal to the average inductor current I_{L1}, which is written as below.

$$I_{DFC} = I_{L1} = \frac{2}{3-4d} I_{out} \tag{3.55}$$

3.5.3 Synchronous Rectification of the QZS Boost Three-Level DC–DC Converter

According to steady-state analysis, the QZS Boost three-level DC–DC converter exhibits advantages such as low voltage stresses across power switches and diodes, which proves beneficial in reducing switching losses. To further mitigate conducting losses, a decrease in the on-resistance of semiconductors is anticipated. Diodes typically have higher conducting resistance and voltage drop, leading to increased conducting losses compared to power switches.

By means of Kirchhoff's Current Law, when Q_1 is turned OFF and Q_2 is turned ON, the currents i_{D1} and i_{D2} that flows through diodes D_1 and D_2 are obtained as follows, where i_{L1} and i_{L2} are the inductor currents of L_1 and L_2, respectively.

$$\begin{cases} i_{\mathrm{D1}} = i_{\mathrm{L1}} + i_{\mathrm{L2}} - i_{\mathrm{D2}} \\ i_{\mathrm{D2}} = i_{\mathrm{L1}} - i_{\mathrm{C2}} \end{cases} \tag{3.56}$$

In addition, when Q_1 is turned ON and Q_2 is turned OFF, the currents i_{D1} and i_{D3} that flows through diodes D_1 and D_3 are derived as follows, where $i_{\mathrm{C2}} > 0$, and i_{L1} and i_{L2} are equal.

$$\begin{cases} i_{\mathrm{D1}} = i_{\mathrm{L1}} + i_{\mathrm{L2}} - i_{\mathrm{D3}} \\ i_{\mathrm{D3}} = i_{\mathrm{L1}} - i_{\mathrm{C2}} \end{cases} \tag{3.57}$$

When Q_1 and Q_2 are both turned OFF, all diodes are OFF. Combining (3.56) and (3.57), the relationships between diode currents and inductor currents can be deduced as below.

$$\begin{cases} i_{\mathrm{D2}} < i_{\mathrm{L1}} \quad \text{when} \quad S_1 S_2 = 01 \\ i_{\mathrm{D3}} < i_{\mathrm{L1}} \quad \text{when} \quad S_1 S_2 = 10 \\ i_{\mathrm{D1}} > i_{\mathrm{L1}} \quad \text{when} \quad S_1 S_2 = [01, \quad 10] \end{cases} \tag{3.58}$$

While the diode currents of D_2 and D_3 are smaller than the inductor current, the diode current of D_1 in the QZS cell exceeds the inductor current. Consequently, the conducting loss of D_1 inevitably surpasses that of D_2 and D_3. To mitigate the diode conducting loss and improve converter efficiency, D_1 is replaced with the synchronous rectifier Q_{SR}, with D_{SR} serving as the antiparallel diode. The topology of the QZS Boost three-level DC–DC converter with synchronous rectifier is depicted in Fig. 3.23. Due to the lower conducting resistance of MOSFET compared to diodes, synchronous rectification control for Q_{SR} effectively reduces conducting losses. It's worth noting that the voltage stress across Q_{SR} remains identical to that of the original diode D_1.

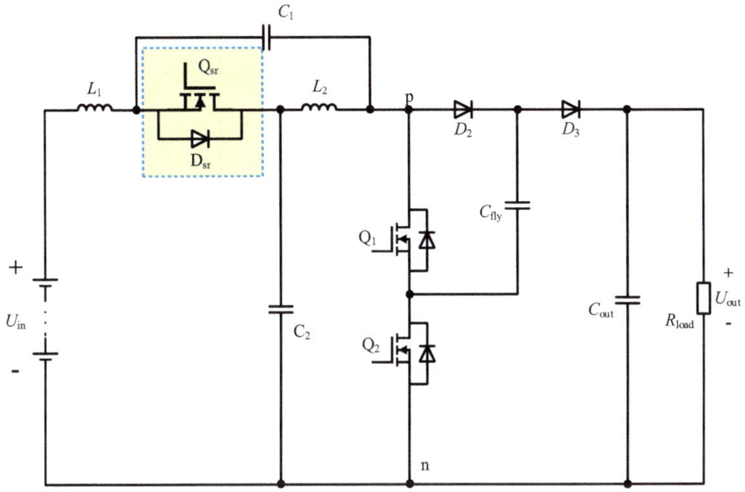

Fig. 3.23 Topology of QZS Boost three-level DC–DC converter with synchronous rectifier

$$U_{\mathrm{QSR}} = U_{C1} + U_{C2} = \frac{U_{\mathrm{out}}}{2} \tag{3.59}$$

The synchronous rectifier Q_{SR} is driven by the complementary gate signal derived from S_1 and S_2, as depicted in Fig. 3.24. The dead time t_{d} is introduced between the synchronous rectification gate signal S_{SR} and S_1S_2, so as to avoid the short circuit fault during switching state transition. As a result, the turn-on instance of Q_{SR} lag behind the turn-off instances Q_1 and Q_2, and vice versa. Furthermore, the duration of dead time depends on the dead time modulation index m_{d}, which is written as below, where T_{s} is the switching period.

$$t_{\mathrm{d}} = m_{\mathrm{d}} \times \frac{T_{\mathrm{s}}}{2} \tag{3.60}$$

Furthermore, owing to the fact that Q_{SR} is turned ON after D_{SR} is conducted, the ZVS ON operation is attained for Q_{SR}. The turn-on voltage of Q_{SR} is the forward voltage drop of D_{SR}, which is relatively small. Conversely, as Q_{SR} is turned OFF ahead of D_{SR}, following which D_{SR} continues conducting. Thus, the turn-off of Q_{SR} is also considered as ZVS.

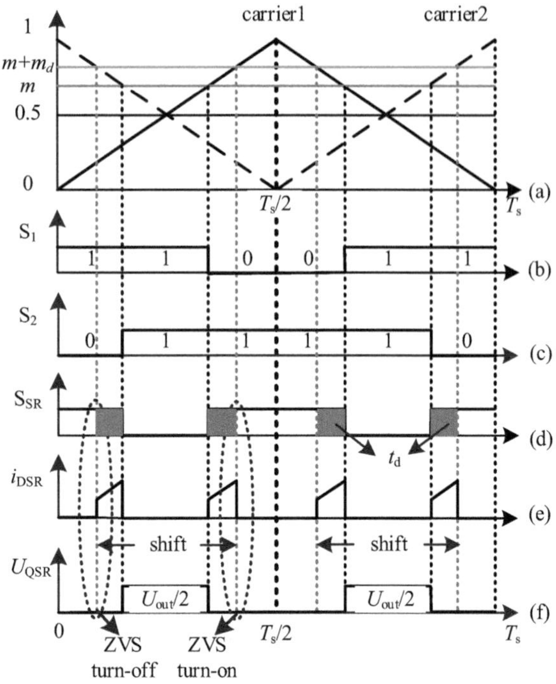

Fig. 3.24 Synchronous rectification control for quasi-Z-source Boost three-level DC–DC convert

3.5.4 Enhanced Hybrid Switching-Frequency Modulation Strategy of the QZS Boost Three-Level DC–DC Converter

In the operation of the QZS Boost three-level DC–DC converter, which utilizes two power switches, there exist four carrier modulation modes. These modes pertain to the modulation strategy involving two carriers: one with the same frequency but phase-shifted, one with both the same frequency and phase, another with different frequencies and phases, and finally, one with different frequencies but the same phase. Additionally, there are alternative methods for deriving modulation indexes. Consequently, the modulation strategy for the QZS Boost three-level DC–DC converter demonstrates considerable flexibility.

The preceding discussion in this section employed equal duty cycles with a 180° phase-shift for power switches Q_1 and Q_2. However, it omitted consideration of the switching state $S_1S_2 = 00$. To address this absence, the QZS Boost three-level DC–DC converter utilizes a hybrid switching-frequency modulation strategy in this context. This strategy is employed to enhance performance by incorporating the previously omitted switching state. Under the hybrid modulation strategy, the carrier frequency of S_2 is twice that of S_1, and both S_1 and S_2 adopt the same modulation index.

The hybrid switching-frequency modulation strategy for QZS Boost three-level DC–DC converter is illustrated in Fig. 3.25, where each switching period contains six switching states, which are $S_1S_2 = [11, 10, 11, 01, 00, 01]$. Thus, the equivalent frequency of inductor current is twice the switching frequency of Q_1. When Q_1 is turned OFF and Q_2 is turned ON, the flying-capacitor voltage is equal to the sum of capacitor voltages of C_1 and C_2, as presented below.

$$U_{\text{Cfly}} = U_{C1} + U_{C2} \tag{3.61}$$

whereas, when both Q_1 and Q_2 are turned OFF, the flying-capacitor voltage can be written as follows.

$$U_{\text{Q2}} = U_{\text{C1}} + U_{\text{C2}} - U_{\text{Cfly}} \tag{3.62}$$

According to (3.61) and (3.62), the voltage stress across Q_2 is zero when $S_1S_2 = 00$. Thus, Q_2 is ZVS OFF from $S_1S_2 = 01$ to $S_1S_2 = 00$. Furthermore, Q_2 is also ZVS ON from $S_1S_2 = 00$ to $S_1S_2 = 10$. As a result, by applying the hybrid

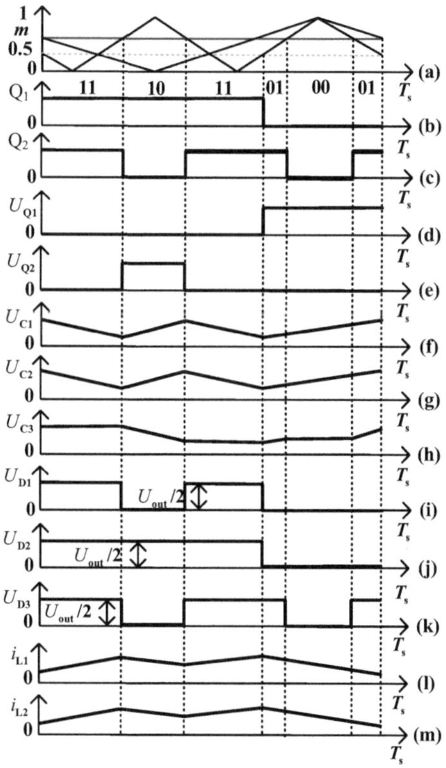

Fig. 3.25 Hybrid switching-frequency modulation strategy for quasi-Z-source Boost three-level DC–DC converter

switching-frequency modulation strategy, the switching losses of the converter can be significantly reduced by achieving ZVS.

3.6 Efficiency and Application Analysis

3.6.1 *Efficiency Analysis*

To verify the operational efficiency of the various three-level DC–DC converter topologies introduced in this chapter, experimental prototypes were constructed for performance analysis. The efficiency of the transformerless Buck three-level DC–DC converter was measured at different power levels and plotted in Fig. 3.26, with an input voltage of 500 V and an output voltage of 68 V. As illustrated, within the output power range of 400–1000 W, the prototype's efficiency increases with higher output power. However, beyond 1000 W, the conversion efficiency of the prototype begins to decline. The maximum efficiency reaches ~ 91.4% at an output power of 1000 W, while the minimum efficiency is around 88.1% at an output power of 400 W.

Figure 3.27 shows the measured efficiency of the hybrid Boost three-level DC–DC converter across the full load range, with an input voltage of 50 V and an output voltage of 600 V. The prototype's switching frequency is designed at 11.5 kHz, and the rated power level is 1 kW. As depicted in the figure, the maximum operational efficiency is 93.1% at an output power of 360 W. The conversion efficiency decreases as the output power decreases from 360 W to light load. However, when the output power increases to the rated level, the operational efficiency declines to 90.3% at full load. Thus, the hybrid Boost three-level DC–DC converter demonstrates an overall high efficiency across a wide voltage gain range.

The prototype of the bidirectional three-level DC–DC converter is designed with a rated power level of 1.2 kW, and the voltage on the high voltage side is controlled at 400 V. The DC source on the low voltage side varies within the range of 60–160 V.

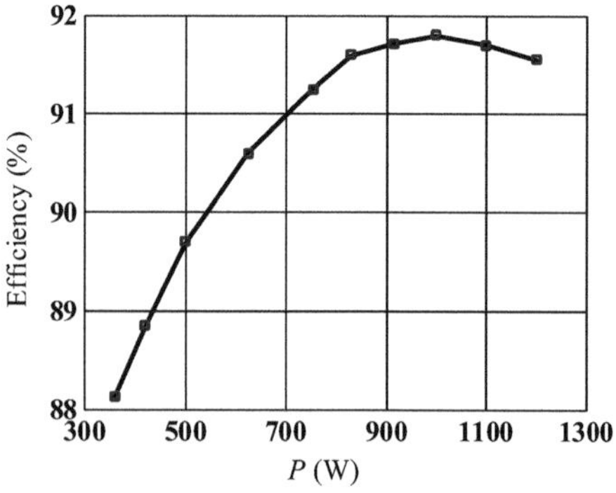

Fig. 3.26 Efficiency of the transformerless Buck DC–DC converter

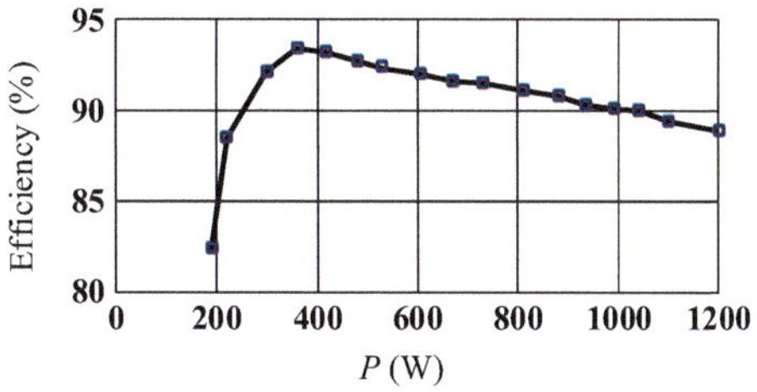

Fig. 3.27 Efficiency of the hybrid Boost three-level DC–DC converter

Moreover, the switching frequency is set at 10 kHz. The conversion efficiency of the prototype in both step-down and step-up modes is plotted in Fig. 3.28a and b, respectively.

As depicted in Fig. 3.28, the conversion efficiency of the converter increases in both step-down and step-up modes as the voltage on the low voltage side increases. Additionally, when the low voltage side is within the range of 120–160 V, the overall efficiency is higher at heavier loads. However, when the low voltage side is lower than 120 V, the overall efficiency is higher at lighter loads. The maximum and minimum operation efficiency in step-down mode is 94.9 and 83.4%, respectively. In contrast,

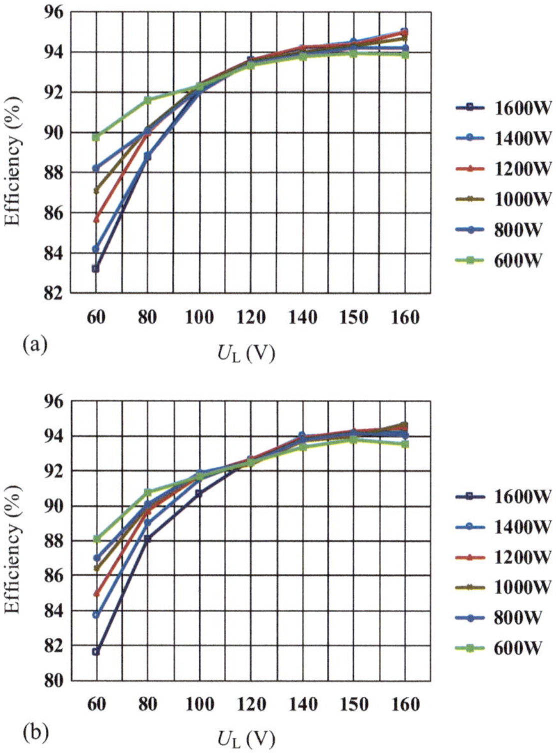

Fig. 3.28 Efficiency of the bidirectional DC–DC converter **a** in step-down mode **b** in step-up mode

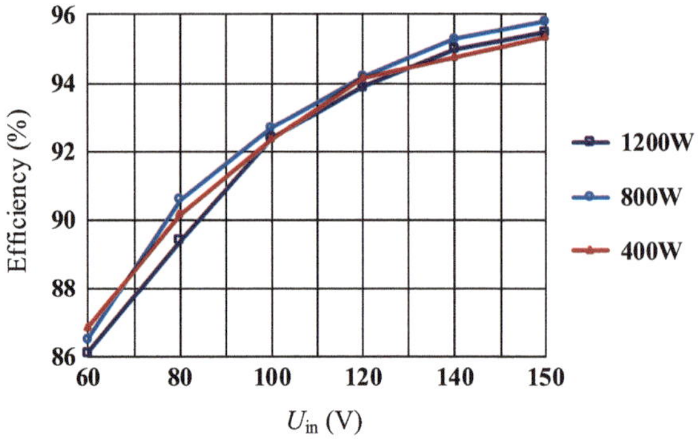

Fig. 3.29 Efficiency of quasi-Z-source Boost three-level DC–DC converter

the maximum and minimum operation efficiency in step-up mode are 94.5 and 81.2%, respectively.

The operation performance of the quasi-Z-source Boost three-level DC–DC converter is demonstrated by a prototype with a rated power level of 1.2 kW. The output voltage is controlled at 400 V through a closed voltage loop, while the input voltage varies within the range of 60–150 V. The switching frequency is set at 10 kHz.

The conversion efficiency of the quasi-Z-source Boost three-level DC–DC converter at various power levels is depicted in Fig. 3.29. It is evident that as the input voltage increases from 60 to 150 V, resulting in a decrease in voltage gain, the operation efficiency correspondingly improves. The utilization of synchronous rectification leads to a significant reduction in conduction losses, contributing to the overall high efficiency of the converter. Specifically, the minimum efficiency is 85% with an input voltage of 60 V and an output power of 12 kW. In contrast, the maximum efficiency reaches 95.66% with an input voltage of 150 V and an output power of 800 W.

3.6.2 Application Analysis

The structures of the transformerless Buck three-level DC–DC converter and the hybrid Boost three-level DC–DC converter are symmetrical. Furthermore, the transformerless Buck three-level topology operates in step-down mode, with energy flowing from the high voltage side to the low voltage side. On the other hand, the hybrid Boost three-level topology converts energy from the low voltage side to the high voltage side. The unidirectional power flow confines these converters to applications in energy storage systems, where energy storage devices require charging and discharging. Nevertheless, they are suitable for photovoltaic systems and hybrid energy systems with fuel cells.

The bidirectional three-level DC–DC converter introduced in this chapter is derived from the unidirectional topologies, inheriting the same voltage gain range and voltage stresses across power switches. Because the converter can achieve high voltage gain and low voltage stresses, it is suitable for high voltage level applications. However, due to the large number of semiconductor components, the three-level DC–DC converter mentioned above has relatively higher costs and a larger volume. Therefore, this family of topologies is more suitable for ground applications than mobile equipments.

Regarding the quasi-Z-source Boost three-level DC–DC converter, it features a common ground structure and can achieve high operation efficiency within a wide load range. However, the converter can only operate in step-up mode, transferring energy from the low voltage side to the high voltage side. Consequently, the converter is applicable to photovoltaic systems and fuel cell vehicles.

3.7 Conclusion

This chapter presents and analyzes a variety of three-level DC–DC converters. The transformerless Buck three-level DC–DC converter demonstrates the ability to operate over a wide input voltage range. Similarly, the hybrid Boost three-level DC–DC converter can achieve a high voltage gain with non-extreme duty cycles. Building upon the principles of the Buck and Boost three-level DC–DC converters, the bidirectional DC–DC converter allows for power flow in both operation modes.

Concerning three-level DC–DC converters, the voltages of series-connected capacitors can self-balance, and the voltage stresses across power switches are half of the high-side voltage. Additionally, this chapter introduces the quasi-Z-source Boost three-level DC–DC converter, which boasts advantages such as low voltage stresses and a common ground between the negative terminals of the input and output. The flying-capacitor voltage is clamped at half the output voltage via the quasi-Z-source cell. Moreover, operational efficiency is significantly enhanced by applying synchronous rectification. This approach enables both ZVS ON and OFF for the synchronous rectifiers. Consequently, the three-level DC–DC converters discussed in this chapter are well-suited for applications involving renewable energies.

Chapter 4
Quasi-Z-Source DC–DC Converters

4.1 Introduction

In the field of three-level DC–DC converters, various challenges and limitations have been observed. While the voltage stresses on power switches are significantly reduced, the practical voltage ratios in both step-up and step-down modes are limited due to parasitic parameters [71, 72]. Multilevel converters can achieve a wider voltage gain range. However, they require additional power switches, hardware circuits, and control strategies to maintain voltage stress balance on the switches [73]. The Cuk/Sepic/Zeta converters exhibit a wider voltage gain range, but their efficiency is constrained by the cascaded structures [74, 75]. Coupled-inductor converters, despite their high voltage gain potential by adjusting the turns ratio, suffer from increased complexity due to the need for more power switches and the challenge of leakage inductance [76–78].

Alternatively, the Z-source, quasi-Z-source, and switched-capacitor DC–DC converters offer simpler structures and control schemes, allowing for easy expansion and high voltage gain by utilizing capacitors to deliver energy through different paths during charge and discharge processes [79]. Switched-inductor bidirectional converters also provide a wide voltage gain range and low voltage stress while avoiding extreme duty cycles, but they sacrifice power density due to the requirement of additional inductors [80].

Recent research efforts have proposed novel solutions to address these limitations. One such solution is the non-isolated single capacitor bidirectional converter, which achieves a wide voltage gain range but at the expense of increased voltage stress on power switches [81]. A bidirectional switched-capacitor converter enhances efficiency, but it necessitates additional power switches [82]. Additionally, a novel coupled-inductor bidirectional converter with increased voltage gain has been proposed, but concerns regarding leakage inductance, dv/dt problems between input and output grounds, and high voltage stresses on power switches near the high voltage side remain [83].

Y. Zhang and S. Gao, *High Efficiency Non-isolated DC-DC Converters with Wide Voltage Gain Range for Renewable Energies*, CPSS Power Electronics Series,
https://doi.org/10.1007/978-981-97-0648-8_4

To address the need for higher and wider voltage gain ranges, several techniques and approaches have been proposed in the field of DC–DC converters. Switched-capacitor configurations have been widely utilized to elevate the output voltage [84, 85]. In [86], a novel bidirectional Buck-Boost converter is introduced, capable of achieving a wide voltage gain range while enabling ZVS operation. Quasi-Z-source networks are adopted in [87] to reduce voltage stresses across components.

In the pursuit of soft-switching operation, various techniques have been explored, such as incorporating auxiliary circuits [88] or coupled-inductors [89, 90]. By combining CIs and auxiliary circuits, the proposed converter in [91] achieves soft-switching operation for all power switches within a specific voltage range. However, this soft-switching capability is not maintained for lower voltages on the low voltage side or at lower power levels. In [92], the bidirectional DC–DC converter achieves soft-switching operation over a wide voltage gain range through the use of an auxiliary circuit. Nevertheless, there are voltage spikes on the power switches resulting from leakage inductance. In [93], a solution is presented that achieves full-load range ZVS for all power switches with low current stresses, while ensuring ZCS OFF for all diodes. However, power switches in the auxiliary circuit do not achieve ZVS, and the output voltage is subject to limitations.

In order to achieve higher voltage gains and wider operation ranges, various modifications have been made to conventional converters. These modifications include the use of three-level, four-level [94], or multilevel converters [95], which require additional capacitors and power switches. However, these multilevel converters often come with the drawback of increased complexity, requiring more power switches, additional hardware circuits, and control strategies to maintain voltage balance and ensure efficient operation. Additionally, converters based on Cuk/Sepic/Zeta topologies can be adapted for bidirectional operation, but their unique structures can limit their overall efficiency.

Another approach involves retrofitting simple structure converters, such as buck-boost converters, with additional switched capacitors or inductors to create non-isolated hybrid DC–DC converters capable of achieving higher voltage gains [96, 97]. On the other hand, Z-source, quasi-Z-source, and switched-capacitor DC–DC converters offer simple and expandable structures. These converters utilize capacitors to deliver energy through different paths during charging and discharging processes, allowing for high voltage gains [98–100]. Switched-inductor DC–DC converters can achieve wide voltage gain ranges while maintaining lower voltage stresses on the power switches, albeit at the expense of lower power densities.

In this chapter, a novel switched-quasi-Z-source bidirectional DC–DC converter is presented. It is based on the conventional two-level quasi-Z-source converter, with the main power switch position modified. The converter offers a simple structure, wide voltage gain range, and low voltage stress on power switches. It enables the selection of power switches with lower rated voltage and lower on-state resistance, resulting in improved conversion efficiency. Furthermore, the converter features a common ground, eliminating potential EMI issues.

Furthermore, an optimization design method for inductance parameters and a variable switching frequency control strategy for the switched-quasi-Z-source bidirectional converter is introduced. Through the realization of zero-voltage switching (ZVS) and quasi-zero-current switching, the converter achieves a wide voltage gain range, low voltage stresses, and reduced electromagnetic noise. This leads to high conversion efficiency without the need for additional components.

Moreover, a hybrid switched-capacitor switched-quasi-Z-source bidirectional DC–DC converter is discussed. It serves as a power interface between the high voltage DC bus and the low voltage super-capacitor bank in EVs. The hybrid energy sources enable the exchange of high-frequency instantaneous power during various EV operations, reducing battery degradation impact and improving the dynamic response of the hybrid energy sources system.

4.2 Switched-Quasi-Z-Source Bidirectional DC–DC Converter

The conventional quasi-Z-source DC–DC converter, as depicted in Fig. 4.1, is derived from quasi-Z-source inverters known for their ability to achieve high voltage gain with a small size and low cost. However, the drawback is that the voltage stress across the main power switch is equal to that of the high voltage side, limiting its potential in high voltage applications.

To address this limitation and maintain a wide voltage gain range, the topology of the switched-quasi-Z-source bidirectional DC–DC converter, introduced in this section, is derived from the conventional quasi-Z-source DC–DC converter by modifying the position of the main power switch, as illustrated in Fig. 4.2 [101]. Additionally, the diodes in the conventional quasi-Z-source DC–DC converter are replaced by power switches, resulting in the derived converter featuring three power switches.

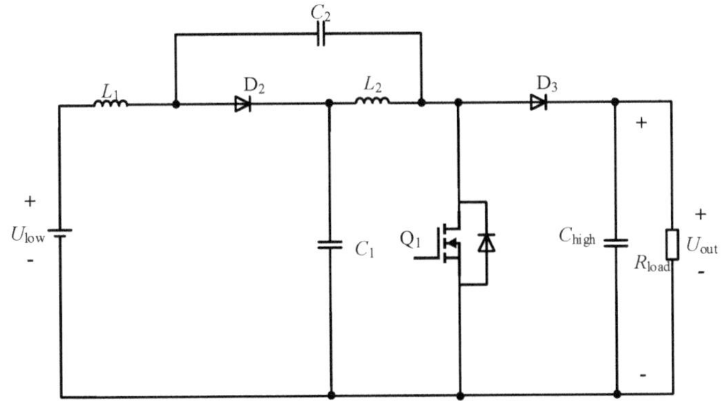

Fig. 4.1 Topology of the quasi-Z-source DC–DC converter

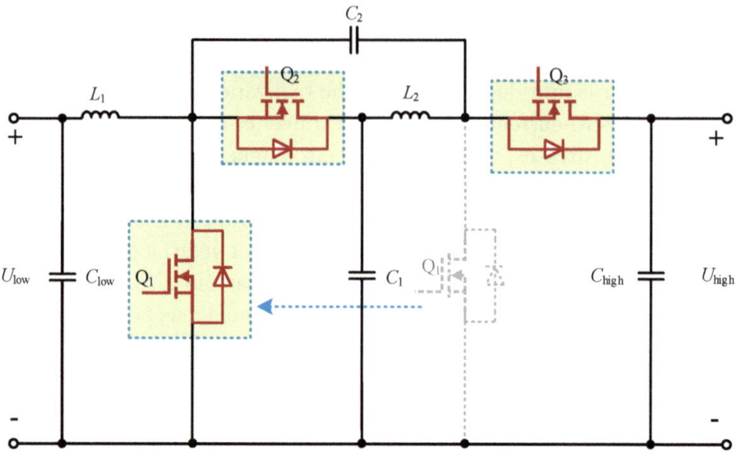

Fig. 4.2 Derivation of the switched-quasi-Z-source bidirectional DC–DC converter

This modification simplifies the structure while enabling a wide voltage gain range and reducing voltage stresses. Moreover, the common ground between the negative terminals of the input and output helps mitigate EMI issues in the derived converter. This feature contributes to achieving high conversion efficiency.

The topology of the switched-quasi-Z-source bidirectional DC–DC converter is shown in Fig. 4.3, where the switched-quasi-Z-source cell consists of power switch Q_2, inductors L_1 and L_2, and capacitors C_1 and C_2. The converter is composed by the switched-quasi-Z-source cell and power switches Q_1 and Q_3, and filter capacitors C_{low} and C_{high}. The power switches Q_1–Q_3 are driven by the gate signals S_1-S_3, while S_2 and S_3 are synchronous and complementary to S_1. The switched-quasi-Z-source converter is capable of bidirectional operation, transferring energy in either step-up or step-down mode. As a result, the converter can be applied in renewable energies and energy storage devices.

This section introduces and examines the switched-quasi-Z-source bidirectional DC–DC converter for application in hybrid energy sources. The topology is derived from the conventional two-level quasi-Z-source bidirectional DC–DC converter through modifications to the main power switch's position and the replacement of diodes with power switches. This converter exhibits a simplified structure, enabling the attainment of a wide voltage gain range and reduced voltage stresses. Notably, the incorporation of a common ground between the negative terminals of the input and output can avoid extra EMI issues, leading to enhanced conversion efficiency.

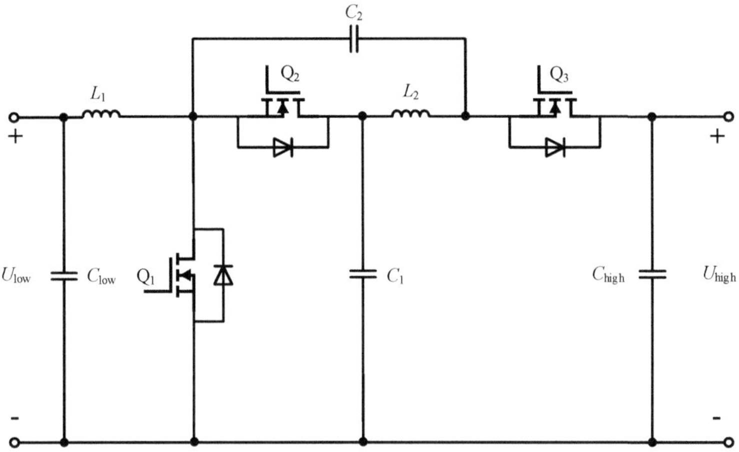

Fig. 4.3 Topology of switched-quasi-Z-source bidirectional DC–DC converter

4.2.1 Operating Principle of the Switched-Quasi-Z-Source Bidirectional DC–DC Converter

Since the gate signals of Q_2 and Q_3 are complementary to that of Q_3, the operation of the switched-quasi-Z-source bidirectional DC–DC converter in step-up and step-down mode consists of two switching states. To simplify the analysis, it is assumed that:

(1) All components are considered as ideal, with the conducting resistance of the power switches and the equivalent series resistance of the inductors and capacitors being ignored.
(2) The variation in inductor currents and capacitor voltages is regarded as linear.
(3) The average inductor currents and capacitor voltages are regarded as constant during steady state.

The operating analysis is based on the continuous current mode. The key waveforms of the switched-quasi-Z-source bidirectional DC–DC converter in step-up mode are depicted in Fig. 4.4, where i_{Q1}–i_{Q3} are the respective currents flow through power switches Q_1–Q_3, and the capacitor voltage of C_{high} is denoted as u_{high}, while i_{L1}, i_{L2} and u_{C1}, u_{C2} are the inductor currents of L_1, L_2 and capacitor voltages of C_1, C_2, respectively. When the converter operates in step-up mode, the energy is converted from the low voltage side to the high voltage side. Meanwhile, Q_1 acts as the main power switch, whereas Q_2 and Q_3 are the synchronous rectifiers that operate complementary to Q_1. The respective duty cycles of power switches d_1-d_3 are described as below, where d_{Boost} is the duty cycle of main power switch in step-up mode.

$$d_1 = 1 - d_2 = 1 - d_3 = d_{\text{Boost}} \tag{4.1}$$

The switching states of the switched-quasi-Z-source bidirectional DC–DC converter in step-up mode are presented in Fig. 4.5, which are analyzed as follows.

State 1) $S_1S_2S_3 = 100$

When Q_1 is turned ON while Q_2 and Q_3 are both turned OFF, the inductor L_1 is charged by the input source U_{low} through Q_1. Meanwhile, the capacitor C_2 and inductor L_2 are charged by C_1. The load R_{load} is powered by the high voltage side capacitor C_{high}. During this state, the following equations can be obtained.

$$\begin{cases} U_{\mathrm{L1}} = U_{\mathrm{low}} \\ U_{\mathrm{L2}} = U_{\mathrm{C1}} - U_{\mathrm{C2}} \end{cases} \tag{4.2}$$

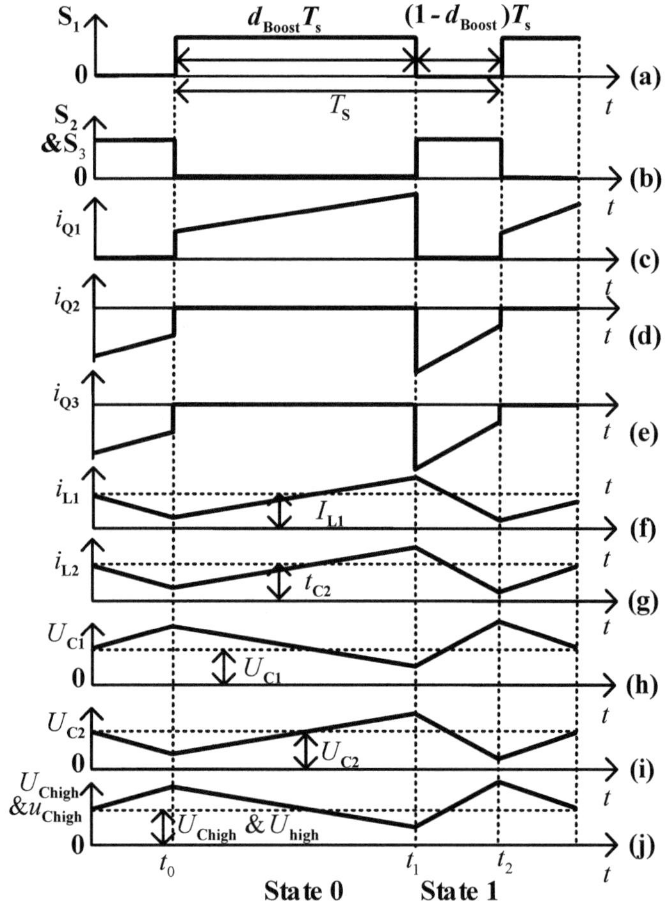

Fig. 4.4 Key waveforms of the switched-quasi-Z-source bidirectional DC–DC converter in step-up mode

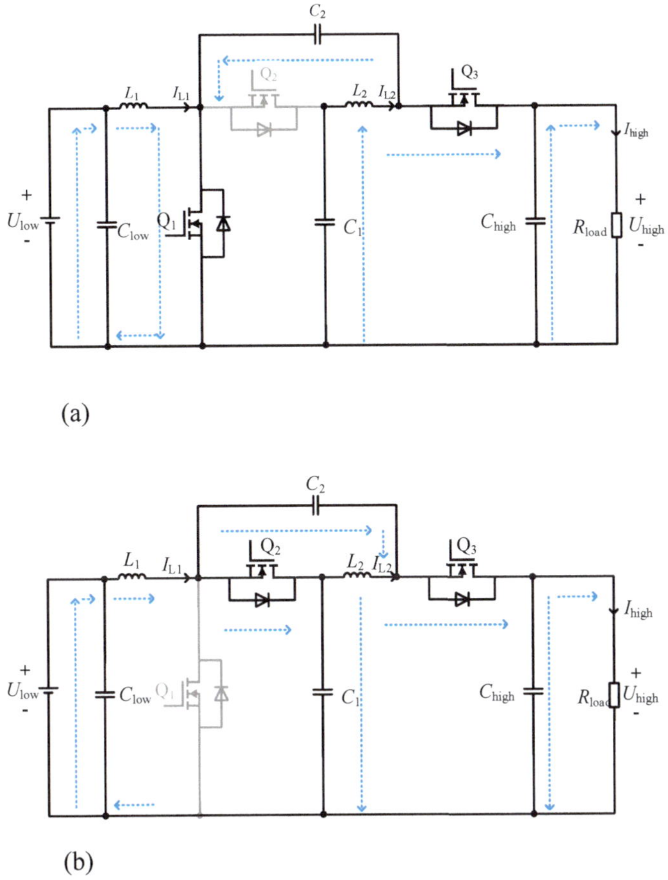

Fig. 4.5 Switching states of the switched-quasi-Z-source bidirectional DC–DC converter in step-up mode during **a** State 1 and **b** State 2

$$\begin{cases} i_{\mathrm{C1}} = -I_{\mathrm{L2}} \\ i_{\mathrm{C2}} = I_{\mathrm{L2}} \\ i_{\mathrm{Chigh}} = -I_{\mathrm{high}} \end{cases} \tag{4.3}$$

State 2) $S_1S_2S_3 = 011$

When Q_1 is turned OFF while Q_2 and Q_3 are turned ON, both inductors L_1 and L_2 are discharged. The capacitor C_2 is paralleled with L_2 and discharged to power the load through Q_2. At the same time, the capacitor C_1 is charged by the input source U_{low} and L_1. During this state, the following equations can be derived.

$$\begin{cases} U_{L1} = U_{low} - U_{C1} \\ U_{L2} = U_{C1} - U_{high} \\ U_{high} = U_{C1} + U_{C2} \end{cases} \tag{4.4}$$

$$\begin{cases} i_{C1} = i_{L1} + i_{C2} - I_{L2} \\ i_{Chigh} = I_{L2} - i_{C2} - I_{high} \end{cases} \tag{4.5}$$

According to the volt-second balance principle, the voltage gain M_{Boost} of the switched-quasi-Z-source bidirectional DC–DC converter in step-up mode can be obtained from (4.2) and (4.4) as follows.

$$M_{Boost} = \frac{1 + d_{Boost}}{1 - d_{Boost}} \tag{4.6}$$

Furthermore, the capacitor voltages of C_1 and C_2 can be written as below.

$$\begin{cases} U_{C1} = \dfrac{U_{low}}{1 - d_{Boost}} = \dfrac{U_{high}}{1 + d_{Boost}} \\ U_{C2} = \dfrac{d_{Boost} U_{low}}{1 - d_{Boost}} = \dfrac{d_{Boost} U_{high}}{1 + d_{Boost}} \end{cases} \tag{4.7}$$

In terms of the inductor currents, the following equation can be derived from (4.3) and (4.5) on the basis of ampere-second balance principle.

$$\begin{cases} I_{L1} = \dfrac{1 + d_{Boost}}{1 - d_{Boost}} I_{high} \\ I_{L2} = I_{high} \end{cases} \tag{4.8}$$

The key waveforms and switching states of the switched-quasi-Z-source bidirectional DC–DC converter in step-down mode are presented in Figs. 4.6 and 4.7, respectively. The energy is converted from the high voltage side to the low voltage side, while Q_2 and Q_3 act as the main power switches and Q_1 is the synchronous rectifier. The duty cycles d_1–d_3 for power switches Q_1–Q_3 can be described as follows, where d_{Buck} is the duty cycle of main power switches in step-down mode.

$$d_2 = d_3 = 1 - d_1 = d_{Buck} \tag{4.9}$$

Symmetric to that in step-up mode, there are two switching states in step-down mode, where S_1 and S_2, S_3 are complementary.

State 1) $S_1S_2S_3 = 011$

When Q_1 is turned OFF while Q_2 and Q_3 are turned ON, the capacitor C_1 is paralleled with L_2. Both C_1 and L_2 are charged by the input source U_{high}. The inducotr L_1 is also charged. Meanwhile, the capacitor C_1 is discharged to power the load R_{load} through Q_2. During this state, the following equations can be obtained.

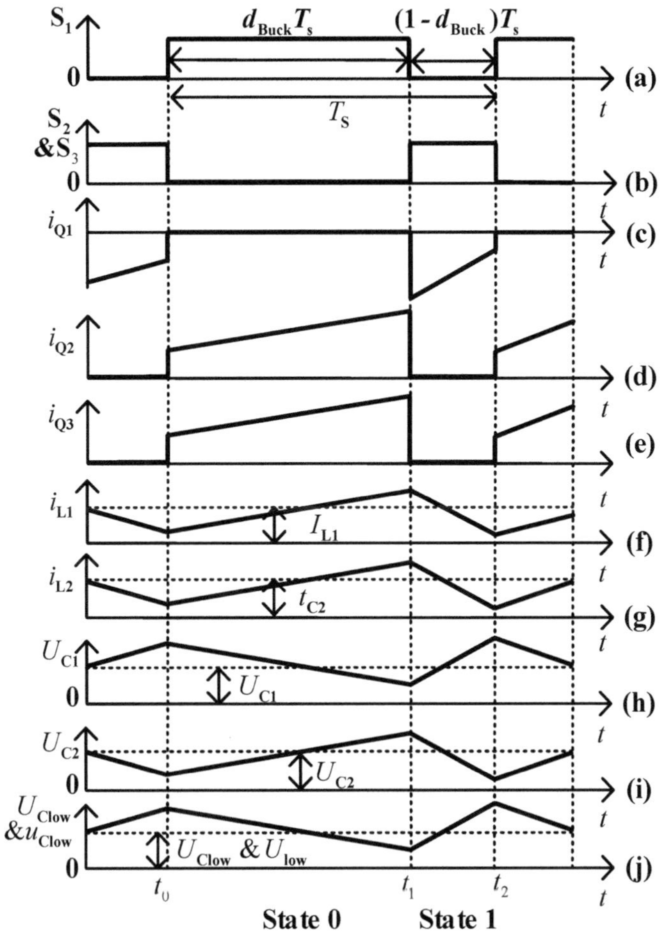

Fig. 4.6 Key waveforms of the switched-quasi-Z-source bidirectional DC–DC converter in step-down mode

$$\begin{cases} U_{L2} = U_{high} - U_{C1} \\ U_{L1} = U_{C1} - U_{low} \\ U_{C1} + U_{C2} = U_{high} \end{cases} \tag{4.10}$$

$$\begin{cases} i_{C1} = i_{C2} - I_{L1} + I_{L2} \\ i_{C2} = I_{high} - i_{Chigh} - I_{L2} \\ I_{L1} = I_{low} + i_{Clow} \end{cases} \tag{4.11}$$

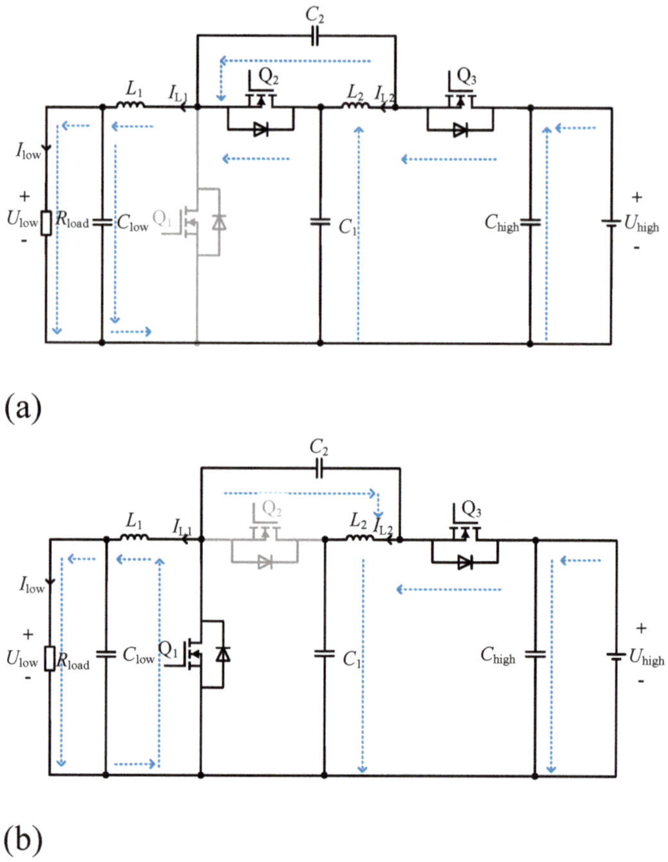

Fig. 4.7 Switching states of the switched-quasi-Z-source bidirectional DC–DC converter in step-down mode during **a** State 1 and **b** State 2

State 2) $S_1S_2S_3 = 100$

When only Q_1 is turned ON, while Q_2 and Q_3 are both turned OFF, the inductor L_1 is discharged to power the load R_{load}. Meanwhile, the capacitor C_2 and inductor L_2 are series connected and discharged. And the capacitor C_1 is charged. During this state, the following equations can be obtained.

$$\begin{cases} U_{\text{L1}} = U_{\text{C2}} - U_{\text{C1}} \\ U_{\text{L2}} = -U_{\text{low}} \\ U_{\text{high}} = U_{\text{C3}} \end{cases} \tag{4.12}$$

$$\begin{cases} i_{\mathrm{C1}} = I_{\mathrm{L2}} \\ i_{\mathrm{C2}} = -I_{\mathrm{L2}} \\ i_{\mathrm{Chigh}} = I_{\mathrm{high}} \\ I_{\mathrm{L1}} = I_{\mathrm{low}} + i_{\mathrm{Clow}} \end{cases} \tag{4.13}$$

According to the volt-second balance principle, the voltage gain M_{Buck} of the switched-quasi-Z-source bidirectional DC–DC converter in step-down mode can be derived from (4.10) and (4.12) as below.

$$M_{\mathrm{Buck}} = \frac{d_{\mathrm{Buck}}}{2 - d_{\mathrm{Buck}}} \tag{4.14}$$

Moreover, the capacitor voltages in step-down mode can be expressed by the duty cycle as follows.

$$\begin{cases} U_{\mathrm{C1}} = \dfrac{U_{\mathrm{low}}}{d_{\mathrm{Buck}}} = \dfrac{U_{\mathrm{high}}}{2 - d_{\mathrm{Buck}}} \\ U_{\mathrm{C2}} = \dfrac{(1 - d_{\mathrm{Buck}})U_{\mathrm{low}}}{d_{\mathrm{Buck}}} = \dfrac{(1 - d_{\mathrm{Buck}})U_{\mathrm{high}}}{2 - d_{\mathrm{Buck}}} \end{cases} \tag{4.15}$$

The inductor currents can be derived from (4.11) and (4.13) by applying the ampere-second balance principle, which are expressed as follows.

$$\begin{cases} I_{\mathrm{L1}} = I_{\mathrm{low}} \\ I_{\mathrm{L2}} = \dfrac{d_{\mathrm{Buck}}}{2 - d_{\mathrm{Buck}}} I_{\mathrm{low}} \end{cases} \tag{4.16}$$

4.2.2 *Bidirectional Control Strategy of the Switched-Quasi-Z-Source Bidirectional DC–DC Converter*

The switched-quasi-Z-source bidirectional DC–DC converter is capable of operating in both step-up/step-down modes, thereby is suitable for application in hybrid energy systems with energy storage devices. The bidirectional power flow of battery or super-capacitors can be achieved by adopting the switched-quasi-Z-source bidirectional DC–DC converter as a charging/discharging regulator. As a result, the energy storage devices can be interfaced with the DC bus through dynamic voltage matching.

In order to realize the mode transition between step-up and step-down modes, the bidirectional control strategy is utilized, where a mode index I_{sgn} is adopted for indicating the operation mode. The control diagram of bidirectional power flow is illustrated in Fig. 4.8.

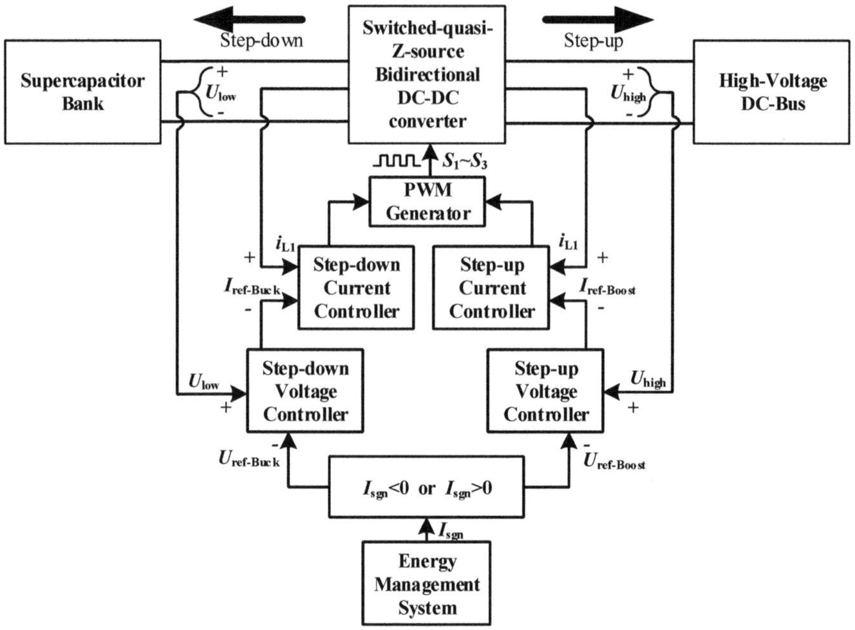

Fig. 4.8 Bidirectional control strategy for the switched-quasi-z-source DC–DC converter

When the mode index I_{sgn} is positive, namely $I_{sgn} > 0$, the converter operates in step-up mode. The high-side voltage is regarded as the controlled object, sampled as the feedback voltage to the closed voltage loop. The voltage error is derived from the sampled voltage and the reference voltage U_{ref_Boost}, ensuring that the high voltage side is maintained constant at the desired DC bus voltage. Under this condition, the output of control loop is the duty cycle of the main power switch Q_1, from which the gate signals are obtained via the PWM generator. The gate signals S_2 and S_3 of synchronous rectifiers Q_2 and Q_3 are derived complementally from S_1.

On the contrary, when the mode index I_{sgn} is negative ($I_{sgn} < 0$), the converter operates in step-down mode. The voltage across the energy storage device on the low voltage side is regarded as the controlled parameter, allowing for the regulation of the charging state. The low-side voltage is sampled and compared with the reference voltage U_{ref_Buck}. The voltage error is generated by the outer voltage loop and output to the inner current loop. The current loop regulates the inductor current of L_1 (i.e., the charging current of the energy storage device) based on the feedback current i_{L1} and reference current I_{ref_Buck}. The duty cycle of the main power switches Q_2 and Q_3 in step-down mode is generated by the control loop. The gate signals S_2 and S_3 are derived sequentially, and S_1 is complementary to S_2 and S_3.

4.2.3 Steady-State Characteristics of the Switched-Quasi-Z-Source Bidirectional DC–DC Converter

It is assumed that the conducting resistance of power switches is ignored, thereby the voltage stress across power switches can be obtained from the above switching state analysis by Kirchhoff's voltage law. In step-up mode, the voltage stresses across power switches can be expressed by the duty cycle d_{Boost} as follows, where $U_{\text{Q1\_Boost}}$-$U_{\text{Q1\_Boost}}$ are the voltage stresses across Q_1–Q_3 in step-up mode, respectively.

$$U_{\text{Q1\_ Boost}} = U_{\text{Q2\_ Boost}} = U_{\text{Q3\_ Boost}} = \frac{U_{\text{low}}}{1 - d_{\text{Boost}}} = \frac{U_{\text{high}}}{1 + d_{\text{Boost}}} \tag{4.17}$$

Similarly, the voltage stresses across powers switches in step-down mode can be expressed by the duty cycle d_{Buck} as follows, where $U_{\text{Q1\_Buck}}$-$U_{\text{Q3\_Buck}}$ are the respective voltage stresses across power switches Q_1–Q_3.

$$U_{\text{Q1\_ Buck}} = U_{\text{Q2\_ Buck}} = U_{\text{Q3\_ Buck}} = \frac{U_{\text{low}}}{d_{\text{Buck}}} = \frac{U_{\text{high}}}{2 - d_{\text{Boost}}} \tag{4.18}$$

Since the duty cycles in step-up and step-down modes are complementary, which can be written as below.

$$d_{\text{Boost}} + d_{\text{Buck}} = 1 \tag{4.19}$$

Therefore, according to (4.17) and (4.18), the voltage stresses across power switches in both modes are consistent.

In terms of the current stresses on power switches, the following equation can be obtained according to the Kirchhoff's current law, where $I_{\text{Q1\_Boost}}$-$I_{\text{Q1\_Boost}}$ are the respective current stresses on Q_1–Q_3 in step-up mode, and $I_{\text{Q1\_Buck}}$-$I_{\text{Q1\_Buck}}$ are the respective current stresses on Q_1–Q_3 in step-down mode.

$$\begin{cases} I_{\text{Q1\_ Boost}} = \dfrac{2}{1 - d_{\text{Boost}}} I_{\text{high}} \\ I_{\text{Q2\_ Boost}} = I_{\text{Q3\_ Boost}} = -\dfrac{1}{1 - d_{\text{Boost}}} I_{\text{high}} \end{cases} \tag{4.20}$$

$$\begin{cases} I_{\text{Q1\_ Buck}} = -\dfrac{2}{2 - d_{\text{Buck}}} I_{\text{low}} \\ I_{\text{Q2\_ Buck}} = I_{\text{Q3\_ Buck}} = \dfrac{1}{2 - d_{\text{Buck}}} I_{\text{low}} \end{cases} \tag{4.21}$$

Based on the complementary duty cycles in (4.19), it is concluded that the current stresses on power switches of the switched-quasi-Z-source bidirectional DC–DC converter are consistent in step-down and step-up modes.

4.2.4 Small-Signal Analysis of the Switched-Quasi-Z-Source Bidirectional DC–DC Converter

In the small-signal model of the switched-quasi-Z-source bidirectional DC–DC converter, the state variables include the inductor currents i_{L1} and i_{L2} and the capacitor voltages u_{C1}, u_{C2}, u_{Clow}, and u_{Chigh}. To decouple the capacitor voltages, the small-signal model introduces the parasitic resistance r of C_{high}, preventing the inclusion of an invalid state variable. Additionally, both inductor current ripple and capacitor voltage ripple are neglected. Consequently, the small-signal model of the switched-quasi-Z-source bidirectional DC–DC converter in step-up mode is derived using the state-space averaging method, presented as follows.

$$
\begin{bmatrix} \frac{d\hat{i}_{L1}(t)}{dt} \\ \frac{d\hat{i}_{L2}(t)}{dt} \\ \frac{d\hat{u}_{C1}(t)}{dt} \\ \frac{d\hat{u}_{C2}(t)}{dt} \\ \frac{d\hat{u}_{Chigh}(t)}{dt} \end{bmatrix}
=
\begin{bmatrix}
0 & 0 & -\frac{1-D}{L_1} & 0 & 0 \\
0 & -\frac{rD}{L_2} & \frac{1}{L_2} & -\frac{D}{L_2} & -\frac{1-D}{L_2} \\
\frac{1-D}{C_1} & -\frac{1}{C_1} & -\frac{1-D}{C_1 r} & -\frac{1-D}{C_1 r} & \frac{1-D}{C_1 r} \\
0 & \frac{D}{C_2} & -\frac{1-D}{C_2 r} & -\frac{1-D}{C_2 r} & \frac{1-D}{C_2 r} \\
0 & \frac{1-D}{C_{high}} & \frac{1-D}{C_{high} r} & \frac{1-D}{C_{high} r} & -\frac{1}{C_{high} R} - \frac{1-D}{C_{high} r}
\end{bmatrix}
\begin{bmatrix} \hat{i}_{L1}(t) \\ \hat{i}_{L2}(t) \\ \hat{u}_{C1}(t) \\ \hat{u}_{C2}(t) \\ \hat{u}_{Chigh}(t) \end{bmatrix}
+
\begin{bmatrix} \frac{1}{L_1} \\ 0 \\ 0 \\ 0 \\ 0 \end{bmatrix} \hat{u}_{in}(t)
$$

$$
+
\begin{bmatrix}
0 & 0 & \frac{1}{L_1} & 0 & 0 \\
0 & -\frac{r}{L_2} & 0 & -\frac{1}{L_2} & \frac{1}{L_2} \\
-\frac{1}{C_1} & 0 & \frac{1}{C_1 r} & \frac{1}{C_1 r} & -\frac{1}{C_1 r} \\
0 & \frac{1}{C_2} & \frac{1}{C_2 r} & \frac{1}{C_2 r} & -\frac{1}{C_2 r} \\
0 & -\frac{1}{C_{high}} & -\frac{1}{C_{high} r} & -\frac{1}{C_{high} r} & \frac{1}{C_{high} r}
\end{bmatrix}
\begin{bmatrix} I_{L1} \\ I_{L2} \\ U_{C1} \\ U_{C2} \\ U_{Chigh} \end{bmatrix} \hat{d}(t)
$$

$$
\hat{u}_{high}(t) = \begin{bmatrix} 0 & 0 & 0 & 0 & 1 \end{bmatrix} \begin{bmatrix} \hat{i}_{L1}(t) & \hat{i}_{L2}(t) & \hat{u}_{C1}(t) & \hat{u}_{C2}(t) & \hat{u}_{Chigh}(t) \end{bmatrix}^{T} \tag{4.22}
$$

The control-to-output transfer function of the converter in step-up mode can be obtained by substituting the parameters of components into (4.19). Similarly, the small-signal model of the converter in the step-down mode can be derived as follows.

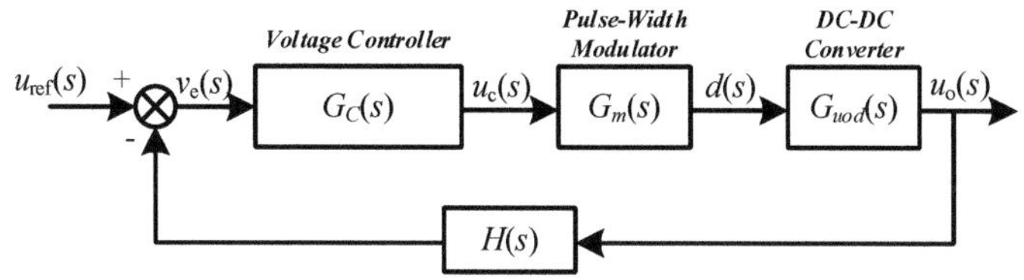

Fig. 4.9 Control strategy for the switched-quasi-Z-source bidirectional DC–DC converter

$$
\begin{bmatrix} \frac{\mathrm{d}\hat{i}_{L1}(t)}{\mathrm{d}t} \\ \frac{\mathrm{d}\hat{i}_{L2}(t)}{\mathrm{d}t} \\ \frac{\mathrm{d}\hat{u}_{C1}(t)}{\mathrm{d}t} \\ \frac{\mathrm{d}\hat{u}_{C2}(t)}{\mathrm{d}t} \\ \frac{\mathrm{d}\hat{u}_{\mathrm{Clow}}(t)}{\mathrm{d}t} \end{bmatrix} = \begin{bmatrix} 0 & 0 & \frac{D}{L_1} & 0 & -\frac{1}{L_1} \\ 0 & -\frac{r(1-D)}{L_2} & -\frac{1}{L_2} & \frac{1-D}{L_2} & 0 \\ -\frac{D}{C_1} & \frac{1}{C_1} & -\frac{D}{C_1 r} & -\frac{D}{C_1 r} & 0 \\ 0 & -\frac{1-D}{C_2} & -\frac{D}{C_2 r} & -\frac{D}{C_2 r} & 0 \\ \frac{1}{C_{\mathrm{low}}} & 0 & 0 & 0 & -\frac{1}{C_{\mathrm{low}} R} \end{bmatrix} \begin{bmatrix} \hat{i}_{L1}(t) \\ \hat{i}_{L2}(t) \\ \hat{u}_{C1}(t) \\ \hat{u}_{C2}(t) \\ \hat{u}_{\mathrm{Clow}}(t) \end{bmatrix}
$$

$$
+ \begin{bmatrix} 0 \\ \frac{D}{L_2} \\ \frac{D}{C_1 r} \\ \frac{D}{C_2 r} \\ 0 \end{bmatrix} \hat{u}_{\mathrm{high}}(t) + \begin{bmatrix} 0 & 0 & \frac{1}{L_1} & 0 & 0 \\ 0 & \frac{r}{L_2} & 0 & -\frac{1}{L_2} & 0 \\ -\frac{1}{C_1} & 0 & -\frac{1}{C_1 r} & -\frac{1}{C_1 r} & 0 \\ 0 & \frac{1}{C_2} & -\frac{1}{C_2 r} & -\frac{1}{C_2 r} & 0 \\ 0 & 0 & 0 & 0 & 0 \end{bmatrix} \begin{bmatrix} I_{L1} \\ I_{L2} \\ U_{C1} \\ U_{C2} \\ U_{\mathrm{Clow}} \end{bmatrix} \hat{d}(t) + \begin{bmatrix} 0 \\ \frac{\hat{d}(t)}{L_2} \\ \frac{\hat{d}(t)}{C_1 r} \\ \frac{\hat{d}(t)}{C_2 r} \\ 0 \end{bmatrix} U_{\mathrm{high}}
$$

$$
\hat{u}_{\mathrm{low}}(t) = \begin{bmatrix} 0 & 0 & 0 & 0 & 1 \end{bmatrix} \begin{bmatrix} \hat{i}_{L1}(t) & \hat{i}_{L2}(t) & \hat{u}_{C1}(t) & \hat{u}_{C2}(t) & \hat{u}_{\mathrm{Clow}}(t) \end{bmatrix}^{\mathrm{T}} \tag{4.23}
$$

On the basis of (4.23), the control-to-output transfer function of the converter in step-down mode can be obtained.

The diagram of the control strategy for the switched-quasi-Z-source bidirectional DC–DC converter is illustrated in Fig. 4.9, where $G_{\mathrm{uod}}(s)$ and $G_{\mathrm{m}}(s)$ are the transfer functions of the converter and pulse-width modulator, respectively. $H(s)$ is the feedback transfer function, and $G_{\mathrm{c}}(s)$ is the voltage controller transfer function. In addition, the value of $H(s)$ and $G_{\mathrm{c}}(s)$ are regard as 1. By means of the closed voltage loop, the output voltage on the high voltage side can be maintained at the reference voltage u_{ref}.

4.2.5 *Soft-Switching Operation of the Switched-Quasi-Z-Source Bidirectional DC–DC Converter*

As analyzed above, the switched-quasi-Z-source bidirectional DC–DC converter exhibits advantages of a common ground, simple structure, a wide voltage gain range and low voltage stresses. In order to realize soft-switching operation for the converter, a variable switching frequency (VSF) control strategy is presented, so that the electromagnetic noise can be effectively suppressed by realizing ZVS ON and quasi-ZCS off. In addition, the turn-on losses of power switches as well as recovery losses of body diodes can be eliminated. As a result, high conversion efficiency within a wide gain range can be achieved without auxiliary components.

A. Soft-switching operating principle

The key waveforms and switching states of the switched-quasi-Z-source bidirectional DC–DC converter in step-up mode are shown in Figs. 4.10 and 4.11, respectively. The beginning and end instances of the $S_1S_2S_3 = 011$ interval are denoted as t_3 and t_0, while the beginning and end instances of the $S_1S_2S_3 = 100$ interval are denoted as t_1 and t_2, respectively. In comparison with the conventional hard-switching condition, the dead time interval is analyzed in detail, which is critical to realizing soft-switching for the converter.

State 1) t_0–t_1: During the dead time, the power switches Q_2 and Q_3 are both turned OFF. The parasitic capacitors C_{s2} and C_{s3} are charged by the inductor currents. Meanwhile, the parasitic capacitor C_{s1} is discharged. The capacitor voltages during this interval are obtained as follows.

$$\begin{cases} u_{Cs1} + u_{Cs2} = U_{C1} \\ u_{Cs1} + u_{Cs3} + U_{C2} = U_{Chigh} \end{cases} \tag{4.24}$$

The capacitor voltages of C_1, C_2 and C_{high} are considered as constant, thus the discharging of C_{s1} is reversely synchronous with the charging of C_{s2} and C_{s3}. The power switches are completely turned OFF when C_{s2} and C_{s3} are fully charged. The freewheeling diode of Q_1 is conduced afterward.

State 2) t_1–t_2: Since the voltage across the main powers switch Q_1 is clamped at zero by its body diode, Q_1 is ZVS ON. Then the inductor L_1 is charged by the input source, and L_2 is charged by the capacitor C_1. The capacitor C_2 connected in series with L_2, and is charged by C. At the same time, the output load R_{load} is powered by the high voltage side capacitor C_{high}. The conducting current of Q_1 is obtained as the sum of inductor currents i_{L1} and i_{L2}.

$$i_{Q1} = i_{L1} + i_{L2} \tag{4.25}$$

As observed in Fig. 4.10, the inductor currents are initially negative during this interval. Consequently, power switch Q_1 is conducted in reverse at the same time. Subsequently, the inductor is charged, leading to current increase for i_{L1} and i_{L2}. Thus, i_{Q1} becoming positive, indicating that Q_1 is forward conducted.

State 3) t_2–t_3: After the main power switch Q_1 is turned OFF at t_2, the parasitic capacitor C_{s1} is charged by C_{s2} and C_{s3}. The capacitor voltages of C_1 and C_2 in the loop are assumed as constant, thus the changing of C_{s1} and discharging of C_{s2}

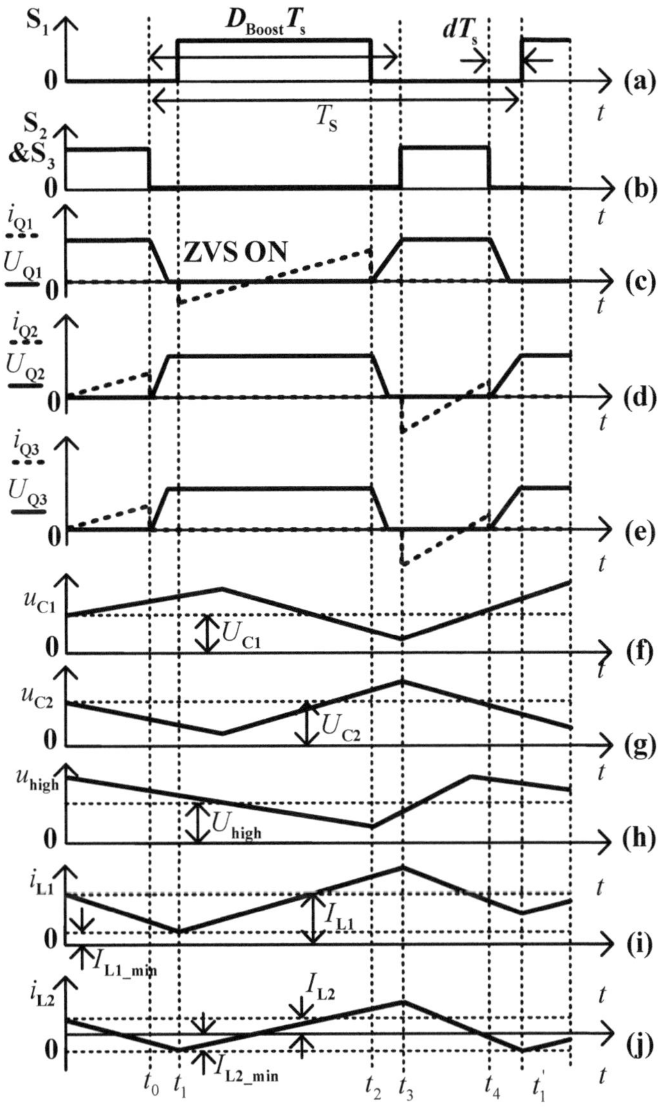

Fig. 4.10 Key waveforms of the switched-quasi-Z-source bidirectional DC–DC converter with soft-switching operation in step-up mode

and C_{s3} are reversely synchronous. Q_1 is completely turned OFF when C_{s1} is fully charged. Afterward, the body diodes of Q_2 and Q_3 are conducted.

State 4) t_3–t_4: The synchronous rectifiers Q_2 and Q_3 are reversely conducted at t_3. During this switching state, the inductor L_1 and capacitor C_1 are charged by the

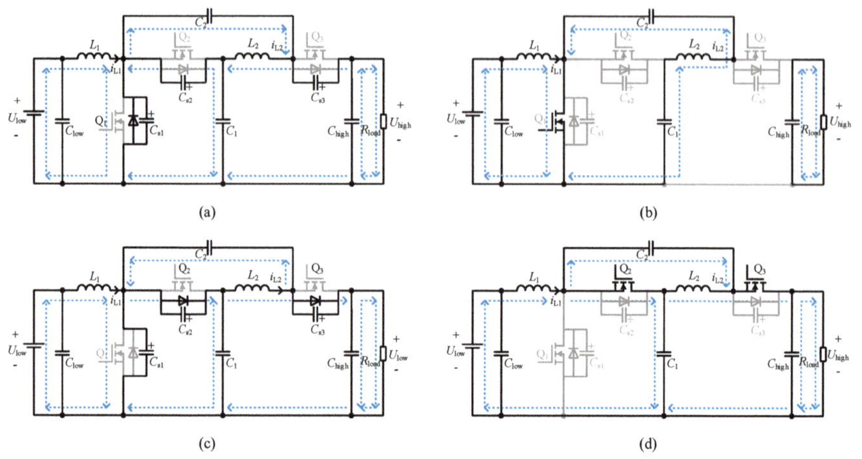

Fig. 4.11 Switching states of switched-quasi-Z-source bidirectional DC–DC converter during **a** State 1 **b** State 2 **c** State 3, and **d** State 4

input source U_{low}. Meanwhile, the capacitor C_2 is discharged to power the load on the high voltage side. The inductor L_2 is connected with the capacitor C_2 in parallel via Q_2. According to Kirchhoff's current law, the respective currents i_{Q2} and i_{Q3} flow through Q_2 and Q_3 can be written as follows.

$$i_{\mathrm{Q2}} = i_{\mathrm{Q3}} = -\frac{i_{\mathrm{L1}} + i_{\mathrm{L2}}}{2} \tag{4.26}$$

Both the inductor currents i_{L1} and i_{L2} decrease linearly during this state. As a result, the current on Q_2 and Q_3 at t_4 can be quasi-ZCS OFF when the sum of inductor currents is close to zero, as shown in Fig. 4.10.

When the switched-quasi-Z-source bidirectional DC–DC converter operates in step-down mode, the energy is converted from the input source U_{high} on the high voltage side to the load R_{load} on the low voltage side. The key waveforms and switching states of the converter in step-down mode are illustrated in Figs. 4.12 and 4.13, respectively. The beginning and end instances of the $S_1S_2S_3 = 011$ state are denoted as t_1 and t_2, while the beginning and end instances of the $S_1S_2S_3 = 100$ state are denoted as t_3 and t_0 (i.e., t_4), respectively. The main power switches are Q_2 and Q_3, whereas Q_1 acts as the synchronous rectifier to reduce the conducting losses. There are also four switching states include two dead time intervals within each switching period. The operating principle in step-down mode are analyzed as follows.

State 1) t_0–t_1: After Q_1 is turned OFF at t_0, the parasitic capacitor C_{s1} is charged, while C_{s2} and C_{s3} are discharged. Due to the constant voltages of the energy storage capacitors, (4.24) can also be established in step-down mode. This implies that the charging and discharging of parasitic capacitors are synchronously reversed. When

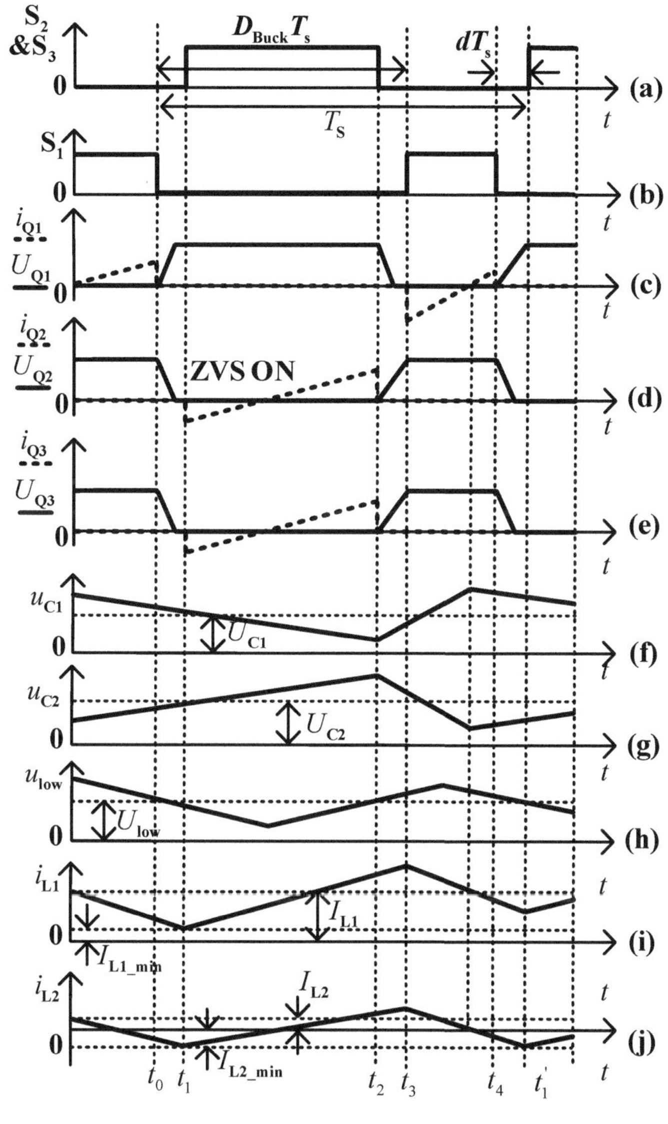

Fig. 4.12 Key waveforms of switched-quasi-Z-source bidirectional DC–DC converter with soft-switching operation in step-down mode

C_{s1} is fully charged, Q_1 is completely turned OFF, and the body diodes of Q_2 and Q_3 are conducted in turn.

State 2) t_1–t_2: At t_1, the main power switches Q_2 and Q_3 are ZVS ON, as the voltage is clamped by the conducted body diodes. The currents flow through Q_2 and Q_3 can be obtained as the sum of inductor currents, as shown below.

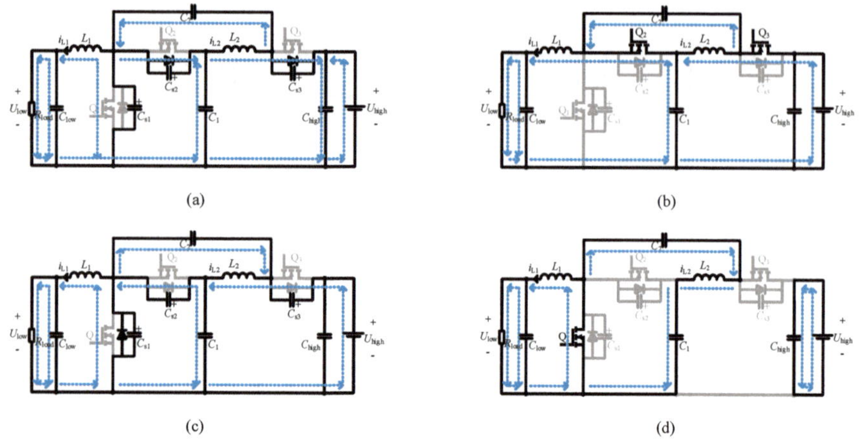

Fig. 4.13 Switching states of switched-quasi-Z-source bidirectional DC–DC converter with soft-switching operation in step-down mode during **a** State 1, **b** State 2, **c** State 3, and **d** State 4

$$i_{\mathrm{Q2}} = i_{\mathrm{Q3}} = \frac{i_{\mathrm{L1}} + i_{\mathrm{L2}}}{2} \tag{4.27}$$

Meanwhile, both the inductors L_1 and L_2 are charged, that the inductor currents i_{L} and i_{L2} increase linearly. The capacitor C_1 is discharged to power the load R_{load}, whereas C_2 is charged by the input source U_{high} on the high voltage side.

State 3) t_2**–**t_3: When the main switches Q_2 and Q_3 are turned OFF at t_2, the parasitic capacitors C_{s2} and C_{s3} are charged, while C_{s1} is discharged synchronously. The body diode of the synchronous rectifier Q_1 is conducted after the parasitic capacitors are fully charged/discharged, indicating Q_2 and Q_3 are completely turned OFF.

State 4) t_3**–**t_4: As the body diode of Q_1 is conducted in advance so that the voltage across Q_1 is clamped, Q_1 is ZVS ON at t_3 for synchronous rectification. The capacitor C_1 is charged while C_2 is discharged. The conducting current of Q_1 is composed of the inductor currents, which is written as follows, where the inductor currents i_{L1} and i_{L2} decrease linearly as the inductors are discharged.

$$i_{\mathrm{Q1}} = -(i_{\mathrm{L1}} + i_{\mathrm{L2}}) \tag{4.28}$$

As a result, the conducting current of Q_1 decreases, so shat the quasi-ZCS OFF of Q_1 can be achieved.

B. Condition of soft-switching

Based on the operating principle analysis, the synchronous rectifiers (i.e., Q_1 in step-down mode, Q_2 and Q_3 in step-up mode) consistently operate with ZVS ON. Consequently, the focus shifts to studying the soft-switching condition of the main switches in terms of ZVS operation. To prevent voltage spikes across the power switch, it is crucial to ensure that the parasitic capacitor is fully discharged by the reversed discharging current within the dead time interval. This enables the body diode to conduct before the power switch is turned ON.

The relationship among capacitor voltages is described in (4.24), where the voltages of energy storage capacitors are regarded as constant during each switching interval. Thereby, the voltage across power switches can be written as

$$\frac{du_{Q1}}{dt} = -\frac{du_{Q2}}{dt} = -\frac{du_{Q3}}{dt} \tag{4.29}$$

In addition, it is assumed that all parasitic capacitances are equal.

$$C_{s1} = C_{s2} = C_{s3} \tag{4.30}$$

By combining (4.29) and (4.30), the currents i_{Cs1}-i_{Cs3} of parasitic capacitors C_{s1}–C_{s3} can be obtained as

$$i_{Cs1} = -i_{Cs2} = -i_{Cs3} \tag{4.31}$$

Hence, the charging/discharging of C_{s1} is reverse synchronous with that of C_{s2} and C_{s3}. Moreover, the current of parasitic capacitor C_{s1} in step-up mode can be derived based on KCL, which is expressed by the inductor currents as follows.

$$i_{Cs1} = \frac{i_{L1} + i_{L2}}{3} \tag{4.32}$$

Similarly, when the converter operates in step-down mode, the currents i_{Cs2} and i_{Cs3} of parasitic capacitors can be obtained as below.

$$i_{Cs2} = i_{Cs3} = \frac{i_{L1} + i_{L2}}{3} \tag{4.33}$$

In order to fully discharge the parasitic capacitors within dead time, the following equation must be satisfied to achieve ZVS ON for the main power switches.

$$\int_0^{D_{dead}T_s} i_{Csi}(t)dt > C_{si}\Delta u_{dsi}, (i = 1, 2, 3) \tag{4.34}$$

In both step-up and step-down modes, the inductor currents at t_1 are the minimum values. Hence, the boundary condition of ZVS ON for main power switches can be deduced from (4.32)–(4.34), which is described as follows, where δ is the reserved current margin to ensure that the parasitic capacitors can be fully charged/discharged by the inductor currents within the dead time.

$$I_{L1\_\min} + I_{L2\_\min} = -\delta \tag{4.35}$$

The expression of δ is displayed below.

$$\delta \geq \frac{3C_s(U_{\text{low}} + U_{\text{high}})}{2D_{\text{dead}}T_s} \tag{4.36}$$

The minimum values of inductor currents depend on the operation condition of the converter. Thus, the ZVS condition can be expressed by the output power P, high-side voltage and low-side voltage as follows, where d_1 is duty cycle of power switch Q_1, and L is the equivalent inductance of L_1 and L_2 in parallel.

$$\frac{U_{\text{low}}d_1T_s}{2L\cdot[P(\frac{1}{U_{\text{high}}}+\frac{1}{U_{\text{low}}})+\delta]} \geq 1 \tag{4.37}$$

The value of L is written as below.

$$L = \frac{L_1L_2}{L_1 + L_2} \tag{4.38}$$

As a result, as long as (4.37) is satisfied for the switched-quasi-Z-source bidirectional DC–DC converter, the ZVS ON for all power switches can be achieved in both step-up and step-down modes.

In terms of the synchronous rectifiers, the turn-off current of Q_1 in step-down mode, as well as the turn-off currents of Q_2 and Q_3 is step-up mode can be derived as (4.39) and (4.40), respectively.

$$i_{\text{ds}1}(t_4) = \delta \tag{4.39}$$

$$i_{\text{ds}2}(t_4) = i_{\text{ds}3}(t_4) = \frac{\delta}{2} \tag{4.40}$$

According to (4.36), the value of δ is relatively small, namely the turn-off currents of synchronous rectifiers are near zero. Therefore, under the soft-switching operation, the synchronous rectifiers of the switched-quasi-Z-source bidirectional DC–DC converter are quasi-ZCS OFF. As a result, the switching losses of power switches can be significantly reduced.

C. Design of inductances

As can be observed from the soft-switching condition in (4.37), smaller inductance of L can realized a wider soft-switching operation range for the switched-quasi-Z-source bidirectional DC–DC converter. However, small inductance results large inductor current ripple, leading to considerable conducting losses at light load condition. Hence, the inductances of L_1 and L_2 should be designed carefully on the basis of the boundary soft-switching condition, so as to achieve optimal operation performance for the converter.

The value of L that satisfies the boundary soft-switching condition is represented by L^*, which can be calculated from (4.37) by substituting the parameters of rated operation condition. As long as the inductances of L_1 and L_2 meet the following inequality, ZVS can be implemented within the rated operation range.

$$\frac{L_1 L_2}{L_1 + L_2} < L^* \tag{4.41}$$

Thus, the current ripple θ_1 of L_1 at the boundary condition can be obtained, which is presented as below.

$$\theta_1 = \frac{\Delta i_{L1}}{I_{L1}} = \frac{2}{1 + L_1/L_2}\left(1 + \frac{1+\delta}{M_{\text{Boost}}}\right) \tag{4.42}$$

Considering fixed inductances, the current ripple is negatively correlated with the voltage gain as the operating condition of the converter varies. To ensure the converter operates with a positive continuous inductor current for L_1, the inductances are designed based on the minimum voltage gain $M_{\text{Boost\_min}}$.

$$L_1/L_2 > \frac{1+\delta}{M_{\text{Boost\_ min}}} \tag{4.43}$$

D. VSF control scheme

When the converter operates under the boundary soft-switching conditions, the main power switches exhibit ZVS ON operation, while the synchronous rectifiers are quasi-ZCS OFF. This configuration minimizes switching losses during each switching cycle. Consequently, a variable switching frequency control scheme is adopted to adjust the converter, ensuring it operates precisely at the boundary soft-switching condition.

The main idea of the VSF control scheme is to adjust the switching frequency on the basis of the boundary soft-switching condition, namely the equation presented in (4.37). The block diagram of VSF control is illustrated in Fig. 4.14, where the PI controller is utilized in the closed voltage loop to maintain the high-side voltage U_{high} at the reference voltage U_{ref}. The carrier frequency of PWM modulator is calculated from the sample operation conditions include low-side voltage U_{low} and current I_{high} on the high voltage side, which is described as follows.

$$f_s = \frac{U_{\text{low}} D_{\text{Boost}}}{2L \cdot \left[P\left(\frac{1}{U_{\text{high}}} + \frac{1}{U_{\text{low}}}\right) + \delta\right]} \tag{4.44}$$

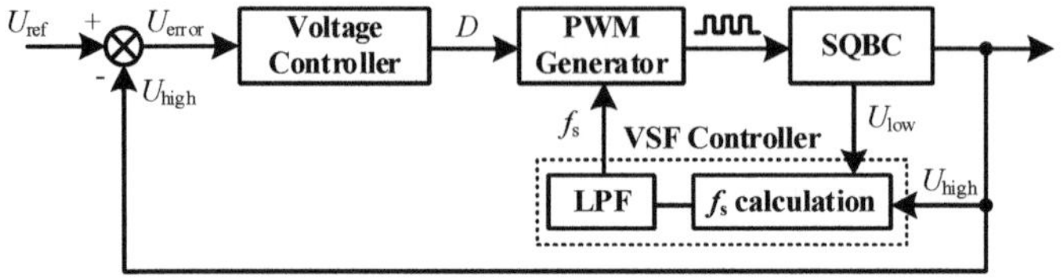

Fig. 4.14 Variable switching frequency control scheme for the switched-quasi-Z-source bidirectional DC–DC converter

As a result, the switching frequency is adjusted to track the boundary soft-switching condition, thereby the conversion efficiency of the switched-quasi-Z-source bidirectional DC–DC converter can be improved.

4.3 Hybrid Switched-Capacitor/Switched-Quasi-Z-Source Bidirectional DC–DC Converter

Built upon the switched-quasi-Z-source bidirectional DC–DC converter illustrated in Fig. 4.15, an additional switched-capacitor cell is introduced to further expand the voltage gain range and reduce voltage stresses across the power switches. As depicted in Fig. 4.16, a novel hybrid switched-capacitor/switched-quasi-Z-source bidirectional DC–DC converter is obtained and analyzed in this section [102].

The topology of the hybrid switched-capacitor/switched-quasi-Z-source bidirectional DC–DC converter is shown in Fig. 4.17, where L_1 and L_2 are energy storage inductors, and C_1–C_1 are energy storage capacitors. There are five power switches Q_1–Q_5 employed in the converter. Then hybrid converter consists of a switched-quasi-Z-source cell and a switched-capacitor cell. In addition, the power switches

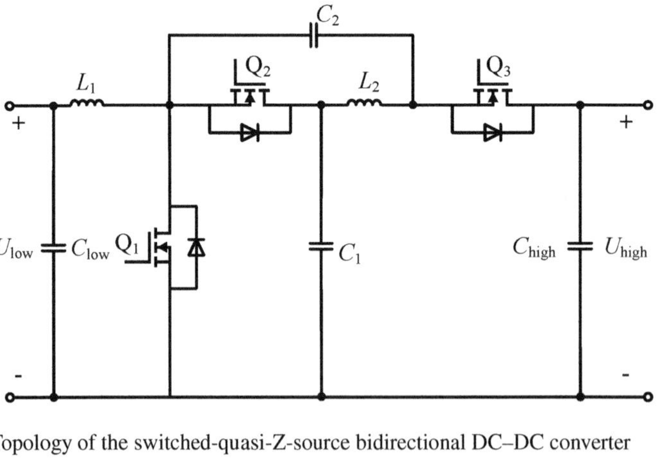

Fig. 4.15 Topology of the switched-quasi-Z-source bidirectional DC–DC converter

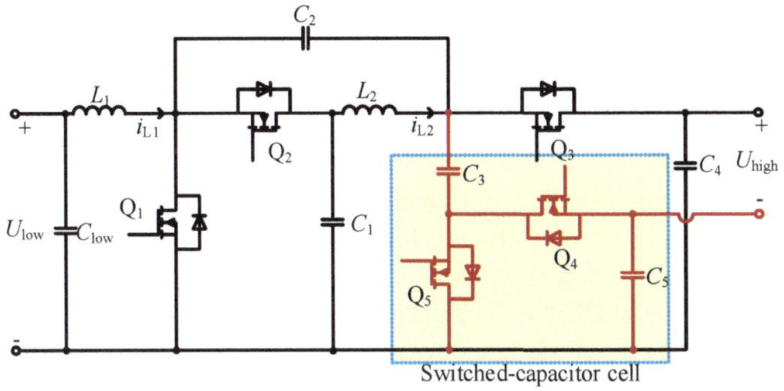

Fig. 4.16 Derivation of the hybrid switched-capacitor/switched-quasi-Z-source bidirectional DC–DC converter

Q_1 and Q_4 act as main power switches in step-up mode, whereas Q_2, Q_3, and Q_5 act as main power switches in step-down mode. In order to achieve synchronous rectification to reduce the conducting losses of freewheeling diodes, the gate signals S_1 and S_4 are complementary to S_2, S_3, and S_5.

In comparison to the switched-quasi-Z-source converter, the hybrid converter incorporates two additional power switches and two more capacitors, allowing for a wider voltage gain range and lower voltage stresses. Applicability of this converter is found in renewable energy systems. Furthermore, due to its bidirectional power flow capability, the converter can serve as a power interface between energy storage units and the DC bus in hybrid power systems.

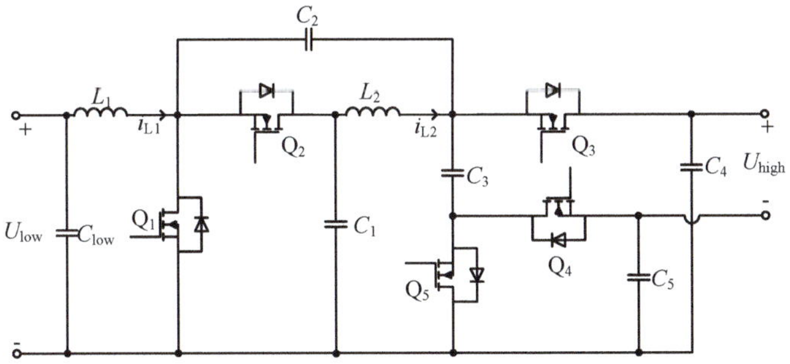

Fig. 4.17 Topology of the switched-capacitor/switched-quasi-Z-source bidirectional DC–DC converter

4.3.1 Operating Principle of the Hybrid Switched-Capacitor/ Switched-Quasi-Z-Source Bidirectional DC–DC Converter

When the hybrid switched-capacitor/switched-quasi-Z-source bidirectional DC–DC converter is applied in DC grids, the low voltage side is connected with energy storage devices such as batteries and super-capacitors. Whereas, the high voltage side is connected to the DC bus. The energy is converted in bidirection for charging and discharging the energy storage devices.

Assuming that the inductance is sufficiently large for continuous current mode operation in the converter, Q_1 and Q_4 serve as the main power switches in step-up mode, with Q_2, Q_3, and Q_5 acting as synchronous rectifiers. The key waveforms and switching states of the converter in step-up mode are illustrated in Figs. 4.18 and 4.19, respectively. The operating principles of the switching states are analyzed as follows.

State 1) $S_1S_2S_3S_4S_5 = 10010$

When the main switches Q_1 and Q_4 are turned ON and the synchronous rectifiers are turned OFF, the inductor L_1 is charged by the input source U_{low} via Q_1. Meanwhile, the capacitor C_2 and inductor L_2 are charged by C_1, while the capacitor C_5 are charged by C_3. The output load R_{load} is powered by C_4.

State 2) $S_1S_2S_3S_4S_5 = 01101$

When the main switches Q_1 and Q_4 are turned OFF and the synchronous rectifiers are reversely turned ON, the capacitor C_1 is charged by the input source U_{low} and inductor L_1. The capacitor C_2 and inductor L_2 are connected in parallel through Q_2 and discharged. The capacitors C_3 and C_4 are both charged, while C_5 is discharged to power the load R_{load}.

When the converter operates in step-down mode, it functions as the charging regulator for energy devices, converting power from the DC bus on the high voltage side to the low voltage side. In step-down mode, the hybrid converter exhibits two switching states, where Q_2, Q_3, and Q_5 act as the main power switches, while Q_1 and Q_4 function as the synchronous rectifiers. The key waveforms and switching states of the switched-capacitor/switched-quasi-Z-source bidirectional DC–DC converter in step-down mode are presented in Figs. 4.20 and 4.21, respectively.

State 1) $S_1S_2S_3S_4S_5 = 01101$

When the main power switches Q_2, Q_3, and Q_5 are turned ON and the synchronous rectifiers are turned OFF, both the inductors L_1 and L_2 are charged. The capacitor C_2 is charged by C_3, C_4, and the input source U_{high} on the high voltage side. Meanwhile, C_5 is charged by U_{high}, and the load R_{load} on the low voltage side is powered by the capacitors C_1 and C_4.

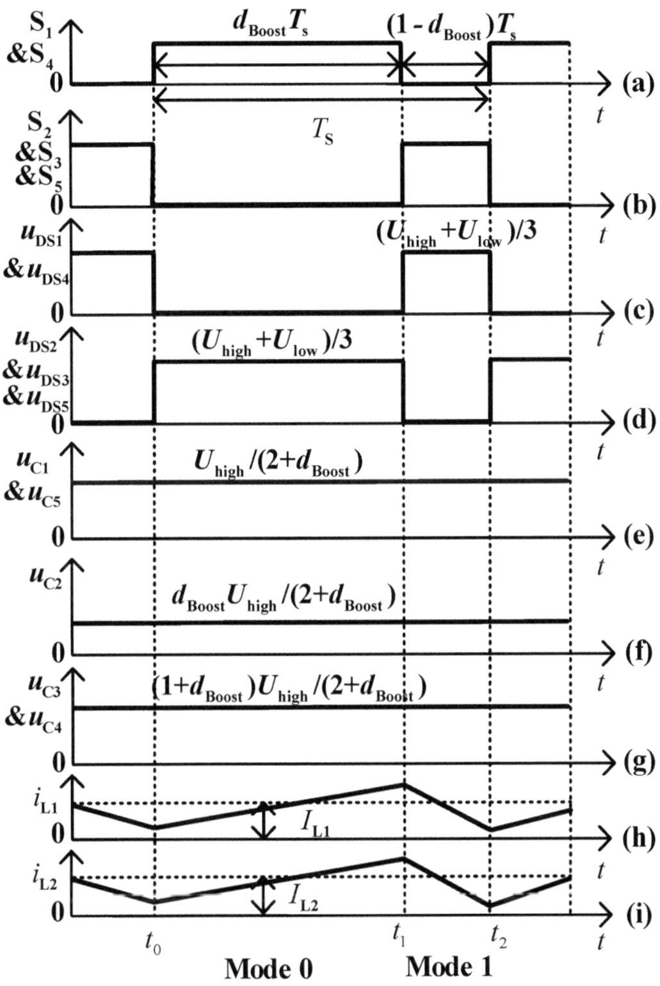

Fig. 4.18 Key waveforms of the switched-capacitor/switched-quasi-Z-source bidirectional DC–DC converter in step-up mode

State 2) $S_1S_2S_3S_4S_5 = 10010$

When the main power switches Q_2, Q_3, and Q_5 are turned OFF and the synchronous rectifiers are reversely conducted, the capacitors C_3 and C_4 are charged by the input source U_{high} on the high voltage side. Meanwhile, the capacitor C_1 is charged by C_2 and L_2 through v_1, and C_3 is charged by C_2 and C_5. Moreover, the load R_{load} is powered by the inductor L_1 through Q_1.

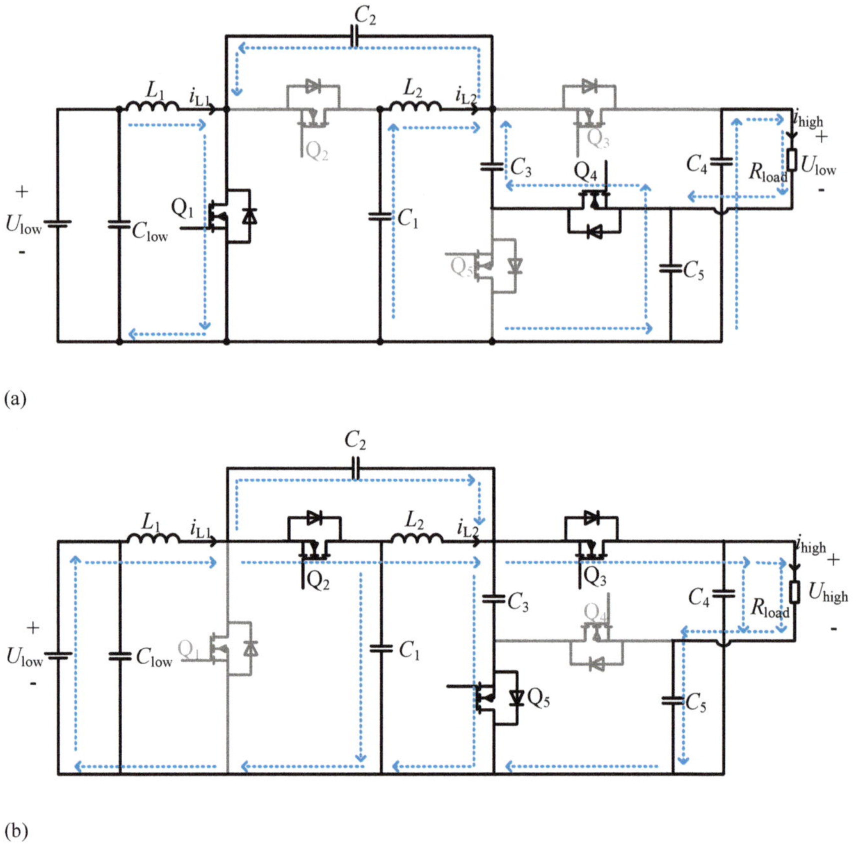

Fig. 4.19 Switching states of the switched-capacitor/switched-quasi-Z-source bidirectional DC–DC converter in step-up mode during **a** State 1 and **b** State 2

4.3.2 Steady-State Characteristics of the Hybrid Switched-Capacitor/Switched-Quasi-Z-Source Bidirectional DC–DC Converter

In the context of the steady-state characteristics of the switched-capacitor/switched-quasi-Z-source bidirectional DC–DC converter in step-up mode, the following equation can be derived by applying the volt-second balance principle to inductors L_1 and L_2, where d_{Boost} is the duty cycle of the main power switches (i.e., Q_1 and Q_4) in the switched-capacitor/switched-quasi-Z-source bidirectional DC–DC converter during step-up mode.

$$\begin{cases} d_{\text{Boost}}U_{\text{low}} + (1 - d_{\text{Boost}})(U_{\text{low}} - U_{\text{C1}}) = 0 \\ d_{\text{Boost}}(U_{\text{C1}} - U_{\text{C2}}) + (1 - d_{\text{Boost}})(U_{\text{C1}} - U_{\text{C4}}) = 0 \end{cases} \tag{4.45}$$

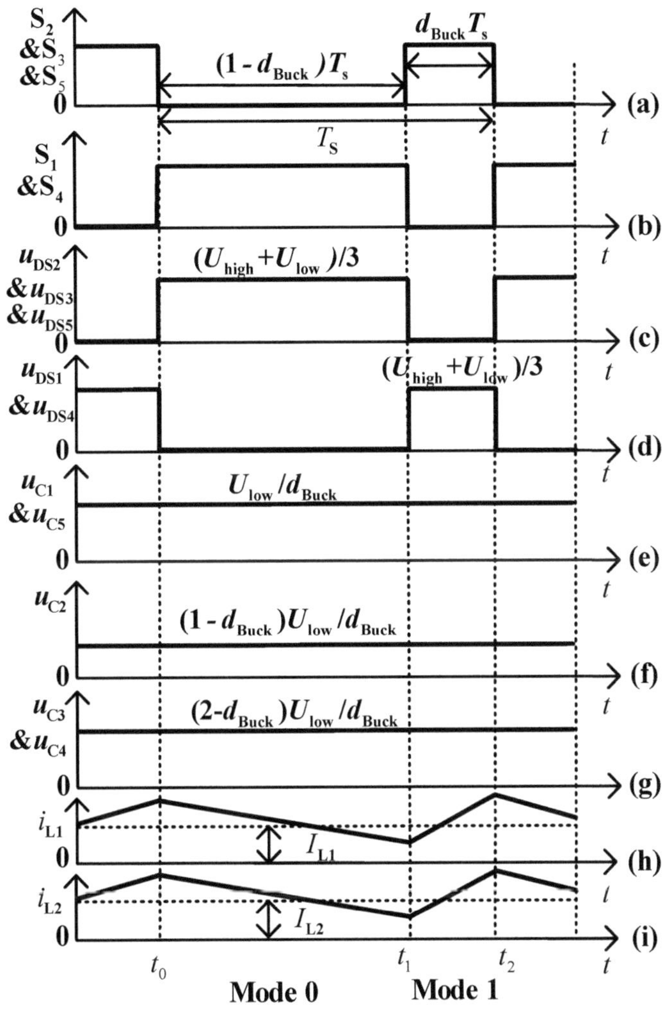

Fig. 4.20 Key waveforms of the switched-capacitor/switched-quasi-Z-source bidirectional DC–DC converter in step-down mode

$$\begin{cases} U_{\text{high}} = U_{\text{C4}} + U_{\text{C5}} \\ U_{\text{C5}} + U_{\text{C2}} = U_{\text{C3}} \\ U_{\text{C5}} = U_{\text{C1}} \\ U_{\text{C3}} = U_{\text{C4}} \end{cases} \tag{4.46}$$

According to (4.45) and (4.46), the capacitor voltages can be obtained, which are written as follows.

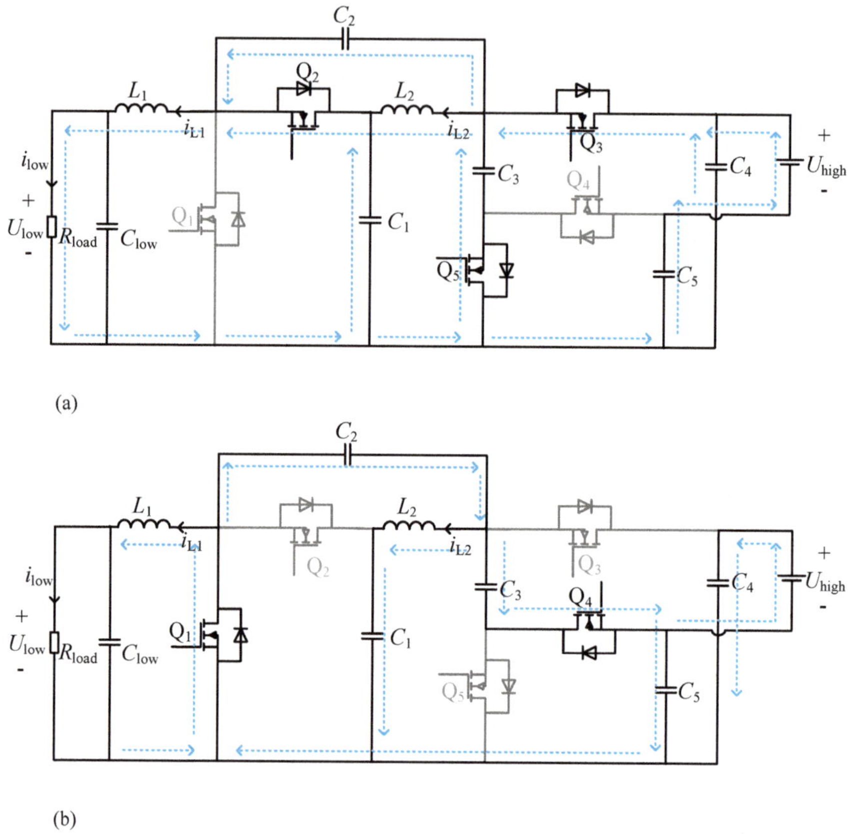

Fig. 4.21 Switching states of the switched-capacitor/switched-quasi-Z-source bidirectional DC–DC converter in step-down mode during **a** State 1 and **b** State 2

$$\begin{cases} U_{C1} = U_{C5} = \dfrac{U_{low}}{1 - d_{Boost}} = \dfrac{U_{high}}{2 + d_{Boost}} \\ U_{C2} = \dfrac{d_{Boost} U_{low}}{1 - d_{Boost}} = \dfrac{d_{Boost} U_{high}}{2 + d_{Boost}} \\ U_{C3} = U_{C4} = \dfrac{(1 + d_{Boost}) U_{low}}{1 - d_{Boost}} = \dfrac{(1 + d_{Boost}) U_{high}}{2 + d_{Boost}} \end{cases} \tag{4.47}$$

Furthermore, the voltage stresses across power switches are derived as follows.

$$U_{Q1} = U_{Q2} = U_{Q3} = U_{Q4} = U_{Q5} = \frac{U_{high}}{2 + d_{Boost}} = \frac{U_{high} + U_{low}}{3} \tag{4.48}$$

The voltage gain M_{Boost} of the converter in step-up mode can be obtained by combining (4.46) and (4.47), which is described as below.

$$M_{\text{Boost}} = \frac{2 + d_{\text{Boost}}}{1 - d_{\text{Boost}}} \tag{4.49}$$

The steady-state behavior of the converter in step-down mode aligns with that in step-up mode. Applying the volt-second balance principle, the following equations can be derived based on the previously discussed operating principle analysis, where d_{Buck} is the duty cycle of the main power switches in step-down mode (i.e., Q_2, Q_3, and Q_5) of the switched-capacitor/switched-quasi-Z-source bidirectional DC–DC converter.

$$\begin{cases} d_{\text{Buck}}(U_{\text{C4}} - U_{\text{C2}} - U_{\text{low}}) + (1 - d_{\text{Buck}})(-U_{\text{low}}) = 0 \\ d_{\text{Buck}}(U_{\text{C4}} - U_{\text{C1}}) + (1 - d_{\text{Buck}})(-(U_{\text{C5}} + U_{\text{C1}}) + U_{\text{C3}}) = 0 \end{cases} \tag{4.50}$$

In terms of the relationships among capacitor voltages, (4.46) can also be established in step-down mode. On the basis of (4.46) and (4.50), the voltage stresses across capacitors and power switches of the converter in step-down mode can be derived as follows.

$$\begin{cases} U_{\text{C1}} = U_{\text{C5}} = \dfrac{U_{\text{low}}}{d_{\text{Buck}}} = \dfrac{U_{\text{high}}}{3 - d_{\text{Buck}}} \\ U_{\text{C2}} = \dfrac{(1 - d_{\text{Buck}})U_{\text{low}}}{d_{\text{Buck}}} = \dfrac{(1 - d_{\text{Buck}})U_{\text{high}}}{3 - d_{\text{Boost}}} \\ U_{\text{C3}} = U_{\text{C4}} = \dfrac{(2 - d_{\text{Buck}})U_{\text{low}}}{d_{\text{Buck}}} = \dfrac{(2 - d_{\text{Buck}})U_{\text{high}}}{3 - d_{\text{Boost}}} \end{cases} \tag{4.51}$$

$$U_{\text{Q1}} = U_{\text{Q2}} = U_{\text{Q3}} = U_{\text{Q4}} = U_{\text{Q5}} = \frac{U_{\text{high}}}{3 - d_{\text{Buck}}} = \frac{U_{\text{high}} + U_{\text{low}}}{3} \tag{4.52}$$

As a result, the voltage gain of the converter in step-down mode can be obtained.

$$M_{\text{Buck}} = \frac{d_{\text{Buck}}}{3 - d_{\text{Buck}}} \tag{4.53}$$

4.3.3 *Small-Signal Modeling of the Hybrid Switched-Capacitor/Switched-Quasi-Z-Source Bidirectional DC–DC Converter*

To decouple the capacitor currents i_{C1}-i_{C5}, equivalent series resistances are introduced to the switched-capacitor/switched-quasi-Z-source bidirectional DC–DC converter, including R_{C1} for C_1 and R_{C3} for C_3. This approach prevents invalid state variables in the small-signal modeling. Assuming that inductor current ripples and capacitor voltage ripples are neglected, the small-signal model of the converter in step-up mode through the state-space averaging method can be obtained as (4.54).

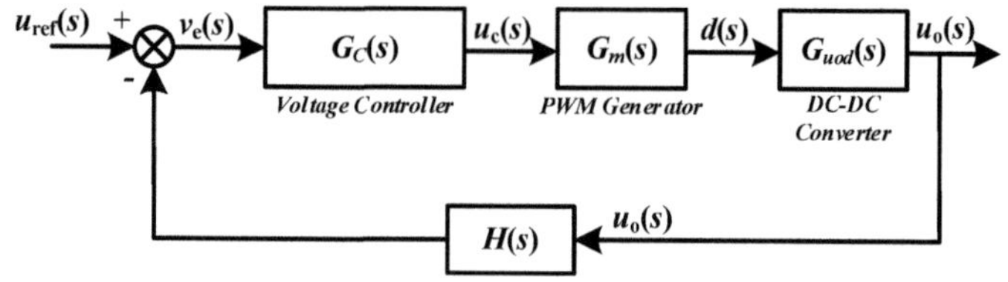

Fig. 4.22 Control strategy for the switched-capacitor/switched-quasi-Z-source bidirectional DC–DC converter

Figure 4.22 presents the control strategy for the switched-capacitor/switched-quasi-Z-source bidirectional DC–DC converter, which is adopted to regulate the DC bus voltage through closed voltage loop. The transfer function of the converter is denoted as $G_{\text{uod}}(\text{s})$, and $G_{\text{m}}(\text{s})$ is the transfer function of pulse-width modulator, while $G_{\text{c}}(\text{s})$ and $H(\text{s})$ are the PI controller and feedback transfer functions, respectively. In addition, $G_{\text{c}}(\text{s})$ and $H(\text{s})$ are considered as 1. Through the voltage control strategy, the DC bus voltage on the high voltage side of the converter can be regulated at the reference voltage u_{ref}.

$$
\begin{aligned}
\begin{bmatrix} \frac{\mathrm{d}\hat{i}_{\mathrm{L1}}(t)}{\mathrm{d}t} \\ \frac{\mathrm{d}\hat{i}_{\mathrm{L2}}(t)}{\mathrm{d}t} \\ \frac{\mathrm{d}\hat{u}_{\mathrm{C1}}(t)}{\mathrm{d}t} \\ \frac{\mathrm{d}\hat{u}_{\mathrm{C2}}(t)}{\mathrm{d}t} \\ \frac{\mathrm{d}\hat{u}_{\mathrm{C3}}(t)}{\mathrm{d}t} \\ \frac{\mathrm{d}\hat{u}_{\mathrm{C4}}(t)}{\mathrm{d}t} \\ \frac{\mathrm{d}\hat{u}_{\mathrm{C5}}(t)}{\mathrm{d}t} \end{bmatrix}
&= \begin{bmatrix}
0 & 0 & 0 & \frac{1-D}{L_1} & 0 & -\frac{1-D}{L_1} & 0 \\
0 & -\frac{DR_{\mathrm{C1}}}{L_2} & \frac{D}{L_2} & -\frac{1}{L_2} & 0 & 0 & 0 \\
0 & -\frac{D}{C_1} & -\frac{1-D}{C_1R_{\mathrm{C1}}} & -\frac{1-D}{C_1R_{\mathrm{C1}}} & 0 & \frac{1-D}{C_1R_{\mathrm{C1}}} & 0 \\
-\frac{1-D}{C_2} & \frac{1}{C} & -\frac{1-D}{C_2R_{\mathrm{C1}}} & -\frac{1}{C_2}\left(\frac{D}{R_{\mathrm{C3}}}+\frac{1-D}{R_{\mathrm{C1}}}\right) & \frac{D}{C_2R_{\mathrm{C3}}} & \frac{1-D}{C_2R_{\mathrm{C1}}} & -\frac{D}{C_2R_{\mathrm{C3}}} \\
0 & 0 & 0 & \frac{D}{C_3R_{\mathrm{C3}}} & -\frac{1}{C_3R_{\mathrm{C3}}} & \frac{1-D}{C_3R_{\mathrm{C3}}} & \frac{D}{C_3R_{\mathrm{C3}}} \\
\frac{1-D}{C_4} & 0 & \frac{1-D}{C_4R_{\mathrm{C1}}} & \frac{1-D}{C_4R_{\mathrm{C1}}} & \frac{1-D}{C_4R_{\mathrm{C3}}} & -\frac{1}{C_4R}-\frac{1-D}{C_4}\left(\frac{1}{R_{\mathrm{C1}}}+\frac{1}{R_{\mathrm{C3}}}\right) & -\frac{1}{C_4R} \\
0 & 0 & 0 & -\frac{D}{C_5R_{\mathrm{C3}}} & \frac{D}{C_5R_{\mathrm{C3}}} & -\frac{1}{C_5R} & -\frac{1}{C_5}\left(\frac{1}{R}+\frac{D}{R_{\mathrm{C3}}}\right)
\end{bmatrix}
\begin{bmatrix} \hat{i}_{\mathrm{L1}}(t) \\ \hat{i}_{\mathrm{L2}}(t) \\ \hat{u}_{\mathrm{C1}}(t) \\ \hat{u}_{\mathrm{C2}}(t) \\ \hat{u}_{\mathrm{C3}}(t) \\ \hat{u}_{\mathrm{C4}}(t) \\ \hat{u}_{\mathrm{C5}}(t) \end{bmatrix} \\
&\quad + \begin{bmatrix} \frac{1}{L_1} \\ 0 \\ 0 \\ 0 \\ 0 \\ 0 \\ 0 \end{bmatrix} \hat{u}_{\mathrm{low}}(t) + \begin{bmatrix}
0 & 0 & 0 & -\frac{1}{L_1} & 0 & \frac{1}{L_1} & 0 \\
0 & -\frac{R_{\mathrm{C1}}}{L_2} & \frac{1}{L_2} & 0 & 0 & 0 & 0 \\
0 & -\frac{1}{C_1} & \frac{1}{C_1R_{\mathrm{C1}}} & \frac{1}{C_1R_{\mathrm{C1}}} & 0 & -\frac{1}{C_1R_{\mathrm{C1}}} & 0 \\
\frac{1}{C_2} & 0 & \frac{1}{C_2R_{\mathrm{C1}}} & \frac{1}{C_2}\left(\frac{1}{R_{\mathrm{C1}}}-\frac{1}{R_{\mathrm{C3}}}\right) & \frac{1}{C_2R_{\mathrm{C3}}} & -\frac{1}{C_2R_{\mathrm{C1}}} & -\frac{1}{C_2R_{\mathrm{C3}}} \\
0 & 0 & 0 & \frac{1}{C_3R_{\mathrm{C3}}} & 0 & -\frac{1}{C_3R_{\mathrm{C3}}} & \frac{1}{C_3R_{\mathrm{C3}}} \\
-\frac{1}{C_4} & 0 & -\frac{1}{C_4R_{\mathrm{C1}}} & -\frac{1}{C_4R_{\mathrm{C1}}} & -\frac{1}{C_4R_{\mathrm{C3}}} & \frac{1}{C_4}\left(\frac{1}{R_{\mathrm{C1}}}+\frac{1}{R_{\mathrm{C3}}}\right) & 0 \\
0 & 0 & 0 & -\frac{1}{C_5R_{\mathrm{C3}}} & \frac{1}{C_5R_{\mathrm{C3}}} & 0 & -\frac{1}{C_5R_{\mathrm{C3}}}
\end{bmatrix}
\begin{bmatrix} I_{\mathrm{L1}} \\ I_{\mathrm{L2}} \\ U_{\mathrm{C1}} \\ U_{\mathrm{C2}} \\ U_{\mathrm{C3}} \\ U_{\mathrm{C4}} \\ U_{\mathrm{C5}} \end{bmatrix} \hat{d}(t) \\
\hat{u}_{\mathrm{high}}(t) &= \begin{bmatrix} 0 & 0 & 0 & 0 & 0 & 1 & 1 \end{bmatrix} \begin{bmatrix} \hat{i}_{\mathrm{L1}}(t) & \hat{i}_{\mathrm{L2}}(t) & \hat{u}_{\mathrm{C1}}(t) & \hat{u}_{\mathrm{C2}}(t) & \hat{u}_{\mathrm{C3}}(t) & \hat{u}_{\mathrm{C4}}(t) & \hat{u}_{\mathrm{C5}}(t) \end{bmatrix}^{\mathrm{T}}
\end{aligned}
\tag{4.54}
$$

4.4 Efficiency and Application Analysis

In order to validate the performance of the switched-quasi-Z-source topologies introduced in this chapter, experimental prototypes were developed for efficiency measurement and application analysis.

4.4.1 Efficiency Analysis

The prototype of the switched-quasi-Z-source bidirectional DC–DC converter is designed with a rated power of 300 W. The converter's conversion efficiency was measured in both step-down and step-up modes at the rated power level. It is noteworthy that the high voltage side is maintained at a constant level of 240 V, while the low voltage side varies within the range of 40–120 V, resulting in a voltage gain range of 2–6.

The measured operation efficiency of the switched-quasi-Z-source bidirectional DC–DC converter, with a fixed switching frequency of 20 kHz, is plotted in Fig. 4.23. Notably, in step-down mode, the maximum and minimum efficiency are 96.24% and 92.31%, respectively. In step-up mode, the maximum and minimum operation efficiency are 96.44% and 88.17%, respectively. In both step-down and step-up modes, the conversion efficiency increases as the low-side voltage increases.

To validate the effectiveness of the variable switching frequency strategy, the comparative results of operation efficiency are presented in Fig. 4.24. The variable switching frequency method can significantly reduce the switching losses of power switches by achieving soft-switching operation. Besides, the volume of inductors required in this converter can be reduced. Thereby, the core losses and copper losses

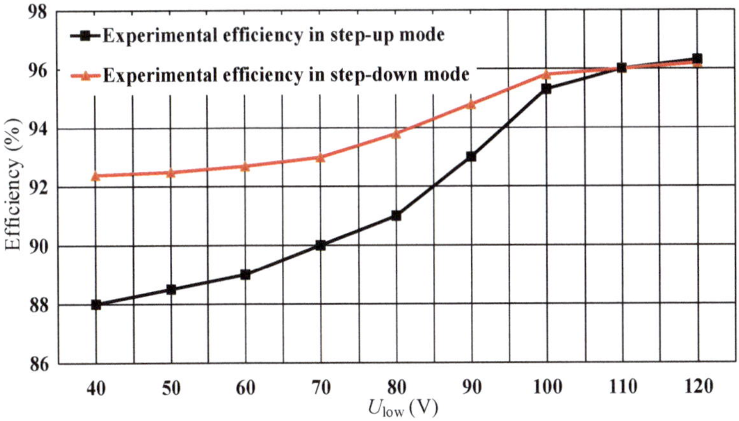

Fig. 4.23 Efficiency of the switched-quasi-Z-source bidirectional DC–DC converter with fixed switching frequency

resulted from the inductors can be suppressed. As can be seen from the figure, when the high voltage side is constant at 300 V, and the low voltage side varies within the range of 50–150 V, the conversion efficiency of the prototype with variable switching frequency is much higher than that with fixed switching frequency at 36 kHz. The maximum efficiency in step-down and step-up modes are 98.28% and 97.78%, respectively.

In the context of the hybrid switched-capacitor/switched-quasi-Z-source bidirectional DC–DC converter, the prototype operates with a constant high voltage side at 400 V, while the low voltage side varies within the range of 40–120 V. The rated power is designed at 400 W. Figure 4.25 illustrates the conversion efficiency of the prototype in both step-down and step-up modes at the rated power level and half-load condition.

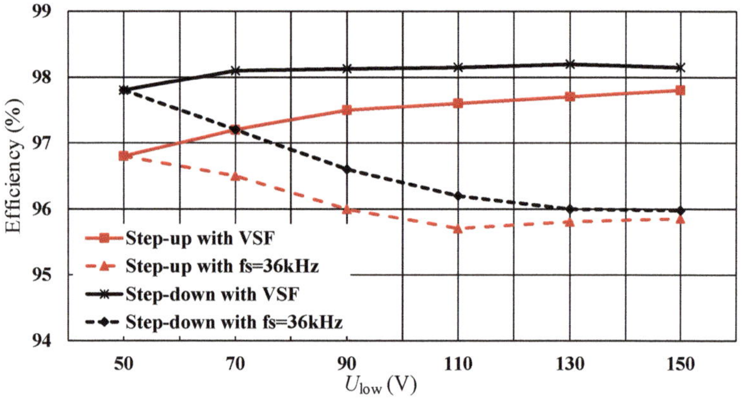

Fig. 4.24 Efficiency of the switched-quasi-Z-source bidirectional DC–DC converter with variable switching frequency

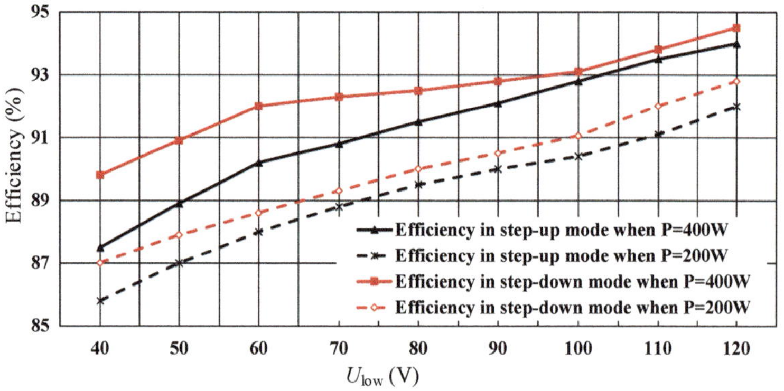

Fig. 4.25 Efficiency of the hybrid switched-capacitor/switched-quasi-Z-source bidirectional DC–DC converter

As depicted in the curves, at the rated power level of 400 W, the prototype achieves its maximum efficiency in step-down mode at 94.41% with the highest output voltage on the low side. Conversely, the minimum efficiency is recorded at 89.78% when the output voltage is at 40 V. In step-up mode, the prototype attains its peak efficiency of 94.09% when the input voltage on the low side is 120 V, while the minimum efficiency drops to 87.37% with the highest voltage gain. As the output power decreases to 200 W, the overall operational efficiency declines, maintaining the same variation trend associated with the low voltage side.

4.4.2 *Application Analysis*

By applying the variable switching frequency method, the switched-quasi-Z-source bidirectional DC–DC converter can achieve high efficiency across nearly the entire operational range. Furthermore, the topology features a common ground structure, resulting in reduced inductor volume and high power density. Consequently, the converter is suitable for applications in photovoltaic systems and energy storage systems, particularly in scenarios with stringent size and weight requirements, such as unmanned aircraft and portable equipment.

In comparison to the switched-quasi-Z-source topology, the hybrid switched-capacitor/switched-quasi-Z-source bidirectional DC–DC converter offers a wider voltage gain range and lower voltage stresses with the same output. However, the converter lacks a common ground and requires additional components, making it more challenging to improve the power density of the system. Therefore, this converter is better suited for applications in fuel cell vehicles and energy storage systems characterized by higher voltage levels and larger system sizes.

4.5 Conclusion

In conclusion, this chapter offers a comprehensive analysis of both the hybrid switched-capacitor/switched-quasi-Z-source converter and the switched-quasi-Z-source converter. Additionally, a novel variable switching frequency control strategy is introduced, specifically tailored for the switched-quasi-Z-source converter, aimed at achieving soft-switching operation. The synergistic combination of these converters and the variable switching frequency scheme presents significant advantages, making them promising solutions for various energy conversion systems.

The hybrid switched-capacitor/switched-quasi-Z-source bidirectional DC–DC converter exhibits a simple structure and lower voltage stresses across the power switches. With its expanded voltage gain range, this converter finds suitability in applications involving hybrid energy sources. On the other hand, the switched-quasi-Z-source bidirectional DC–DC converter offers enhanced efficiency and reduced power losses compared to conventional converters. The introduced variable switching frequency control strategy applied to the switched-quasi-Z-source bidirectional DC–DC converter enables soft-switching operation. By dynamically adjusting the switching frequency, the converter achieves ZVS ON for all power switches and quasi-ZCS OFF for synchronous rectification switches. This results in minimized power losses and improved overall efficiency.

In summary, both the hybrid switched-capacitor/switched-quasi-Z-source bidirectional DC–DC converter and the switched-quasi-Z-source bidirectional DC–DC converter, combined with the variable switching frequency control strategy, provide comprehensive solutions for soft-switching power conversion. These converters offer advantages such as a wider voltage gain range and improved efficiency.

Chapter 5
Switched-Capacitor DC–DC Converters

5.1 Introduction

For the purpose of achieving high voltage gain and low voltage stress in DC–DC converters, numerous topologies have been proposed by researchers worldwide [103–105]. Among these, non-isolated bidirectional DC–DC converters have gained prominence due to their cost-effectiveness and compactness, as they eliminate the need for a transformer and possess a common ground characteristic. Two types of non-isolated bidirectional DC–DC converters can be distinguished: coupled-inductor converters and uncoupled-inductor converters. Uncoupled-inductor converters offer advantages such as simplicity, low cost, and reduced magnetic losses. However, a specific converter discussed in [106] that improves upon the traditional buck/boost bidirectional DC–DC converter to double its voltage gain faces an issue of high-frequency pulse-width modulation (PWM) voltage across the potential difference between the grounds of the input voltage source side and the load side. This issue arises due to the presence of a diode between the two grounds, potentially introducing extra problems related to du/dt and limiting its applications.

To address these concerns, a soft-switching bidirectional DC–DC converter is proposed in [107], which effectively reduces switching losses but maintains the same voltage gain as the traditional buck/boost converter. However, this solution requires a greater number of power devices. Another converter introduced in [108] offers a small number of power devices, enabling high voltage gain and low voltage stress. Nonetheless, it still lacks a common ground structure between the input voltage source side and the load side. In terms of modulation methods, interleaved modulation proves beneficial in reducing current ripple in the low-voltage side. On the other hand, coupled-inductor bidirectional DC–DC converters can achieve arbitrarily high voltage gain by adjusting the turns ratio of the coupled inductor. A high-efficiency bidirectional DC–DC converter with a high voltage gain is presented in [109]. Although the voltage gain is significantly high, the voltage stress across the high-side power switch remains as high as the high-side voltage. Additionally,

Y. Zhang and S. Gao, *High Efficiency Non-isolated DC-DC Converters with Wide Voltage Gain Range for Renewable Energies*, CPSS Power Electronics Series,
https://doi.org/10.1007/978-981-97-0648-8_5

this converter employs three power diodes, increasing the total number of devices. Another converter discussed in [110] demonstrates good performance in terms of voltage gain and voltage stress. However, it suffers from substantial current ripple in the low-voltage side, which can potentially harm the power source's lifespan and reduce conversion efficiency. Phase-shifting control techniques, as presented in [111], can be employed in bidirectional DC–DC converters to achieve zero voltage switching (ZVS). However, the pulse current in the low-voltage side tends to enlarge the current ripple. The voltage stress of the converter specified in [112] is constrained by the turns ratio of the coupled inductor. Similarly, the converter described in [113] shares the same issues as the converter mentioned in [111], as it also lacks a common ground structure.

Another Boost three-level DC–DC converter discussed in [114] addresses the issue of a common ground connection between the output and input sides. However, the voltage gain of this converter is still constrained. Additionally, it requires a complex control scheme to balance the voltage of the flying capacitors, further adding to its complexity. In [115], a multilevel DC–DC Boost converter achieves both high voltage gain and low voltage stress on power semiconductors. Nevertheless, this converter is deemed excessively intricate for automotive applications and necessitates cost and size reductions. A different approach is presented in [116], where a Z-source network is employed to achieve higher voltage gain. However, the output and input sides of this converter do not share a common ground, which can introduce safety concerns and additional EMI-related challenges. Various types of DC–DC converters with coupled inductors are discussed in [117–120]. The converter with coupled inductors in [117] demonstrates high voltage gain and low voltage stress on power semiconductors but suffers from significant input current ripple. Interleaved boost converters with coupled inductors, as mentioned in [118], exhibit high voltage gain and low input current ripple. However, these converters still face high voltage stress on power semiconductors. Stacked high step-up coupled-inductor boost converters discussed in [119] achieve very high voltage gain and low voltage stress on power semiconductors. Nevertheless, these converters exhibit higher input current ripple. The interleaved boost converters with winding-cross-coupling mentioned in [119] provide high voltage gain, low voltage stress on power semiconductors, and low input current ripple. Cascaded boost converters in [120] also achieve high voltage gain. However, these converters suffer from high input current ripple, and their efficiencies are determined by the product of the efficiencies of each stage.

In this chapter, a family of switched-capacitor DC–DC converters are introduced to overcome the limitations associated with traditional converters and provide enhanced performance and efficiency. Compared to conventional converters, these advanced DC–DC converters exhibit benefits such as low input current ripple, low voltage stress on power semiconductors, and high voltage gain. In addition, a coupled-inductor bidirectional DC–DC converter is also presented, which offers a high conversion ratio in both boost and buck modes while reducing voltage stresses on power switches. This improvement enhances the converter's overall efficiency and reliability. Furthermore, the converter effectively minimizes current ripple on the input voltage source side and load side, resulting in improved stability and reduced power losses.

5.2 Switched-Capacitor Bidirectional DC–DC Converter

The conventional Boost converter exhibits a simple common ground structure and requires only few components to achieve voltage lifting. However, as analyzed before, the narrow voltage gain range put limit on its application. To deal with this issue, the switched-capacitor bidirectional DC–DC converter can be comprised by combining the Boost converter with a switched-capacitor cell. As shown in Fig. 5.1, by connecting the switched-capacitor cell between the drain of Q_1 and the positive terminal of diode in the Boost converter, and replacing all diodes with power switches, the topology of switched-capacitor bidirectional DC–DC converter can be obtained [121].

This section presents and analyzes the derived switched-capacitor bidirectional DC–DC converter, exhibits high voltage gain and low voltage stresses across power switches. The converter is suitable for integration into hybrid power systems, enabling

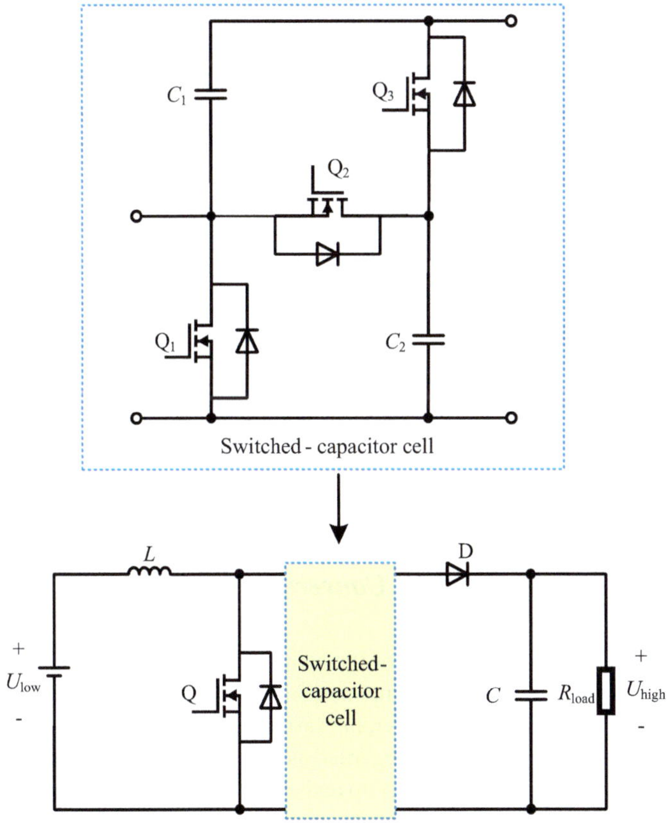

Fig. 5.1 Derivation of switched-capacitor bidirectional DC–DC converter

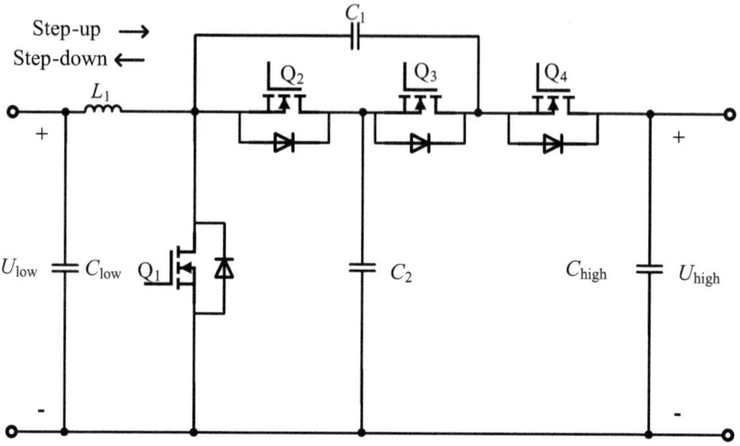

Fig. 5.2 Topology of the switched-capacitor bidirectional DC–DC converter

bidirectional power flow with a wide voltage gain range. Additionally, ZVS operation can be implemented without the need for auxiliary circuits. Furthermore, the adoption of synchronous rectification reduces conducting losses in power switches, enhancing the conversion efficiency of the converter. In contrast to coupled inductors, the utilization of switched-capacitors simplifies the design procedure while allowing for the achievement of continuous inductor current and dynamic balancing of switched-capacitor voltages.

The topology of the switched-capacitor bidirectional DC–DC converter is shown in Fig. 5.2, employing four power switches Q_1–Q_4, and a filtering inductor L. C_1 and C_2 are the energy storage capacitors, while C_{low} and C_{high} are the filtering capacitors on the low voltage side and high voltage side, respectively. Compared with the conventional Boost converter, the switched-capacitor bidirectional DC–DC converter can achieve higher voltage gain range.

5.2.1 Operating Principle of the Switched-Capacitor Bidirectional DC–DC Converter

In terms of the operating principle of the switched-capacitor bidirectional DC–DC converter, it is assumed that all components are ideal, and the inductance is large enough for CCM operation. Moreover, the variation of inductor current is considered linear, while capacitor voltages are regarded as constant during each switching cycle. The converter is capable of both step-up and step-down operation. In step-up mode, power switch Q_1 acts as the main power switch, whereas in step-down mode, the main power switches are Q_2 and Q_4.

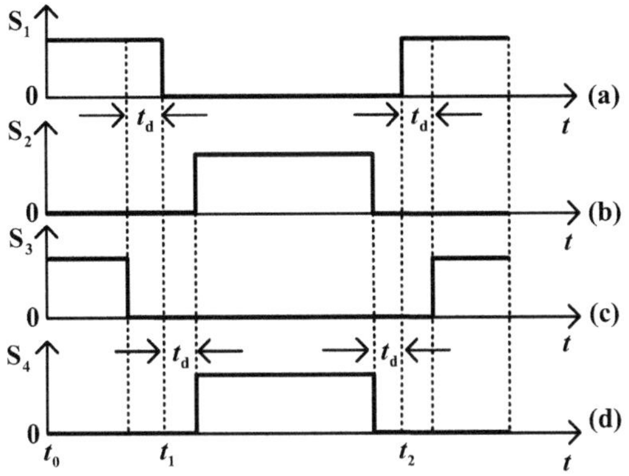

Fig. 5.3 Synchronous rectification for the switched-capacitor bidirectional DC–DC converter

Considering the relatively large on-state resistances of body diodes in power switches, leading to higher conducting losses, the synchronous rectification strategy is employed to enhance conversion efficiency through complementary gate signals, as shown in Fig. 5.3, where S_1–S_4 represent the gate signals for power switches Q_1–Q_4, respectively. In the illustration, gate signals S_1 and S_3 for power switches Q_1 and Q_3 are synchronous, as are S_2 and S_4 for power switches Q_2 and Q_4. Notably, S_1 and S_3 are complementary to S_2 and S_4. Additionally, a small dead time is reserved to prevent short circuit faults.

The switched-capacitor bidirectional DC–DC converter in step-up mode operates through two switching states, enabling the conversion of energy from the low voltage side to the high voltage side. The key waveforms and switching states of the converter in step-up mode are presented in Figs. 5.4 and 5.5.

State 1) $S_1S_2S_3S_4 = 1010$

When the main power switch Q_1 is turned ON, while Q_2 and Q_4 are turned OFF, the inductor L is charged through Q_1. Thus, the inductor current increases linearly, as shown in Fig. 5.4. The output load R_{load} is powered by C_{high}. Meanwhile, the body diodes of Q_3 are conducted. The capacitor C_1 is charged by C_2 through Q_1 and Q_3. Since the voltage across Q_3 is clamped at zero by the body diode, ZVS ON can be achieved.

State 2) $S_1S_2S_3S_4 = 1010$

When power switches Q_1 and Q_3 are turned OFF, the body diodes of Q_2 and Q_4 are conducted. Thus, the synchronous rectifiers Q_2 and Q_4 are turned ON, so as to reduce the on-state resistances. Besides, the voltages across Q_2 and Q_4 are both clamped at zero by the corresponding body diodes, thus ZVS ON can be realized. At the same time, the inductor L is discharged to C_2 through Q_2, thus the inductor current decreases linearly. The capacitor C_1 is discharged to supply energy to the load through Q_4.

Similar to the condition in step-up mode, the operation of the switched-capacitor bidirectional DC–DC converter in step-down mode also consists of two switching

Fig. 5.4 Key waveforms of the switched-capacitor bidirectional DC–DC converter in step-up mode

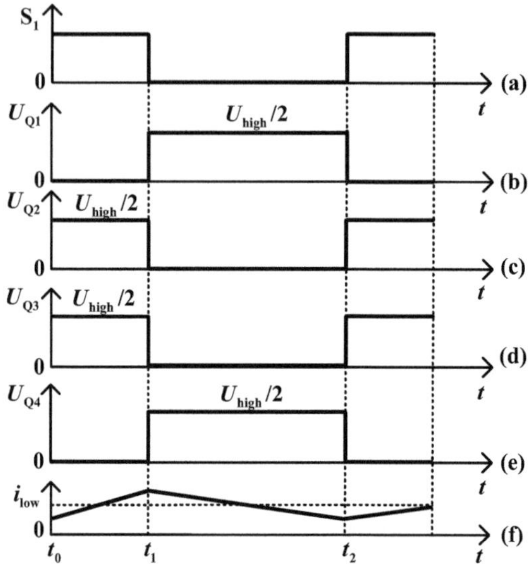

states. The key waveforms and switching states of the converter in step-down mode are presented in Figs. 5.6 and 5.7, respectively.

State 1) $S_1S_2S_3S_4 = 0101$

When the power switches Q_1 and Q_3 are turned OFF, and Q_2 and Q_4 are turned ON, the inductor L is charged by capacitor C_2 through Q_2, that the inductor current increases linearly. Meanwhile, the capacitor C_1 is charged by the input source U_{high} on the high voltage side. The output load R_{load} on the low voltage side is powered by C_2 and U_{high}.

State 2) $S_1S_2S_3S_4 = 1010$

When the power switches Q_2 and Q_4 are turned OFF, while Q_1 and Q_3 are turned ON, the body diode of Q_3 is conducted. Due to the synchronous rectification, Q_1 is ZVS ON and OFF, and the conducting losses of Q_2 is reduced. During this state, the output load is powered by the inductor L. The capacitor C_2 is charged by C_1 through Q_1 and Q_3.

5.2.2 Control Strategy of the Switched-Capacitor Bidirectional DC–DC Converter

The switched-capacitor bidirectional DC–DC converter is applicable to energy storage devices in hybrid power systems, offering the capability to regulate charging

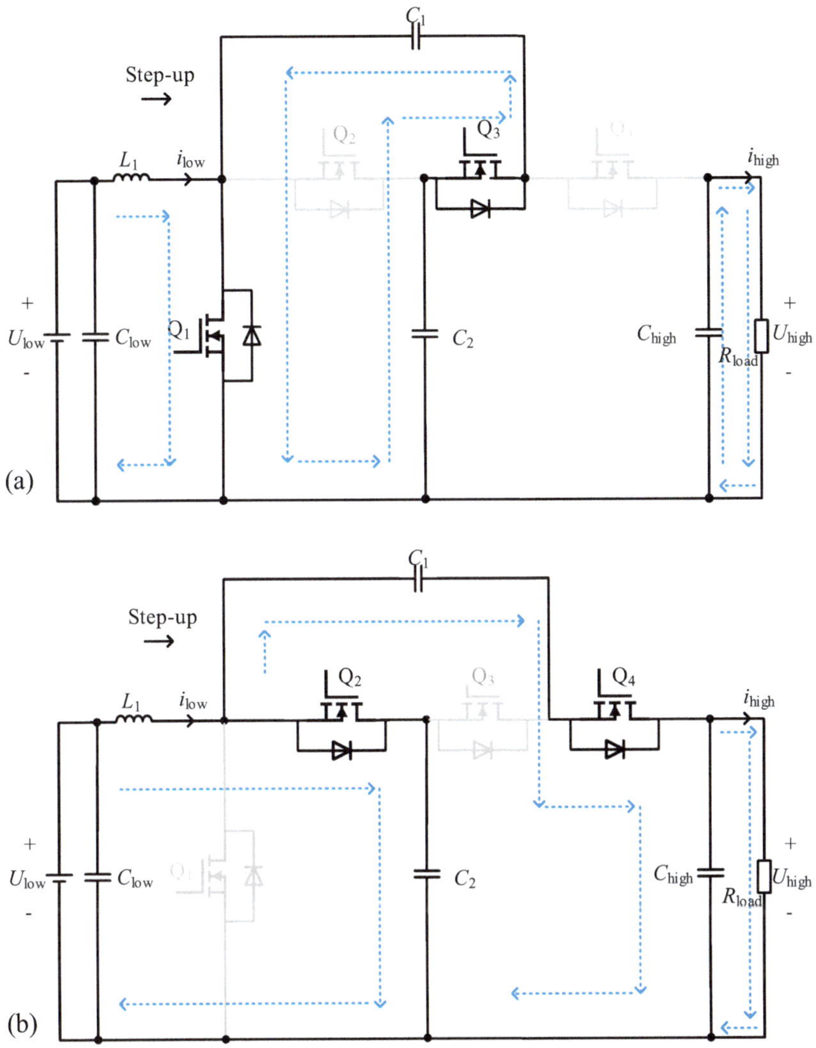

Fig. 5.5 Switching states of the switched-capacitor bidirectional DC–DC converter in step-up mode during **a** State 1 and **b** State 2

and discharging power for batteries and super-capacitors. Achieving high performance with the converter necessitates the practical incorporation of a digital controller and sensing circuit. The experimental configuration for the switched-capacitor bidirectional DC–DC converter is illustrated in Fig. 5.8, where the sensing circuit samples inductor currents and input/output voltages. The sampled signals undergo processing by the digital signal controller (DSP) to generate PWM signals. Beyond data processing, the DSP is responsible for duty cycle calculation and mode selection.

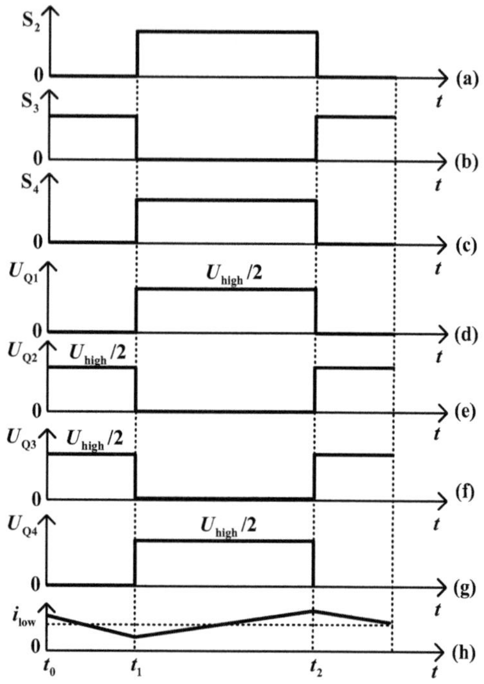

Fig. 5.6 Key waveforms of the switched-capacitor bidirectional DC–DC converter in step-down mode

In order to regulate the voltages on the input and output ports, as well as the inductor current, the double closed-loop control strategy is adopted for the switched-capacitor bidirectional DC–DC converter. As can be observed from Fig. 5.9, the control diagram contains an inner current loop and outer voltage loop. Furthermore, the mode of voltage loop is selected by the mode index U_c, which represents the charging and discharging condition for the energy storage devices according to its SOC. For example, when the step-down mode is selected, the reference for the voltage loop is $U_{\mathrm{ref\_Buck}}$, and the voltage U_{low} is sampled as the feedback signal. The voltage error is calculated, and the reference current I_{ref} is provided through the voltage controller. The inductor current is sampled as the current feedback. By means of the current controller, the duty cycle for Q_2 and Q_4 is obtained, and the PWM signals can be generated.

5.2.3 Steady-State Analysis of the Switched-Capacitor Bidirectional DC–DC Converter

According to the volt-second balance and the above analysis on operating principle of the switched-capacitor bidirectional DC–DC converter, the following equations can be obtained, where d_{Boost} is the duty cycle of main power switch Q_1 in step-up mode.

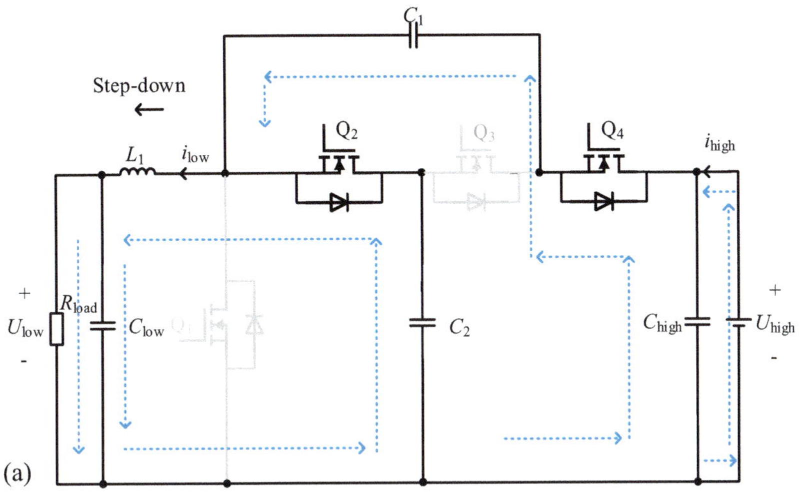

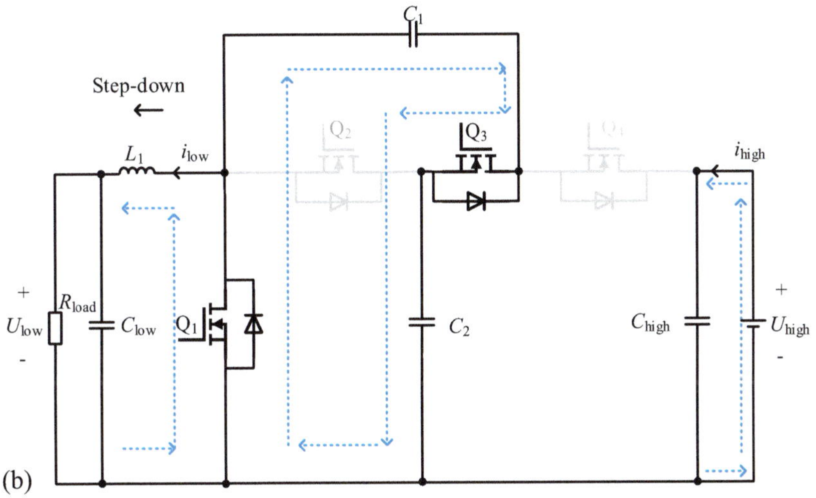

Fig. 5.7 Switching states of the switched-capacitor bidirectional DC–DC converter in step-down mode during **a** State 1 and **b** State 2

$$\begin{cases} d_{\mathrm{Boost}} U_{\mathrm{low}} = (1 - d_{\mathrm{Boot}})(U_{\mathrm{C2}} - U_{\mathrm{low}}) \\ U_{\mathrm{C1}} + U_{\mathrm{C2}} = U_{\mathrm{high}} \\ U_{\mathrm{C1}} = U_{\mathrm{C2}} \end{cases} \tag{5.1}$$

In turn, the relationship of capacitor voltages and duty cycles can be derived as below.

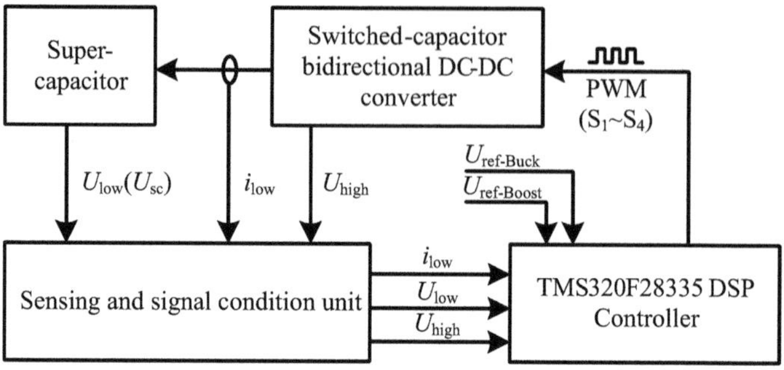

Fig. 5.8 Experimental configuration for the switched-capacitor bidirectional DC–DC converter

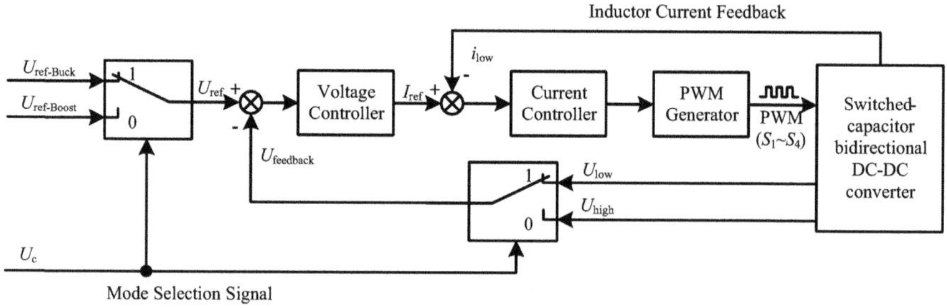

Fig. 5.9 Double closed-loop for the switched-capacitor bidirectional DC–DC converter

$$\begin{cases} U_{C1} = U_{C2} = \dfrac{1}{1 - d_{\text{Boost}}} U_{\text{low}} \\ U_{\text{high}} = \dfrac{2}{1 - d_{\text{Boost}}} U_{\text{low}} \end{cases} \tag{5.2}$$

Hence, the voltage gain of the converter is twice the conventional Boost converter. Besides, the voltages across energy storage capacitors are half the high voltage side of the converter.

Since the input and output power are equal, (5.3) can be obtained.

$$I_{\text{low}} U_{\text{low}} = I_{\text{high}} U_{\text{high}} \tag{5.3}$$

By combining (5.2) and (5.3), the relationship between currents on the low and high voltage sides in step-up mode can be obtained as follows.

$$I_{\text{low}} = \frac{2}{1 - d_{\text{Boost}}} I_{\text{high}} \tag{5.4}$$

Similarly, in step-down mode, the capacitor voltages can be described as follows, where d_{Buck} is the duty cycle of main power switches Q_2 and Q_4 of the converter in step-down mode.

$$\begin{cases} (1 - d_{\text{Buck}})U_{\text{low}} = d_{\text{Buck}}(U_{\text{C2}} - U_{\text{low}}) \\ U_{\text{C1}} + U_{\text{C2}} = U_{\text{high}} \\ U_{\text{C1}} = U_{\text{C2}} \end{cases} \tag{5.5}$$

The relationship of the duty cycle and capacitor voltages can be derived from (5.5) as below.

$$\begin{cases} U_{\text{C1}} = U_{\text{C2}} = \dfrac{1}{2}U_{\text{high}} \\ U_{\text{low}} = \dfrac{d_{\text{Buck}}}{2}U_{\text{high}} \end{cases} \tag{5.6}$$

Thus, the average input and output currents are obtained as follows.

$$I_{\text{high}} = \frac{d_{\text{Buck}}}{2}I_{\text{low}} \tag{5.7}$$

As evident from Eqs. (5.2), (5.4), (5.6), and (5.7), the steady-state characteristics of the switched-capacitor bidirectional DC–DC converter in both step-up and step-down modes remain consistent. The capacitor voltages of C_1 and C_2 are observed to be half the voltage on the high voltage side, resulting in a voltage gain twice that of the conventional Boost converter. In essence, the converter can achieve a higher voltage gain with low voltage stress.

Based on the switching states shown in Figs. 5.5 and 5.7, the voltage stress across Q_1 is equal to the capacitor voltage of C_2. Besides, the voltage stresses across Q_2 and Q_3 are equal to the capacitor voltage of C_1, and the voltage stress across Q_4 is equal to the difference between voltages on the high voltage side and capacitor C_2.

$$\begin{cases} U_{\text{Q1}} = U_{\text{C2}} = \dfrac{U_{\text{high}}}{2} \\ U_{\text{Q2}} = U_{\text{Q3}} = U_{\text{C1}} = \dfrac{U_{\text{high}}}{2} \\ U_{\text{Q4}} = U_{\text{high}} - U_{\text{C2}} = \dfrac{U_{\text{high}}}{2} \end{cases} \tag{5.8}$$

$$\begin{cases} I_{\text{Q1}} = \left(\dfrac{2}{1 - d_{\text{Boost}}} + \dfrac{1}{d_{\text{Boost}}}\right)I_{\text{high}} \\ I_{\text{Q2}} = I_{\text{Q4}} = \dfrac{1}{1 - d_{\text{Boost}}}I_{\text{high}} \\ I_{\text{Q3}} = \dfrac{1}{d_{\text{Boost}}}I_{\text{high}} \end{cases} \tag{5.9}$$

The respective voltage stresses U_{Q1}–U_{Q4} of Q_1–Q_4 are displayed in (5.8), and the current stresses I_{Q1}–I_4 of Q_1–Q_4 are displayed in (5.9). The voltage stresses across all power switches are equal to half of the high-side voltage U_{high}. Moreover, the current stresses of power switches can be expressed by the duty cycle d_{Buck} in step-down mode as follows.

$$\begin{cases} I_{Q1} = \left(1 + \dfrac{d_{Buck}}{2(1 - d_{Buck})}\right) I_{low} \\ I_{Q2} = I_{Q4} = \dfrac{1}{2} I_{low} \\ I_{Q3} = \dfrac{d_{Buck}}{2(1 - d_{Buck})} I_{low} \end{cases} \tag{5.10}$$

According to (5.9) and (5.10), in both step-up and step-down modes, the current stresses of Q_2–Q_4 are relatively low and the current stress of Q_1 is a little bit larger than the average current on the low voltage side. As a result, the component selection for the switched-capacitor bidirectional DC–DC converter would be easier.

5.3 Three-Phase Interleaved Switched-Capacitor Bidirectional DC–DC Converter

Based on the switched-capacitor bidirectional DC–DC converter introduced in the previous section as a basic module, the converter can be further expanded to acquire higher voltage gain and output power level. As illustrated in Fig. 5.10, by connecting the energy storage capacitor to the inductor of the next phase, the multiphase interleaved topology of switched-capacitor bidirectional DC–DC converter can be derived [122].

In this section, a three-phase interleaved switched-capacitor bidirectional DC–DC converter is presented, which can achieve higher power conversion, as well as current ripple reduction on the low voltage side. In additional, the converter has the capability of wide voltage gain range and low voltage stresses across power switches. Furthermore, the common ground and multiphase structure enables lower electromagnetic interferences and current ripple on the low voltage side.

The topology of the three-phase interleaved switched-capacitor bidirectional DC–DC converter is depicted in Fig. 5.11, where Q_1–Q_8 are power switches, L_1–L_3 are filtering inductors, C_1, C_2, C_{low} and C_{h1}–C_{h3} are the energy storage capacitors.

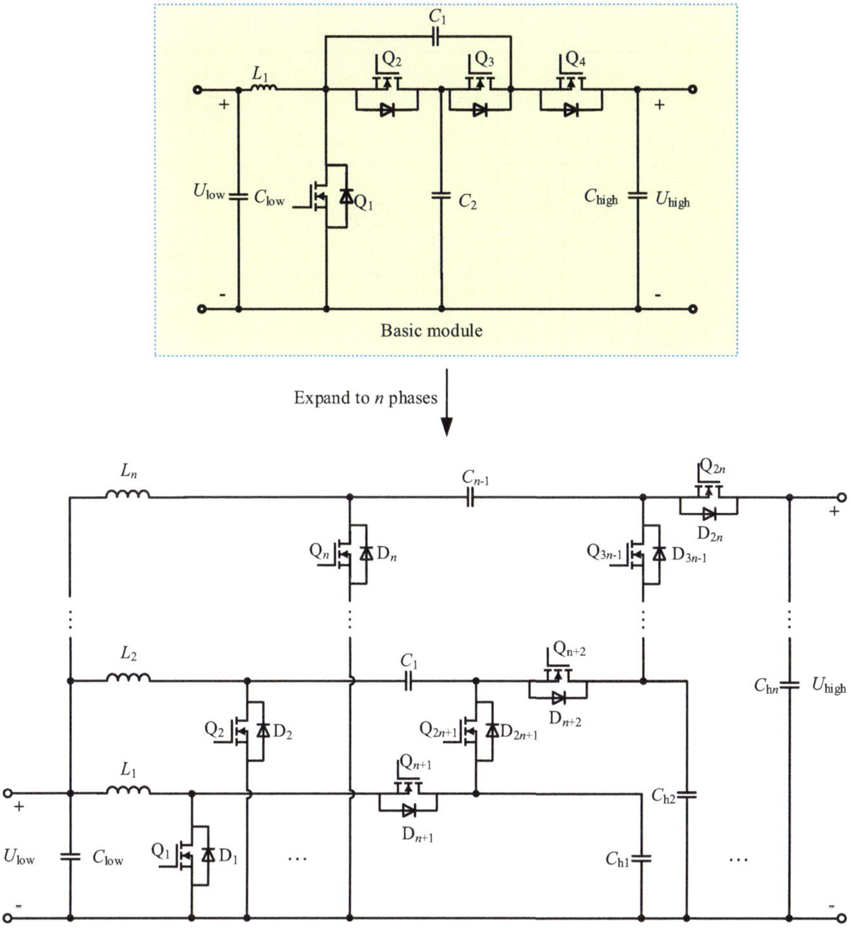

Fig. 5.10 Derivation of multiphase topology based on the switched-capacitor bidirectional DC–DC converter

5.3.1 *Operating Principle of the Three-Phase Interleaved Switched-Capacitor Bidirectional DC–DC Converter*

Due to the capability of bidirectional power flow, the three-phase interleaved switched-capacitor bidirectional DC–DC converter can operate in both step-up and step-down modes. In step-up mode, the discharging of the input source on the low voltage side can be regulated. Meanwhile, Q_1–Q_3 are the main power switches, of which the gate signals S_1–S_3 are 120° shifted, and Q_4–Q_8 act as synchronous rectifiers. Whereas, when the converter operates in step-down mode, the energy is converted from the DC bus on the high voltage side to the low voltage sider for charging the batteries/super-capacitors. The main power switches are Q_4–Q_8, and

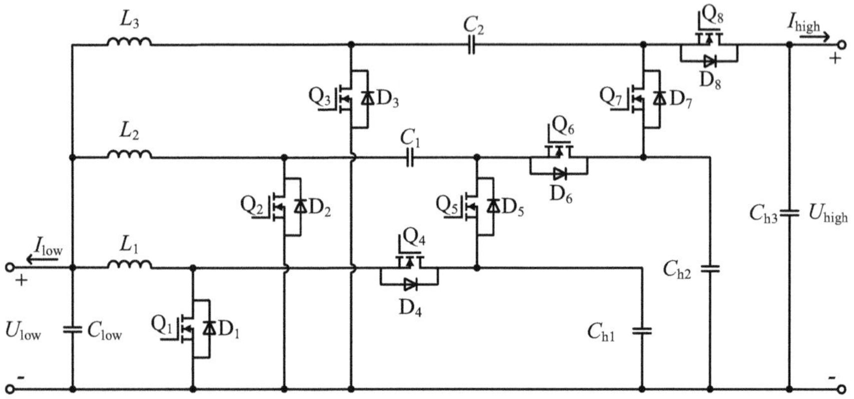

Fig. 5.11 Topology of the three-phase interleaved switched-capacitor bidirectional DC–DC converter

Q_1–Q_3 are the synchronous rectifiers. The gate signals of Q_4, Q_6, and Q_8 are 120° shifted, while Q_5 and Q_7 are complementary to Q_6 and Q_8, respectively. The converter is assumed to operates in continuous current mode, so that the key waveforms of the three-phase interleaved switched-capacitor bidirectional DC–DC converter in step-up mode is illustrated in Fig. 5.12, where U_{Q1}–U_{Q8} are the respective voltage stresses across power switches Q_1–Q_8, and i_{L1}–i_{L3} and i_{low} are the currents of inductor L_1–L_3 and low voltage side, respectively.

The operation of the three-phase interleaved switched-capacitor bidirectional DC–DC converter can be divided into three conditions according to the range of duty cycle. The corresponding duty cycles of the main power switches in step-up and step-down modes are denoted as d_{Boost} and d_{Buck}, respectively. Therefore, there are three conditions: $0 < d_{\mathrm{Boost}}$ (d_{Buck}) $< 1/3$, $1/3 < d_{\mathrm{Boost}}$ (d_{Buck}) $< 2/3$, $2/3 < d_{\mathrm{Boost}}$ (d_{Buck}) < 1. The condition of $1/3 < d_{\mathrm{Boost}}$ (d_{Buck}) $< 2/3$ is discussed in this section, as presented in Fig. 5.12. The other cases follow the similar operating principle.

The switching states of the three-phase interleaved switched-capacitor bidirectional DC–DC converter in step-up mode are illustrated in Fig. 5.13. As can be observed from the illustration, there are six switching states within each switching cycle under the condition that $1/3 < d_{\mathrm{Boost}} < 2/3$, which are explained in detail as below.

State 1) $S_1S_2S_3 = 101$

When the power switches Q_1 and Q_3 are turned ON, whereas Q_3 is turned OFF, the inductors L_1 and L_3 are charged while L_2 is discharged. Meanwhile, the body diodes of Q_6 and Q_7 are conducted. The capacitors C_2 and C_{h2} are both charged by C_1, and the output load on the high voltage side is powered by the capacitor C_{h3}.

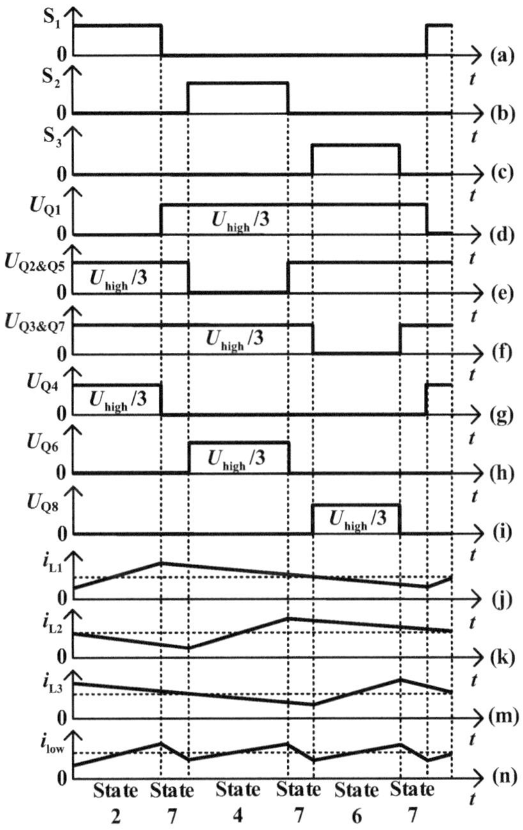

Fig. 5.12 Key waveforms of the three-phase interleaved switched-capacitor bidirectional DC–DC converter in step-up mode

State 2) $S_1S_2S_3 = 100$

When Q_1 is turned ON and Q_2, Q_3 are both turned OFF, the body diodes of Q_6 and Q_8 are conducted. Meanwhile, the inductors L_2 and L_3, as well as the capacitors C_2 and C_3 are discharged, whereas L_1 is charged by the input source on the low voltage side through Q_1. The capacitors $C_{\text{h}2}$ and $C_{\text{h}3}$ are both charged.

State 3) $S_1S_2S_3 = 110$

When Q_2 and Q_3 are turned ON, while Q_3 is turned OFF, the body diodes of Q_5 and Q_8 are conducted. The inductor L_3 is discharged to power the load, whereas L_1 and L_2 are charged by the input source on the low voltage side through Q_1 and Q_2, respectively. The capacitor C_1 is charged by C_{h1}, and C_2 is discharged.

State 4) $S_1S_2S_3 = 010$

When Q_1 and Q_3 are turned OFF, and Q_2 is turned ON, the body diodes of Q_4, Q_5, and Q_8 are conducted. The inductor L_2 is charged via Q_2, while L_1 and L_3 are discharged. Both capacitors C_1 and C_{h1} are charged by the input source. Meanwhile, C_2 is discharged to charge $C_{\text{h}3}$ and power the load.

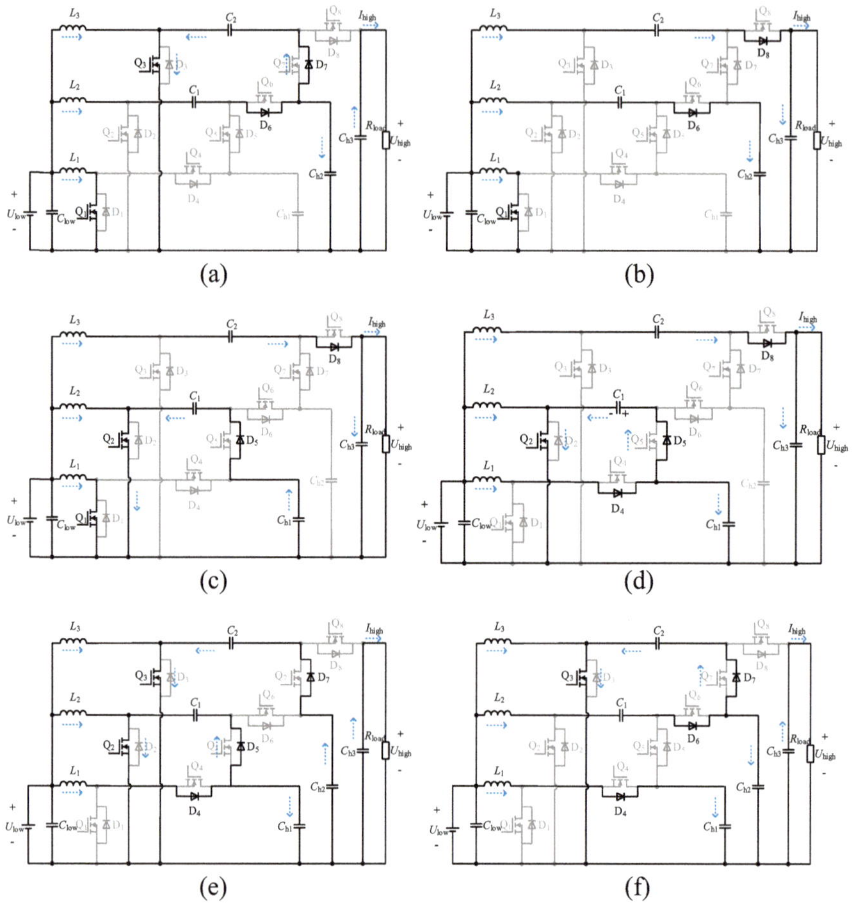

Fig. 5.13 Switching states of the three-phase interleaved switched-capacitor bidirectional DC–DC converter in step-up mode during **a** State 1, **b** State 2, **c** State 3, **d** State 4, **e** State 5, and **f** State 6

State 5) $S_1S_2S_3 = 011$

When Q_1 is turned OFF, whereas Q_2 and Q_3 are turned ON, the inductors L_2 and L_3 are charged while L_1 is discharged. During this state, the body diodes of Q_4, Q_5, and Q_7 are conducted. Meanwhile, the capacitors C_{h1} and C_1 are charged by the input source, and C_2 is charged by C_{h2}. The output load on the high voltage side is powered by C_{h3}.

State 6) $S_1S_2S_3 = 001$

When Q_1 and Q_2 are both turned OFF, and Q_3 is turned ON, the body diodes, only the inductor L_3 is charged by the input source. The inductors L_1 and L_2, as well as the capacitor C_1 are discharged, to charge C_2, C_{h1} and C_{h2}. The output load is still powered by C_{h3}.

In addition to the above six switching states, there are two additional states (i.e., State 7 and State 8) exist in the conditions that $0 < d_{\text{Boost}} < 1/3$ and $2/3 < d_{\text{Boost}} < 1$, which are $S_1S_2S_3 = 000$ and $S_1S_2S_3 = 111$.

When the three-phase interleaved switched-capacitor bidirectional DC–DC converter operates in step-down mode, the energy is transferred from the high voltage side to the low voltage side. The converter can be utilized as the charging regulator for batteries/super-capacitors in the hybrid energy systems. As shown in Fig. 5.14, in step-down mode, there are six switching states during each switching cycle, which are analyzed as follows.

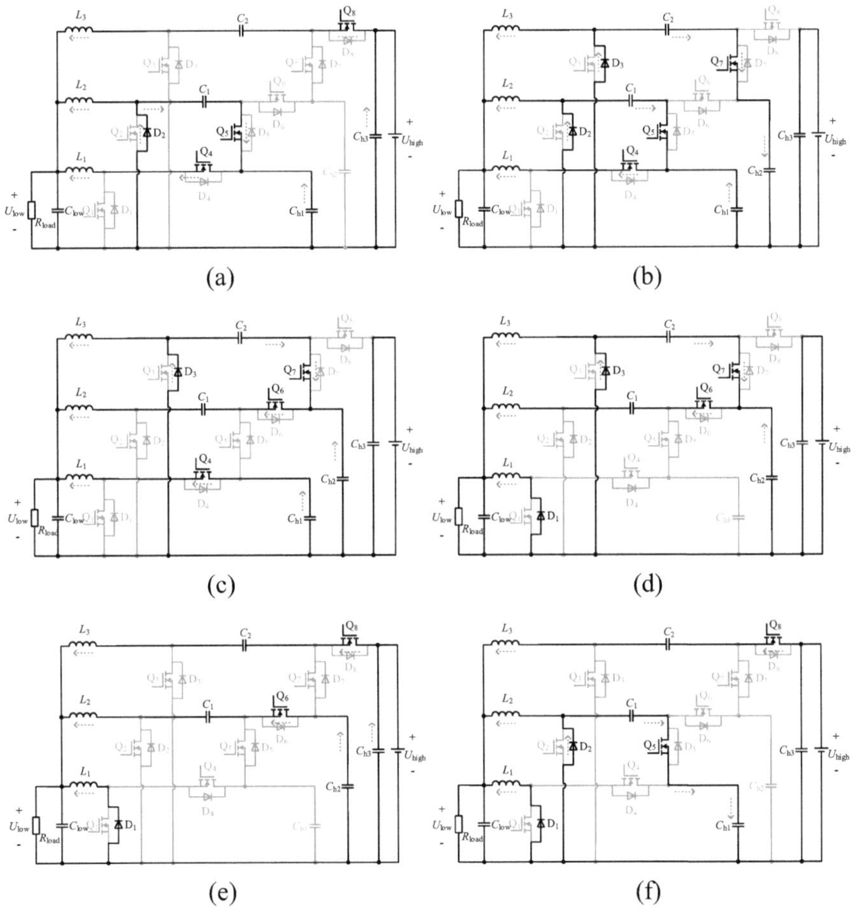

Fig. 5.14 Switching states of the three-phase interleaved switched-capacitor bidirectional DC–DC converter in step-down mode during **a** State 1, **b** State 2, **c** State 3, **d** State 4, **e** State 5, and **f** State 6

State 1) $S_4S_5S_6S_7S_8 = 11001$

When Q_4, Q_5, and Q_8 are turned ON, whereas Q_6 and Q_7 are turned OFF, the body diode of Q_2 is conducted. The inductor L_2 is discharged to power the output load on the low voltage side. Meanwhile, C_{h2} is charged by C_2, while the capacitors C_1 and C_{h1} are discharged to the load.

State 2) $S_4S_5S_6S_7S_8 = 11010$

When Q_4, Q_5, and Q_7 are turned ON, whereas Q_6 and Q_8 are turned OFF, the body diode of Q_2 and Q_3 are both turned ON. Thus, the inductors L_2 and L_3 are discharged to power the load via D_2 and D_3, respectively. The capacitors C_1, C_2 and C_{h1} are all discharged, whereas C_{h2} is charged by C_2.

State 3) $S_4S_5S_6S_7S_8 = 10110$

When Q_4, Q_6 and Q_7 are turned ON, whereas Q_5 and Q_8 are turned OFF, the inductor L_3 is discharged through the conducted body diode D_3 of Q_3. Whereas, the inductors L_1 and L_2 are charged by C_{h1} and C_{h2}, respectively. During this state, C_1 is charged and C_2 is discharged.

State 4) $S_4S_5S_6S_7S_8 = 00110$

When Q_6 and Q_7 are turned ON, whereas Q_4, Q_5, and Q_8 are turned OFF, the body diodes of Q_1 and Q_3 are conducted, through which the inductors L_1 and L_3 are discharged. Meanwhile, the capacitor C_1 is charged, and C_2 and C_{h2} are discharged to power the load.

State 5) $S_4S_5S_6S_7S_8 = 00101$

When Q_6 and Q_8 are turned ON, whereas Q_4, Q_5, and Q_7 are turned OFF, the load on the low voltage side is powered by the inductor L_1 via the body diode D_1 of Q_1. At the same time, the inductors L_2 and L_3, as well as the capacitors C_1 and C_2 are charged by C_{h2} and C_{h3}, respectively.

State 6) $S_4S_5S_6S_7S_8 = 01001$

When Q_5 and Q_8 are turned ON, whereas Q_4, Q_6, and Q_7 are turned OFF, the inductors L_1 and L_2 are discharged through the body diodes D_1 and D_2 of Q_1 and Q_2, respectively. Meanwhile, the capacitor C_{h1} is charged by C_1 through Q_5, and C_2 is charged by C_{h3} through Q_8.

In terms of the operating conditions of $0 < d_{\text{Buck}} < 1/3$ and $2/3 < d_{\text{Buck}} < 1$, there are two extra switching states that $S_4S_5S_6S_7S_8 = 01010$ and $S_4S_5S_6S_7S_8 = 10101$.

5.3.2 Steady-State Characteristics of the Three-Phase Interleaved Switched-Capacitor Bidirectional DC–DC Converter

Based on the volt-second balance principle, the voltages of the input/output and capacitors derived from L_1–L_3, which are described by the following equations.

$$\begin{cases} d_{\text{Boost}}U_{\text{low}} + (1 - d_{\text{Boost}})(U_{\text{low}} - U_{\text{Ch1}}) = 0 \\ d_{\text{Boost}}U_{\text{low}} + (1 - d_{\text{Boost}})(U_{\text{low}} + U_{\text{Ch1}} - U_{\text{Ch2}}) = 0 \\ d_{\text{Boost}}U_{\text{low}} + (1 - d_{\text{Boost}})\left(U_{\text{low}} + U_{C2} - U_{\text{high}}\right) = 0 \end{cases} \tag{5.11}$$

$$\begin{cases} d_{\text{Buck}}(U_{\text{low}} - U_{\text{Ch1}}) + (1 - d_{\text{Buck}})U_{\text{low}} = 0 \\ d_{\text{Buck}}(U_{\text{low}} + U_{\text{Ch1}} - U_{\text{Ch2}}) + (1 - d_{\text{Buck}})U_{\text{low}} = 0 \\ d_{\text{Buck}}\left(U_{\text{low}} + U_{C2} - U_{\text{high}}\right) + (1 - d_{\text{Buck}})U_{\text{low}} = 0 \end{cases} \tag{5.12}$$

The relationship between duty cycle and capacitor voltages in step-up and step-down modes is illustrated in (5.11) and (5.12), respectively. Furthermore, in accordance with the aforementioned analysis of switching states, it is observed that the capacitor voltages U_{C1} and U_{Ch2} are equal, and U_{C2} is equal to U_{Ch2}, as expressed below.

$$\begin{cases} U_{\text{C1}} = U_{\text{Ch1}} \\ U_{\text{C2}} = U_{\text{Ch2}} \end{cases} \tag{5.13}$$

As a result, the capacitor voltages can be expressed by duty cycle and input/output voltages as follows.

$$\begin{cases} U_{\text{C1}} = U_{\text{Ch1}} = \dfrac{1}{3}U_{\text{high}} \\ U_{\text{C2}} = U_{\text{Ch2}} = \dfrac{2}{3}U_{\text{high}} \\ U_{\text{high}} = \dfrac{3}{1 - d_{\text{Boost}}}U_{\text{low}} \\ U_{\text{low}} = \dfrac{d_{\text{Buck}}}{3}U_{\text{high}} \end{cases} \tag{5.14}$$

In terms of the voltage stresses of power switches, the following equation can be obtained according to Kirchhoff's voltage law and switching states shown in Figs. 5.13 and 5.14.

$$U_{\text{Q1}} = U_{\text{Q2}} = U_{\text{Q3}} = U_{\text{Q4}} = U_{\text{Q5}} = U_{\text{Q6}} = U_{\text{Q7}} = U_{\text{Q8}} = \frac{U_{\text{high}}}{3} \tag{5.15}$$

The current stresses of power switches can be derived by applying ampere-second balance principle to the capacitors, as presented below.

$$\begin{cases} I_{Q1} = \dfrac{d_{Boost} I_{high}}{1 - d_{Boost}} \\ I_{Q2} = I_{Q3} = \dfrac{I_{high}}{1 - d_{Boost}} \\ I_{Q4} = I_{Q5} = I_{Q6} = I_{Q7} = I_{Q8} = I_{high} \end{cases} \tag{5.16}$$

The respective average inductor currents I_{L1}–I_{L3} of L_1–L_3 are deduced as follows.

$$I_{L1} = I_{L2} = I_{L3} = \frac{I_{low}}{3} \tag{5.17}$$

Thus, the current on the low voltage side is equally shared by the inductor currents without auxiliary current balancing strategies.

The three-phase interleaved switched-capacitor bidirectional DC–DC converter is compared with its counterparts. In [123], auxiliary inductors and capacitors are utilized to enable soft-switching, but this approach leads to higher voltage stresses across power switches. The converter presented in [124] achieves high voltage gain using parallel capacitors and coupled-inductors. Nevertheless, additional circuits are required to suppress voltage spikes resulting from leakage inductance, and the current ripple on the low voltage side remains relatively large. In [125], series-connected capacitors are adopted to enhance voltage gain, but the design falls short in achieving low voltage stresses across power switches and a common ground between the input and output ports. The converter in [126] exhibits a two-phase structure and achieves of high voltage gain. However, the current stress on the main power switch is significant, equivalent to the sum of the two-phase inductor currents. In addition to the current stress, the voltage stress across the power switch equals the high-side voltage, posing challenges for component selection.

Contrastingly, the three-phase interleaved switched-capacitor bidirectional DC–DC converter, introduced in this section, excels in achieving both a wide voltage gain range and low voltage stress. Additionally, the three-phase structure significantly reduces current ripple on the low voltage side, with inductor currents evenly shared without requiring additional control. The voltage gain curves are depicted in Fig. 5.15.

5.4 Input-Parallel Output-Series DC–DC Converter

On the basis of the conventional Cuk converter, a modified topology can be derived by substituting the inductor on the output side with diode, as displayed in Fig. 5.16. The output voltage is opposite the polarity of the input, while the voltage gain is equal to that of the conventional Boost converter. Combing the modified Cuk topology with the Boost converter, the novel topology of input-parallel output-series DC–DC

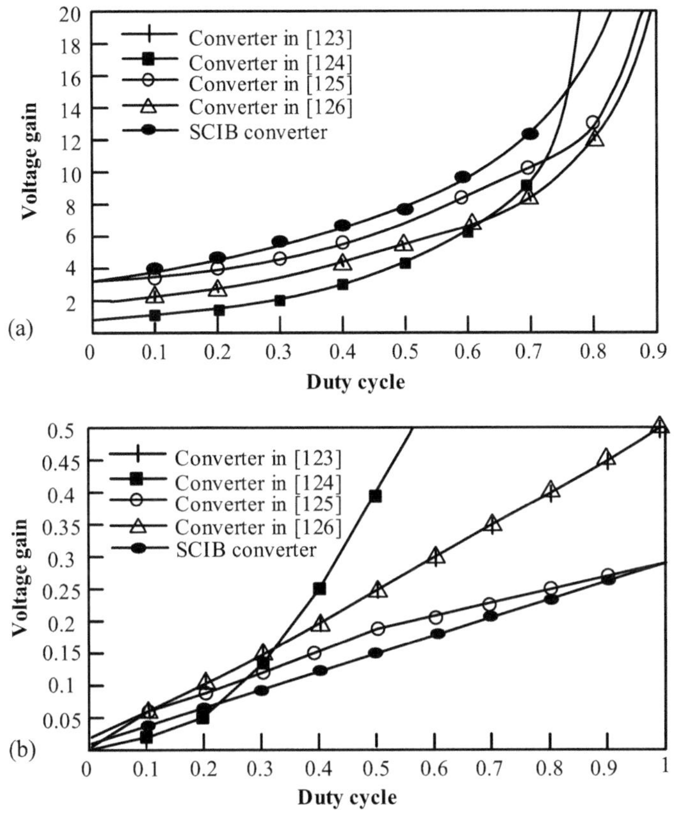

Fig. 5.15 Voltage gain curves of the three-phase interleaved switched-capacitor bidirectional DC–DC converter and the counterparts in **a** step-up and **b** step-down modes

converter can be derived [127]. The low voltage side of the modified topology and the Boost converter are connected to obtain the input-parallel structure. Meanwhile, the positive output terminal of the modified Cuk topology and the ground of the Boost converter are connected to become the new output.

The DC–DC converter with a derived input-parallel output-series structure is introduced in this section, offering advantages such as low current ripple on the low voltage side, high voltage gain, and minimal voltage stresses across power switches. Additionally, the voltage difference between the negative terminals of the input and output manifests as a capacitor voltage rather than a pulse voltage, mitigating the potential for severe electromagnetic interference. In comparison with conventional DC–DC converters, this design has the potential to significantly enhance the utilization depth of capacity for energy storage devices in hybrid power systems.

The topology of the input-parallel output-series Boost DC–DC converter is shown in Fig. 5.17, where Q_1 and Q_2 are power switches, D_1–D_3 are diodes, L_1 and L_2 are filtering inductors, and C_1–C_3 are energy storage capacitors. As can be observed

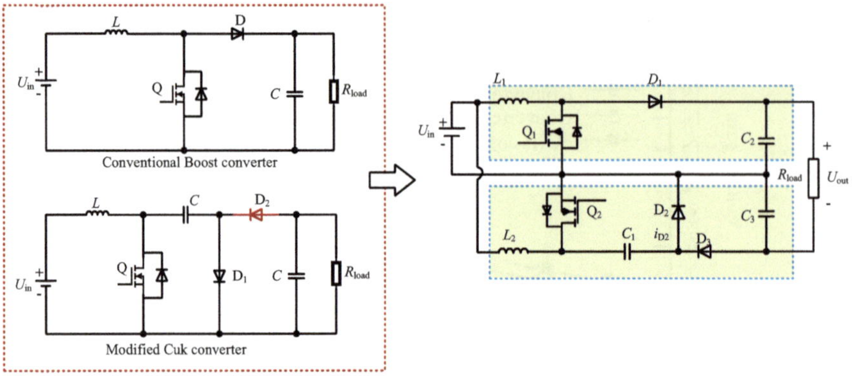

Fig. 5.16 Derivation of input-parallel output-series DC–DC converter

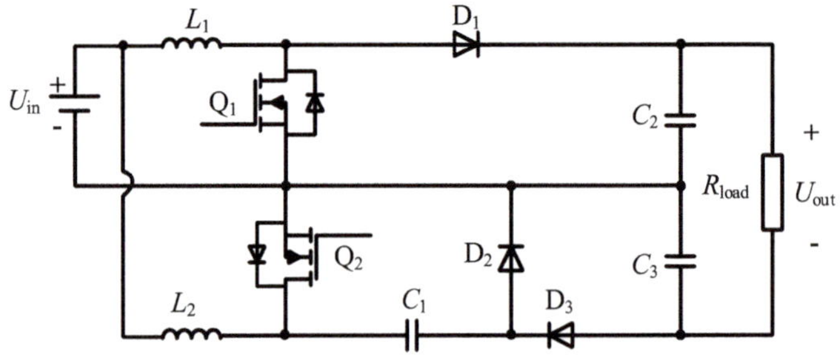

Fig. 5.17 Topology of the input-parallel output-series Boost DC–DC converter

from the figure, the converter contains two conventional Boost DC–DC converters with reversed output voltage polarities. Therefore, high voltage can be achieved by series-connected capacitors C_2 and C_3. Meanwhile, the input current ripple can be suppressed by the interleaved inductor currents of L_1 and L_2.

In order to simplify the operation analysis, the inductances of L_1 and L_2 are designed equally, as well as the capacitances of C_1–C_3. Furthermore, the components are considered as ideal, that is the on-state resistances and forward voltage drop of power switches and diodes are ignored. The parasitic resistances of inductors and capacitors are ignored as well. During each switching cycle, the inductor currents are regard as linear, whereas the capacitor voltages are regard as constant.

5.4.1 *Operating Principle of the Input-Parallel Output-Series DC–DC Converter*

The operation of the inductor currents can be divided in to continuous current mode and discontinuous current mode, depending on the inductances and operating conditions. The key waveforms of the converter while operating in step-up mode under continuous and discontinuous current modes are depicted in Fig. 5.18, where S_1 and S_2 are the respectively gate signals for Q_1 and Q_2. u_{Q1} and u_{Q2}, and u_{D1}–u_{D3} are the voltage stresses across Q_1 and Q_2, and D_1–D_3, respectively. i_{L1} and i_{L2} are the inductor currents of L_1 and L_2. And d_{Boost} is the duty cycle of Q_1 in step-up mode. It can be observed from the illustration that the duty cycles of Q_1 and Q_2 are equal, while the gate signals are phase shifted with half switching cycle.

Due to the utilization of two power switches, there are four switching states of the converter that $S_1S_2 = [00, 01, 10, 11]$. According to the range of duty cycle, the sequence of switching states in continuous current mode can be divided into different conditions as follows.

(1) When the converter operates in continuous current mode and $0 < d_{Boost} < 0.5$, the sequence of switching states during each switching cycle is $S_1S_2 = 10$–00–01–00–10.

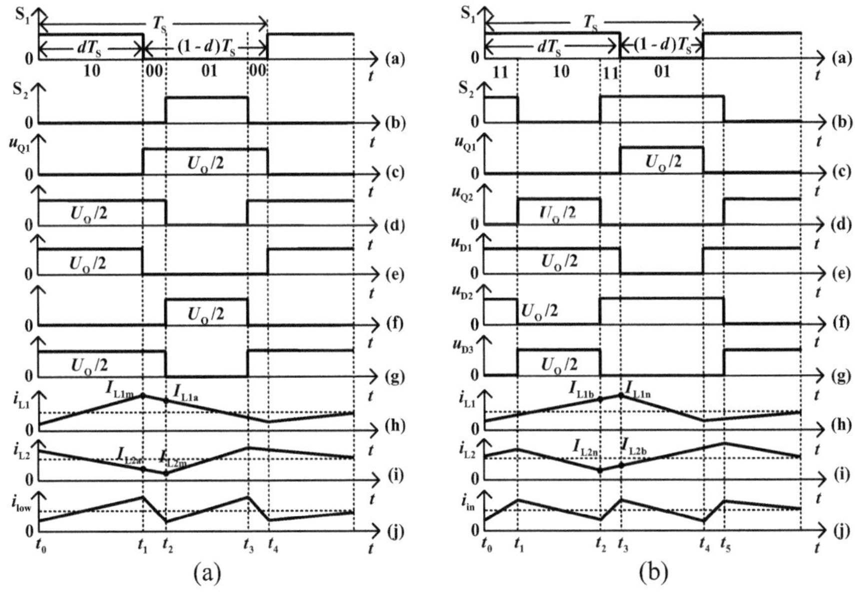

Fig. 5.18 Key waveforms of the input-parallel output-series Boost DC–DC converter in step-up mode under **a** CCM **b** DCM

(2) When the converter operates in continuous current mode and $0.5 < d_{\text{Boost}} < 1$, the sequence of switching states during each switching cycle is $S_1S_2 = 11–10–11–01–11$.

However, when the converter operates in discontinuous current mode, four additional operation states are considered, which are noted as $S_1S_2 = [01_{\text{DCM}}, 10_{\text{DCM}}, 00_{\text{DCM\_1}}, 00_{\text{DCM\_2}}]$. As a result, the sequence of switching states in discontinuous current mode are described as below.

(3) When the converter operates in discontinuous current mode and $0 < d_{\text{Boost}} < 0.5$, the sequence of switching states during each switching cycle is $S_1S_2 = 10–10_{\text{DCM}}–00_{\text{DCM\_2}}–01–01_{\text{DCM}}–00_{\text{DCM\_1}}–10$.
(4) When the converter operates in discontinuous current mode and $0.5 < d_{\text{Boost}} < 1$, the sequence of switching states during each switching cycle is $S_1S_2 = 11–10–10_{\text{DCM}}–11–01–01_{\text{DCM}}–11$.

The switching states are illustrated in Fig. 5.19, which are explained in detail as follows.

State 1) $S_1S_2 = 10$

When the power switch Q_1 is turned ON and Q_2 is turned OFF, while the inductor currents are continuous, the diode D_2 is conducted. The inductor energy of L_2 is transferred to C_1, whereas L_1 is charged by the input source on the low voltage side through Q_1. The capacitors C_2 and C_3 are series connected and both discharged to power the load R_{load} on the high voltage side.

State 2) $S_1S_2 = 00$

When the power switches Q_1 and Q_2 are both turned OFF with continuous inductor currents, the diodes D_1 and D_2 are conducted. Meanwhile, the capacitors C_2 and C_1 are charged by the inductors L_1 and L_2, respectively. The capacitor C_3 is discharged to power the load.

State 3) $S_1S_2 = 01$

When the power switches Q_1 is turned OFF and Q_2 is turned ON, and the inductor currents are continuous, the diodes D_1 and D_3 are conducted. During this state, the inductor L_2 is charged by the input source on the low voltage side through Q_2. And the capacitors C_2 and C_3 are charged by L_1 and C_1, respectively.

State 4) $S_1S_2 = 11$

When the power switches Q_1 and Q_2 are both turned ON with continuous inductor currents, the inductors L_1 and L_2 are charged by the input source through Q_1 and Q_2, respectively. The capacitor C_3 is charged, while the load is powered by C_1 and C_2.

State 5) $S_1S_2 = 10_{\text{DCM}}$

When Q_1 is turned ON and Q_2 is turned OFF, and the inductor current i_{L2} becomes zero, all diodes are turned OFF. The inductor L_1 is still charged by the input source through Q_1, and the output load is powered by the capacitors C_2 and C_3.

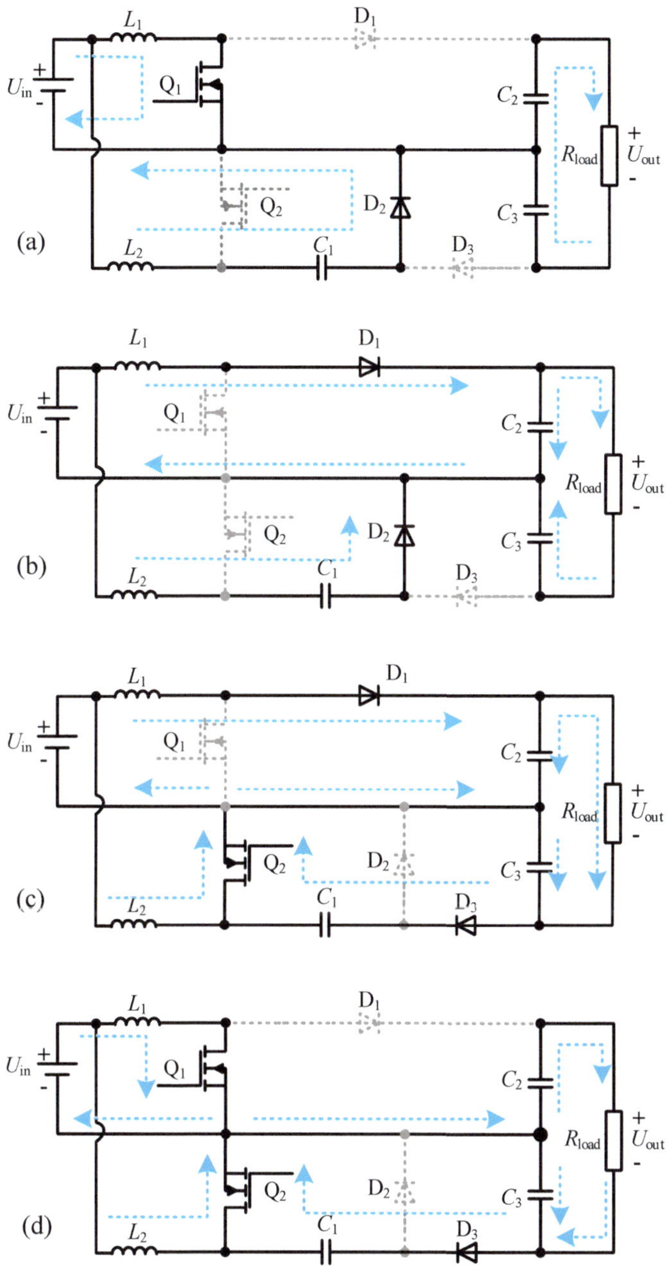

Fig. 5.19 Operation states of the input-parallel output-series Boost DC–DC converter during **a** State 1, **b** State 2, **c** State 3, **d** State 4, **e** State 5, **f** State 6, **g** State7, and **h** State 8

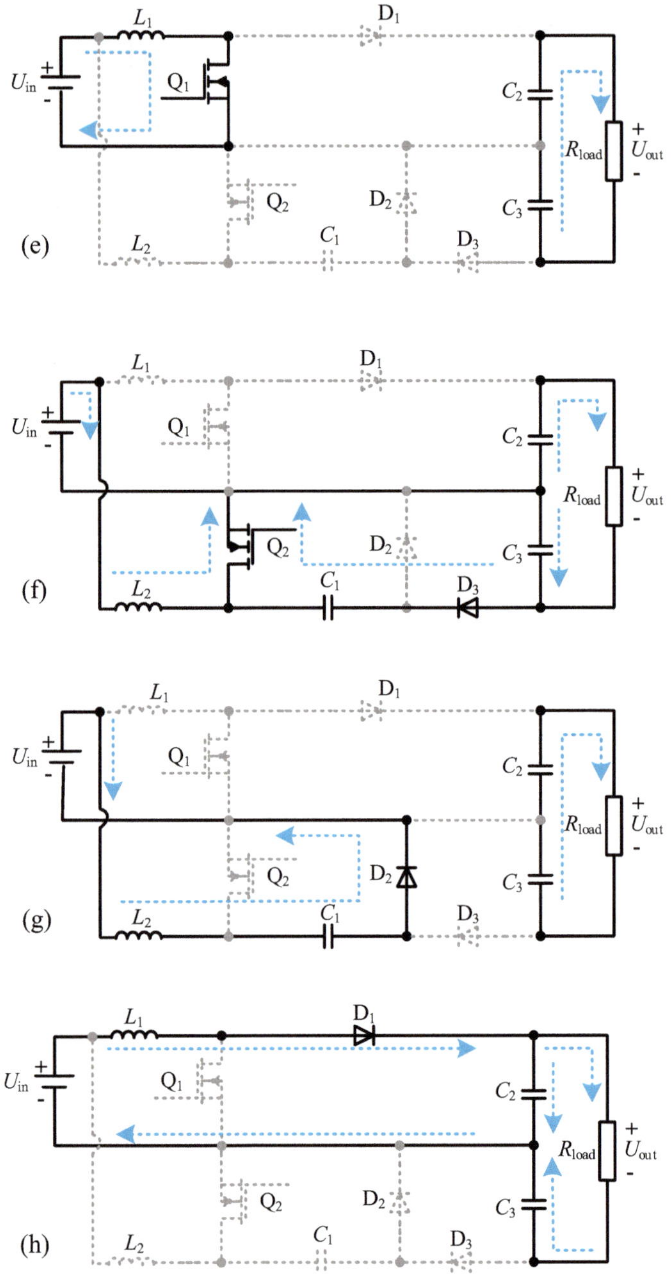

Fig. 5.19 (continued)

State 6) $S_1 S_2 = 01_{\text{DCM}}$

When Q_1 is turned OFF and Q_2 is turned ON, and the inductor current i_{L1} is decreased to zero, only the diode D_3 is conducted. Whereas, L_2 is charged by the input source, and the output load is power by C_1 and C_2.

State 7) $S_1 S_2 = 00_{\text{DCM\_1}}$

When both power switches Q_1 and Q_2 are turned OFF, and the inductor current i_{L1} is discontinuous, only the diode D_2 is conducted. Meanwhile, the capacitor C_1 is charged by input source and L_2, and the output load is powered by C_2 and C_3.

State 8) $S_1 S_2 = 00_{\text{DCM\_2}}$

When both power switches Q_1 and Q_2 are turned OFF, and the inductor current i_{L2} is discontinuous, only the diode D_1 is conducted. The capacitor C_2 is charged by the input source and L_1, while C_3 is discharged.

5.4.2 *Steady-State Characteristics of the Input-Parallel Output-Series DC–DC Converter*

The steady-state characteristics of the input-parallel output-series Boost DC–DC converter depend on the condition of inductor currents. When the converter operates in continuous current mode, as depicted in Fig. 5.19c, capacitors C_1 and C_3 are connected in parallel. Consequently, employing the volt-second balance principle, the relationship among duty cycle and capacitor voltages can be described as follows.

$$\begin{cases} d_{\text{Boost}} U_{\text{low}} = (1 - d_{\text{Boost}})(U_{\text{C2}} - U_{\text{low}}) \\ d_{\text{Boost}} U_{\text{low}} = (1 - d_{\text{Boost}})(U_{\text{C1}} - U_{\text{low}}) \\ U_{\text{C1}} = U_{\text{C3}} \end{cases} \tag{5.18}$$

Thus, the capacitor voltages can be written as below.

$$U_{\text{C1}} = U_{\text{C2}} = U_{\text{C3}} = \frac{1}{1 - d_{\text{Boost}}} U_{\text{low}} \tag{5.19}$$

And the voltage gain M_{Boost} of the input-parallel output-series Boost DC–DC converter can be derived.

$$M_{\text{Boost}} = \frac{U_{\text{high}}}{U_{\text{low}}} = \frac{2}{1 - d_{\text{Boost}}} \tag{5.20}$$

In terms of the inductor currents, the following equations can be obtained through amp-second balance principle applied on C_2.

$$\begin{cases} d_{\text{Boost}} I_{\text{high}} = (1 - d_{\text{Boost}})\left(I_{L1} - I_{\text{high}}\right) \\ I_{\text{low}} U_{\text{low}} = I_{\text{high}} U_{\text{high}} \\ I_{\text{low}} = I_{\text{L1}} + I_{\text{L2}} \end{cases} \tag{5.21}$$

According to (5.21), the inductor currents can be written by the duty cycle and current on the high voltage side.

$$I_{\text{L1}} = I_{\text{L2}} = \frac{1}{1 - d_{\text{Boost}}} I_{\text{high}} \tag{5.22}$$

It is shown in (5.19) and (5.20) that, the input-parallel output-series Boost DC–DC converter exhibits higher voltage gain and lower capacitor voltages than the conventional Boost converter.

Under the discontinuous current mode condition, the peak inductor current I_{L1p} can be calculated from

$$I_{\text{L1p}} = \frac{d_{\text{Boost}} T_{\text{s}} U_{\text{low}}}{L} \tag{5.23}$$

On the other hand, the peak inductor current I_{L1p} can also be expressed in terms of the duty cycle d_{disch}, representing the duration of inductor discharging, as follows.

$$I_{\text{L1p}} = \frac{d_{\text{disch}} T_{\text{s}} (U_{\text{C2}} - U_{\text{low}})}{L} \tag{5.24}$$

Combining (5.23) and (5.24), the discharging duty cycle of inductor currents in discontinuous current mode can be derived as below.

$$d_{\text{disch}} = \frac{d_{\text{Boost}} U_{\text{low}}}{U_{\text{C2}} - U_{\text{low}}} \tag{5.25}$$

In turn, the relationship among duty cycles and capacitor voltages can be described as the following equations by applying the volt-second balance principle on L_1 and L_2.

$$\begin{cases} d_{\text{Boost}} U_{\text{low}} = d_{\text{disch}} (U_{\text{C2}} - U_{\text{low}}) \\ d_{\text{Boost}} U_{\text{low}} = d_{\text{disch}} (U_{\text{C1}} - U_{\text{low}}) \\ U_{\text{C1}} = U_{\text{C3}} \\ U_{\text{high}} = U_{\text{C2}} + U_{\text{C3}} \end{cases} \tag{5.26}$$

Thereby, the capacitor voltages can be written as follows.

$$U_{\text{C1}} = U_{\text{C2}} = U_{\text{C3}} = \frac{U_{\text{high}}}{2} \tag{5.27}$$

According to the amp-second balance principle on C_2, the relationship among currents and discharging duty cycle can be described as below.

$$(1 - d_{\text{disch}})I_{\text{high}} = d_{\text{disch}}\left(\frac{I_{L1p}}{2} - I_{\text{high}}\right) \tag{5.28}$$

On the basis of (5.22)–(5.28), the voltage gain $M_{\text{Boost\_DCM}}$ of the converter in discontinuous current mode can be obtained as below, where τ_L is the normalized inductor time constant.

$$M_{\text{Boost\_DCM}} = \frac{U_{\text{high}}}{U_{\text{low}}} = 1 + \sqrt{1 + \frac{d_{\text{Boost}}^2}{\tau_L}} \tag{5.29}$$

Specifically, the time constant is defined by the following formula.

$$\tau_L = \frac{L}{T_s R_{\text{load}}} \tag{5.30}$$

When the converter operates in the boundary condition between continuous and discontinuous current mode, the boundary normalized inductor time constant τ_{LB} can be obtained as follows.

$$\tau_{\text{LB}} = \frac{d_{\text{Boost}}(1 - d_{\text{Boost}})^2}{4} \tag{5.31}$$

The boundary time constant τ_{LB} is depicted in Fig. 5.20. Thus, the region above and below the boundary time constant refer to the continuous and discontinuous current modes, respectively.

According to the above analysis on operating principles, the voltage stresses across power switches and diodes can be obtained as the following equations.

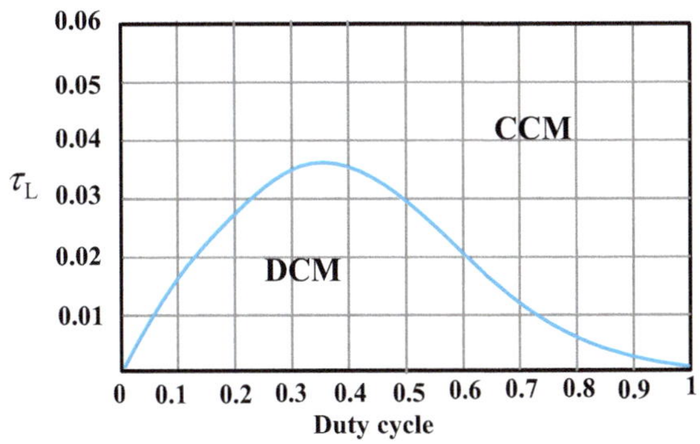

Fig. 5.20 Boundary normalized inductor time constant

$$\begin{cases} U_{Q1} = U_{D1} = U_{C2} = \dfrac{U_{\text{high}}}{2} \\ U_{Q2} = U_{D2} = U_{C1} = \dfrac{U_{\text{high}}}{2} \\ U_{D3} = U_{C3} = \dfrac{U_{\text{high}}}{2} \end{cases} \tag{5.32}$$

Thus, the voltage stresses for all power switches and diodes are half of the output voltage. Moreover, the current stresses of power switches and diodes are obtained as follows.

$$\begin{cases} I_{Q1} = \dfrac{1}{1 - d_{\text{Boost}}} I_{\text{high}} \\ I_{Q2} = \left(\dfrac{1}{1 - d_{\text{Boost}}} + \dfrac{1}{d_{\text{Boost}}} \right) I_{\text{high}} \\ I_{D1} = I_{D2} = \dfrac{1}{1 - d_{\text{Boost}}} I_{\text{high}} \\ I_{D3} = \dfrac{1}{d_{\text{Boost}}} I_{\text{high}} \end{cases} \tag{5.33}$$

It is shown in (5.33) that the current stress of Q_2 is higher than that of the other power semiconductors. Whereas, the current stresses are relatively low, which makes it easier for selection of components.

According to Fig. 5.18, in CCM, the input current ripple Δi_{low} equals to the sum of variations of inductor currents with during $S_1 S_2 = 01$. The variations of inductor currents i_{L1} and i_{L2} are denoted as Δi_{L1} and Δi_{L2}, respectively. Within the duty cycle range of $0 < d_{\text{Boost}} < 0.5$, the inductor current ripples can be obtained as below.

$$\Delta i_{L1} = \Delta i_{L2} = \frac{d_{\text{Boost}} T_s U_{\text{low}}}{L} = \frac{d_{\text{Boost}}(1 - d_{\text{Boost}}) T_s U_{\text{high}}}{2L} \tag{5.34}$$

The input current ripple can be calculated as below.

$$\Delta i_{\text{low}} = \frac{d_{\text{Boost}}(1 - 2d_{\text{Boost}}) T_s U_{\text{low}}}{(1 - d_{\text{Boost}}) L} = \frac{d_{\text{Boost}}(1 - 2d_{\text{Boost}}) T_s U_{\text{high}}}{2L} \tag{5.35}$$

Similarly, within the duty cycle range of $0.5 < d_{\text{Boost}} < 1$, the inductor current ripples are consistent with that in (5.34). Whereas, the input current ripples can be obtained as follows.

$$\Delta i_{\text{low}} = \frac{(2d_{\text{Boost}} - 1) T_s U_{\text{low}}}{L} = \frac{(2d_{\text{Boost}} - 1)(1 - d_{\text{Boost}}) T_s U_{\text{high}}}{2L} \tag{5.36}$$

Tables 5.1 and 5.2 displays the comparative results of the input-parallel output-series Boost DC–DC converter with other previous works. In comparison with the conventional interleaved Boost DC–DC converter, the input-parallel output-series

Table 5.1 Comparison of the input-parallel output-series Boost DC–DC converter with counterparts

	The conventional interleaved Boost DC–DC converter	The Boost three-level DC–DC converter	The typology of the converter in [128]	The input-parallel output-series Boost DC–DC converter
Voltage gain	$1/(1-d)$	$1/(1-d)$	$2/(1-d)$	$2/(1-d)$
Voltage stress for power switches	U_{out}	$U_{out}/2$	$U_{out}/2$	$U_{out}/2$
Voltage stress for diodes	U_{out}	$U_{out}/2$	U_{out}, $U_{out}/2$	$U_{out}/2$
Input current ripple ($d < 0.5$)	$\frac{d(1-2d)U_{out}T_s}{L}$	$\frac{d(1-2d)U_{out}T_s}{2L}$	High	$\frac{d(1-2d)T_sU_{out}}{2L}$
Ripple rate of the input current ($d < 0.5$)	$\frac{d(1-d)(1-2d)R_{load}T_s}{L}$	$\frac{d(1-d)(1-2d)R_{load}T_s}{2L}$	High	$\frac{d(1-d)(1-2d)R_{load}T_s}{4L}$
Input current ripple ($d > 0.5$)	$\frac{(2d-1)(1-d)U_{out}T_s}{L}$	$\frac{(2d-1)(1-d)U_{out}T_s}{2L}$	High	$\frac{(2d-1)(1-d)T_sU_{out}}{2L}$
Ripple rate of the input current ($d > 0.5$)	$\frac{(2d-1)(1-d)^2R_{load}T_s}{L}$	$\frac{(2d-1)(1-d)^2R_{load}T_s}{2L}$	High	$\frac{(2d-1)(1-d)^2R_{load}T_s}{4L}$
The number of inductors	2	1	2	2
The number of power switches	2	2	2	2
The number of diodes	2	2	2	3

Boost DC–DC converter shows higher voltage gain range and lower voltage stresses across power switches and diodes. The three-level Boost DC–DC converter also exhibits low voltage stresses, whereas the voltage gain is the same as the conventional Boost DC–DC converter. The high voltage gain range and low voltage stresses are achieved in [130]. However, the converter in [130] suffers from high current ripples and discontinuous pulse input current.

5.4.3 Operating Principle of the Interleaved Switched-Capacitor Bidirectional DC–DC Converter

By replacing the diodes D_1–D_3 in the input-parallel output-series Boost DC–DC converter with power switches, the interleaved switched-capacitor bidirectional DC–DC converter can be derived. This modification enables energy conversion from the

Table 5.2 Comparison of the input-parallel output-series Boost DC–DC converter with the counterparts

	The converter in [129]	The converter in [130]	The input-parallel output-series Boost DC–DC converter
Voltage gain	$2(N + 1)/(1 - d)$	$(2N + 3 + d)/(1 - d)$	$2/(1 - d)$
Voltage stress for power switches	$U_{out}/2(N + 1)$	$U_{out}/(2N + 3 + d)$	$U_{out}/2$
The maximum voltage stress for diodes	$NU_{out}/(N + 1)$	$(N + 1)U_{out}/(2N + 3 + d)$	$U_{out}/2$
Input current ripple ($d > 0.5$)	$\frac{(1-k)(1-d)(2d-1)U_{out}T_s}{2L_k(N+1)}$	$\frac{(1-k)(1-d)(2d-1)U_{out}T_s}{(2N+3+d)L_k}$	$\frac{(2d-1)(1-d)T_sU_{out}}{2L}$
The number of inductors/coupled inductors	2	2	2
The number of power switches	2	2	2
The number of diodes	4	6	3
The number of capacitors	4	6	3
The potential difference between the output and the input side grounds	A high-frequency pulse voltage	A high-frequency pulse voltage	A constant capacitor voltage

high voltage side to the low voltage side. The topology of the interleaved switched-capacitor bidirectional DC–DC converter is illustrated in Fig. 5.21, featuring five power switches Q_1–Q_5.

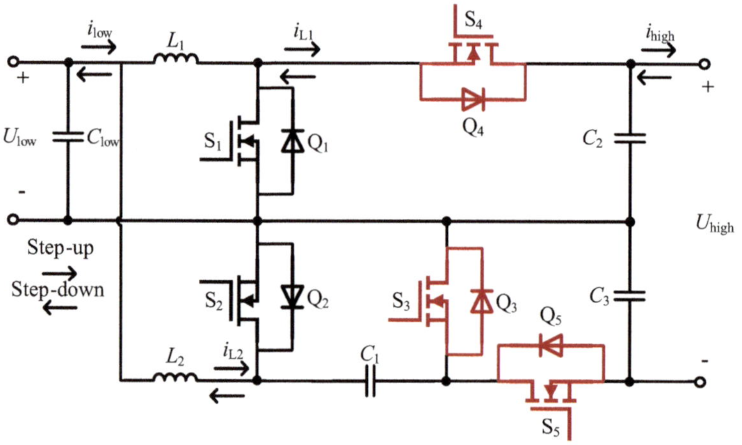

Fig. 5.21 Topology of the interleaved switched-capacitor bidirectional DC–DC converter

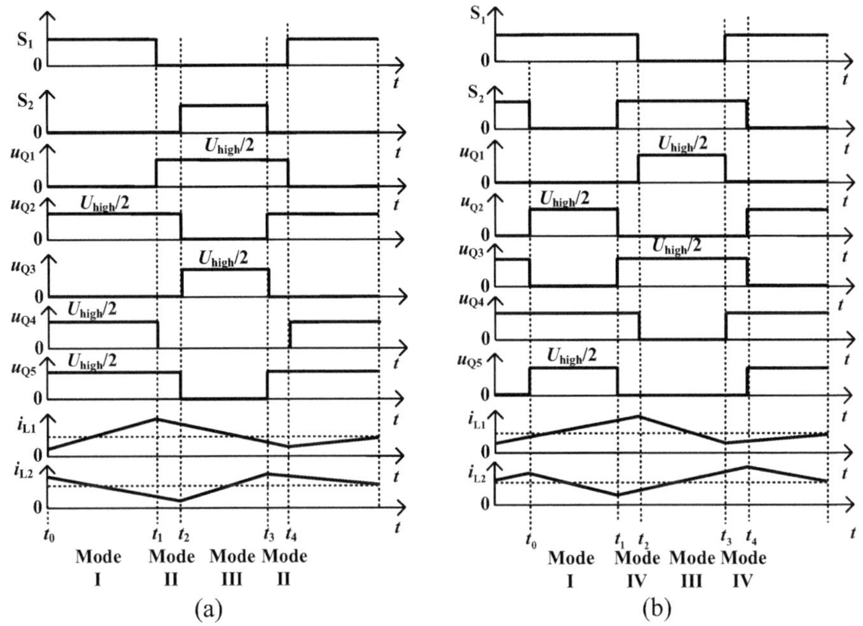

Fig. 5.22 Key waveforms of the interleaved switched-capacitor bidirectional DC–DC converter when **a** $d_{\text{Boost}} < 0.5$ and **b** $d_{\text{Boost}} > 0.5$ in step-up mode

The key waveforms of the interleaved switched-capacitor bidirectional DC–DC converter in step-up and step-down modes are shown in Figs. 5.22 and 5.23, respectively. In step-up mode, Q_1 and Q_2 act as main power switches, and the gate signals S_1 and S_2 of power switches Q_1 and Q_2 are 180° equal duty cycle d_{Boost} with phase-shift. In step-down mode, Q_3–Q_5 are the main power switches, and S_3–S_5 are the corresponding gate signals. Specifically, the duty cycle of S_2 is equal to S_1, and S_2 is 180° shifted relative to S_1. Meanwhile, S_3 is complementary to S_1.

The switching states of the interleaved switched-capacitor bidirectional DC–DC converter in step-up mode is the same as that of the input-parallel output-series Boost DC–DC converter. The switching states of the converter in step-down mode are presented in Fig. 5.24, and the operating principle are explained as below.

State 1) $S_3S_4S_5 = 100$

When the power switch Q_3 is turned ON, while Q_4 and Q_5 are turned OFF, the inductor L_1 is discharged to power the load one the low voltage side through the body diode of Q_1. Meanwhile, the capacitors C_2 and C_3 are charged by the input source on the high voltage side. The inductor L_2 is charged by capacitor C_1 through Q_3.

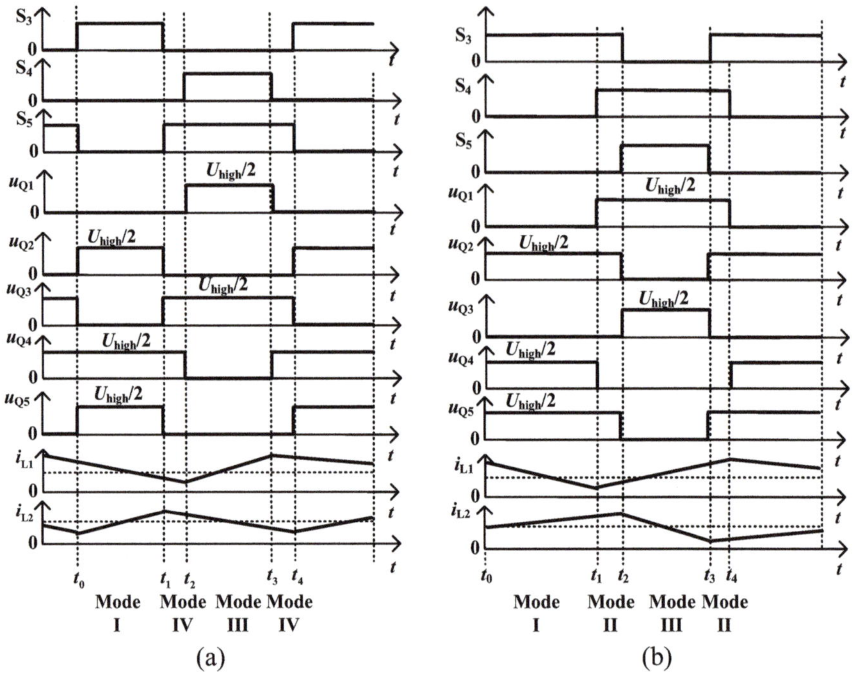

Fig. 5.23 Key waveforms of the interleaved switched-capacitor bidirectional DC–DC converter when **a** $d_{\text{Buck}} < 0.5$ **b** $d_{\text{Buck}} > 0.5$ in step-down mode

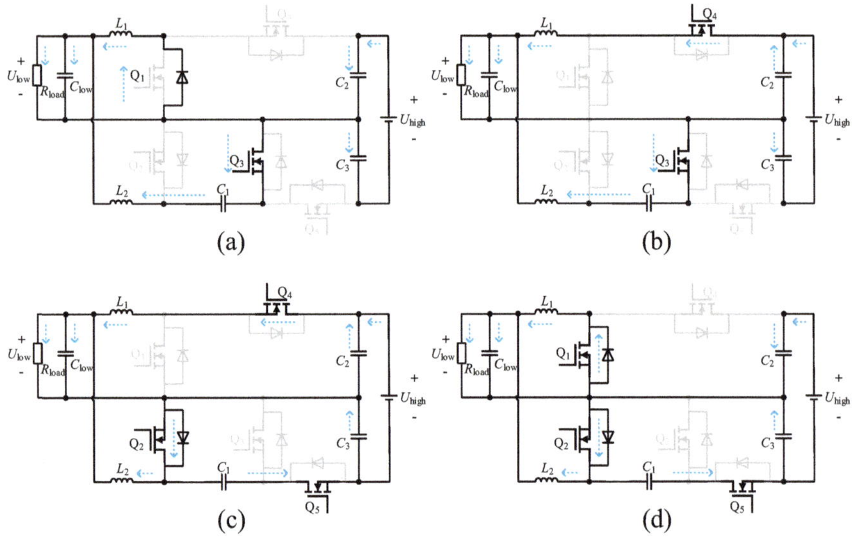

Fig. 5.24 Switching states of the interleaved switched-capacitor bidirectional DC–DC converter in step-down mode during **a** State 1, **b** State 2, **c** State 3, and **d** State 4

State 2) $S_3S_4S_5 = 110$

When the power switches Q_3 and Q_4 are turned ON, while Q_5 is turned OFF, the inductor L_2 is still charged by C_1. The capacitor C_3 is charged by the input source, whereas C_2 is discharged via Q_4 to power the load.

State 3) $S_3S_4S_5 = 011$

When the power switch Q_3 is turned OFF, while Q_4 and Q_5 are both turned ON, the inductor L_1 is charged by C_2 via Q_4, and C_1 is charged by C_3. During this state, the load on the low voltage side is powered by the inductor L_2 via the body diode of Q_2.

State 4) $S_3S_4S_5 = 001$

When the power switches Q_3 and Q_4 are both turned OFF, and Q_5 is turned ON, the inductors L_1 and L_2 are discharged via the body diodes of Q_1 and Q_2, respectively. At the same time, the capacitor C_1 is charged by C_3, and C_2 is charged by the input source.

In order to reduce the conducting losses of the body diodes, the synchronous rectification operation can be implemented for the interleaved switched-capacitor bidirectional DC–DC converter. The gate signals S_1–S_5 with synchronous rectification are depicted in Fig. 5.25, where S_1 is complementary to S_4 and S_2 is complementary to S_3. Furthermore, the dead time is considered between S_1 and S_4, as well as S_2 and S_3. As a results, the body diodes of synchronous rectifiers are conducted within the dead time, thereby the synchronous rectifiers can be ZVS turned ON and OFF.

In terms of the voltage gain of the interleaved switched-capacitor bidirectional DC–DC converter in step-down mode, the following equations can be obtained by applying the volt-second balance principle on L_1 and L_2.

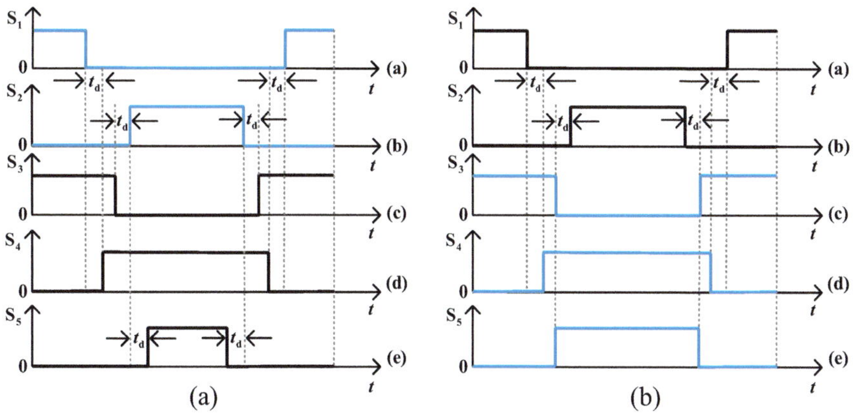

Fig. 5.25 Gate signals of the interleaved switched-capacitor bidirectional DC–DC converter in **a** step-up and **b** step-down modes

$$\begin{cases} d_{\text{Buck}}(U_{C1} - U_{\text{low}}) = (1 - d_{\text{Buck}})U_{\text{low}} \\ d_{\text{Buck}}(U_{C2} - U_{\text{low}}) = (1 - d_{\text{Buck}})U_{\text{low}} \\ U_{C1} = U_{C3} \end{cases} \tag{5.37}$$

The capacitor voltages can be derived in turn, which is consistent with (5.27). Moreover, the step-down voltage gain M_{Buck} can be deduced as below.

$$M_{\text{Buck}} = \frac{U_{\text{low}}}{U_{\text{high}}} = \frac{d_{\text{Buck}}}{2} \tag{5.38}$$

According to KCL, the capacitor currents i_{C1}–i_{C3} of C_1–C_3 in the interleaved switched-capacitor bidirectional DC–DC converter in step-down mode can be described as follows, where I_{L1} and I_{L2} are the average inductor currents of L_1 and L_2, respectively.

$$i_{C1} = \begin{cases} -I_{L2}, \; 0 < t < d_{\text{Buck}}T_s \\ i_{C1}, \; d_{\text{Buck}}T_s < t < d_{\text{Buck}}T_s \end{cases} \tag{5.39}$$

$$i_{C2} = \begin{cases} I_{\text{high}} - I_{L1}, \; 0 < t < d_{\text{Buck}}T_s \\ I_{\text{high}}, \; d_{\text{Buck}}T_s < t < d_{\text{Buck}}T_s \end{cases} \tag{5.40}$$

$$i_{C3} = \begin{cases} I_{\text{high}}, \; 0 < t < d_{\text{Buck}}T_s \\ I_{\text{high}} - i_{C1}, \; d_{\text{Buck}}T_s < t < d_{\text{Buck}}T_s \end{cases} \tag{5.41}$$

Combining (5.39)–(5.41), the average inductor currents and average high voltage side current can be expressed by the average low voltage side current as

$$\begin{cases} I_{L1} = I_{L2} = \dfrac{1}{2}I_{\text{low}} \\ I_{\text{high}} = \dfrac{d_{\text{Buck}}}{2}I_{\text{low}} \end{cases} \tag{5.42}$$

It can be seen from (5.42) that the average inductor currents in step-down mode are equal to half of the low voltage side current, which is consistent with that in step-up mode. Thus, in both step-up and step-down modes, the inductor currents of the interleaved switched-capacitor bidirectional DC–DC converter can be self-balanced without additional control.

The current ripples of inductor currents and low voltage side currents in step-down mode can be derived on the basis of switching states analysis. With the duty cycle range of $0 < d_{\text{Boost}} < 0.5$, the current ripples can be calculated as follows.

$$\begin{cases} \Delta i_{L1} = \Delta i_{L2} = \dfrac{d_{\text{Buck}}(1 - d_{\text{Buck}})T_s U_{\text{high}}}{2L} \\ \Delta i_{\text{low}} = \dfrac{d_{\text{Buck}}(1 - 2d_{\text{Buck}})T_s U_{\text{high}}}{2L} \end{cases} \tag{5.43}$$

whereas the current ripples are derived as with the duty cycle range of $0 < d_{\text{Boost}} < 0.5$.

$$\begin{cases} \Delta i_{L1} = \Delta i_{L2} = \dfrac{d_{\text{Buck}}(1 - d_{\text{Buck}})T_s U_{\text{high}}}{2L} \\ \Delta i_{\text{low}} = \dfrac{(2d_{\text{Buck}} - 1)(1 - d_{\text{Buck}})T_s U_{\text{high}}}{2L} \end{cases} \tag{5.44}$$

5.4.4 Small-Signal Modeling of the Interleaved Switched-Capacitor Bidirectional DC–DC Converter

To avoid the invalid state variables, the equivalent series resistances for capacitors are introduced for decoupling. Then, the average and small-signal model of the interleaved switched-capacitor bidirectional DC–DC converter can be established through state-space averaging method.

When the converter operates in step-up mode within the duty cycle range of $0 < d_{\text{Boost}} < 0.5$, the switching states during each switching cycle consists of $S_1S_2 = [10, 00, 01]$. The respective state-space average models of the switching states are derived as (5.45)–(5.47), where $i_{L1}(t)$, $i_{L2}(t)$, and $u_{C1}(t)$-$u_{C3}(t)$ are the state variables, while $u_{\text{low}}(t)$, $u_{\text{high}}(t)$, and $d_1(t)$ and $d_2(t)$ are input variable, output variable, and the control variables, respectively.

$$\begin{bmatrix} \frac{di_{L1}(t)}{dt} \\ \frac{di_{L2}(t)}{dt} \\ \frac{du_{C1}(t)}{dt} \\ \frac{du_{C2}(t)}{dt} \\ \frac{du_{C3}(t)}{dt} \end{bmatrix} = \begin{bmatrix} 0 & 0 & 0 & 0 & 0 \\ 0 & 0 & -\frac{1}{L_2} & 0 & 0 \\ 0 & \frac{1}{C_1} & 0 & 0 & 0 \\ 0 & 0 & 0 & -\frac{1}{C_2 R_{\text{L - Boost}}} & -\frac{1}{C_2 R_{\text{L - Boost}}} \\ 0 & 0 & 0 & -\frac{1}{C_3 R_{\text{L - Boost}}} & -\frac{1}{C_3 R_{\text{L - Boost}}} \end{bmatrix} \begin{bmatrix} i_{L1}(t) \\ i_{L2}(t) \\ u_{C1}(t) \\ u_{C2}(t) \\ u_{C3}(t) \end{bmatrix} + \begin{bmatrix} \frac{1}{L_1} \\ \frac{1}{L_2} \\ 0 \\ 0 \\ 0 \end{bmatrix} u_{\text{low}}(t)$$
$$u_{\text{high}}(t) = \begin{bmatrix} 0\,0\,0\,1\,1 \end{bmatrix} \begin{bmatrix} i_{L1}(t)\; i_{L2}(t)\; u_{C1}(t)\; u_{C2}(t)\; u_{C3}(t) \end{bmatrix}^T \tag{5.45}$$

$$\begin{bmatrix} \frac{di_{L1}(t)}{dt} \\ \frac{di_{L2}(t)}{dt} \\ \frac{du_{C1}(t)}{dt} \\ \frac{du_{C2}(t)}{dt} \\ \frac{du_{C3}(t)}{dt} \end{bmatrix} = \begin{bmatrix} 0 & 0 & 0 & -\frac{1}{L_1} & 0 \\ 0 & 0 & -\frac{1}{L_2} & 0 & 0 \\ 0 & \frac{1}{C_1} & 0 & 0 & 0 \\ \frac{1}{C_2} & 0 & 0 & -\frac{1}{C_2 R_{\text{L - Boost}}} & -\frac{1}{C_2 R_{\text{L - Boost}}} \\ 0 & 0 & 0 & -\frac{1}{C_3 R_{\text{L - Boost}}} & -\frac{1}{C_3 R_{\text{L - Boost}}} \end{bmatrix} \begin{bmatrix} i_{L1}(t) \\ i_{L2}(t) \\ u_{C1}(t) \\ u_{C2}(t) \\ u_{C3}(t) \end{bmatrix} + \begin{bmatrix} \frac{1}{L_1} \\ \frac{1}{L_2} \\ 0 \\ 0 \\ 0 \end{bmatrix} u_{\text{low}}(t)$$
$$u_{\text{high}}(t) = \begin{bmatrix} 0\,0\,0\,1\,1 \end{bmatrix} \begin{bmatrix} i_{L1}(t)\; i_{L2}(t)\; u_{C1}(t)\; u_{C2}(t)\; u_{C3}(t) \end{bmatrix}^T \tag{5.46}$$

$$\begin{bmatrix} \frac{di_{L1}(t)}{dt} \\ \frac{di_{L2}(t)}{dt} \\ \frac{du_{C1}(t)}{dt} \\ \frac{du_{C2}(t)}{dt} \\ \frac{du_{C3}(t)}{dt} \end{bmatrix} = \begin{bmatrix} 0 & 0 & 0 & -\frac{1}{L_1} & 0 \\ 0 & 0 & 0 & 0 & 0 \\ 0 & 0 & -\frac{1}{C_1 r} & 0 & \frac{1}{C_1 r} \\ \frac{1}{C_2} & 0 & 0 & -\frac{1}{C_2 R_{\text{L - Boost}}} & -\frac{1}{C_2 R_{\text{L - Boost}}} \\ 0 & 0 & \frac{1}{C_3 r} & -\frac{1}{C_3 R_{\text{L - Boost}}} & -\frac{1}{C_3 r} - \frac{1}{C_3 R_{\text{L - Boost}}} \end{bmatrix} \begin{bmatrix} i_{L1}(t) \\ i_{L2}(t) \\ u_{C1}(t) \\ u_{C2}(t) \\ u_{C3}(t) \end{bmatrix} + \begin{bmatrix} \frac{1}{L_1} \\ \frac{1}{L_2} \\ 0 \\ 0 \\ 0 \end{bmatrix} u_{\text{low}}(t)$$

$$u_{\text{high}}(t) = \begin{bmatrix} 0 & 0 & 0 & 1 & 1 \end{bmatrix} \begin{bmatrix} i_{L1}(t) & i_{L2}(t) & u_{C1}(t) & u_{C2}(t) & u_{C3}(t) \end{bmatrix}^T \tag{5.47}$$

Thereby, the average mode of the converter can be obtained as follows.

$$\begin{bmatrix} \frac{di_{L1}(t)}{dt} \\ \frac{di_{L2}(t)}{dt} \\ \frac{du_{C1}(t)}{dt} \\ \frac{du_{C2}(t)}{dt} \\ \frac{du_{C3}(t)}{dt} \end{bmatrix} = \begin{bmatrix} 0 & 0 & 0 & -\frac{1-d_1(t)}{L_1} & 0 \\ 0 & 0 & -\frac{1-d_1(t)}{L_2} & 0 & 0 \\ 0 & \frac{1-d_2(t)}{C_1} & -\frac{d_2(t)}{C_1 r} & 0 & \frac{d_2(t)}{C_1 r} \\ \frac{1-d_1(t)}{C_2} & 0 & 0 & -\frac{1}{C_2 R_{\text{L - Boost}}} & -\frac{1}{C_2 R_{\text{L - Boost}}} \\ 0 & 0 & \frac{d_2(t)}{C_3 r} & -\frac{1}{C_3 R_{\text{L - Boost}}} & -\frac{d_2(t)}{C_3 r} - \frac{1}{C_3 R_{\text{L - Boost}}} \end{bmatrix} \begin{bmatrix} i_{L1}(t) \\ i_{L2}(t) \\ u_{C1}(t) \\ u_{C2}(t) \\ u_{C3}(t) \end{bmatrix}$$

$$+ \begin{bmatrix} \frac{1}{L_1} \\ \frac{1}{L_2} \\ 0 \\ 0 \\ 0 \end{bmatrix} u_{\text{low}}(t)$$

$$u_{\text{high}}(t) = \begin{bmatrix} 0 & 0 & 0 & 1 & 1 \end{bmatrix} \begin{bmatrix} i_{L1}(t) & i_{L2}(t) & u_{C1}(t) & u_{C2}(t) & u_{C3}(t) \end{bmatrix}^T \tag{5.48}$$

Thereby, the small-signal model of the converter can be established as in (5.49).

$$\begin{bmatrix} \frac{d\hat{i}_{L1}(t)}{dt} \\ \frac{d\hat{i}_{L2}(t)}{dt} \\ \frac{d\hat{u}_{C1}(t)}{dt} \\ \frac{d\hat{u}_{C2}(t)}{dt} \\ \frac{d\hat{u}_{C3}(t)}{dt} \end{bmatrix} = \begin{bmatrix} 0 & 0 & 0 & -\frac{1-D_{\text{Boost}}}{L_1} & 0 \\ 0 & 0 & -\frac{1-D_{\text{Boost}}}{L_2} & 0 & 0 \\ 0 & \frac{1-D_{\text{Boost}}}{C_1} & -\frac{D_{\text{Boost}}}{C_1 r} & 0 & \frac{D_{\text{Boost}}}{C_1 r} \\ \frac{1-D_{\text{Boost}}}{C_2} & 0 & 0 & -\frac{1}{C_2 R_{\text{L - Boost}}} & -\frac{1}{C_2 R_{\text{L - Boost}}} \\ 0 & 0 & \frac{D_{\text{Boost}}}{C_3 r} & -\frac{1}{C_3 R_{\text{L - Boost}}} & -\frac{D_{\text{Boost}}}{C_3 r} - \frac{1}{C_3 R_{\text{L - Boost}}} \end{bmatrix} \begin{bmatrix} i_{L1}(t) \\ i_{L2}(t) \\ u_{C1}(t) \\ u_{C2}(t) \\ u_{C3}(t) \end{bmatrix}$$

$$+ \begin{bmatrix} \frac{1}{L_1} \\ \frac{1}{L_2} \\ 0 \\ 0 \\ 0 \end{bmatrix} \hat{u}_{\text{low}}(t)$$

$$+ \begin{bmatrix} 0 & 0 & 0 & \frac{1}{L_1} & 0 \\ 0 & 0 & \frac{1}{L_2} & 0 & 0 \\ 0 & -\frac{1}{C_1} & -\frac{1}{C_1 r} & 0 & \frac{1}{C_1 r} \\ -\frac{1}{C_2} & 0 & 0 & 0 & 0 \\ 0 & 0 & \frac{1}{C_3 r} & 0 & -\frac{1}{C_3 r} \end{bmatrix} \begin{bmatrix} I_{L1}(t) \\ I_{L2}(t) \\ U_{C1}(t) \\ U_{C2}(t) \\ U_{C3}(t) \end{bmatrix} \hat{d}_{\text{Boost}}(t)$$

$$u_{\text{high}}(t) = \begin{bmatrix} 0 & 0 & 0 & 1 & 1 \end{bmatrix} \begin{bmatrix} \hat{i}_{L1}(t) & \hat{i}_{L2}(t) & \hat{u}_{C1}(t) & \hat{u}_{C2}(t) & \hat{u}_{C3}(t) \end{bmatrix}^T \tag{5.49}$$

While the converter operates in step-down mode, and the duty cycle range is $0 < d_{\mathrm{Buck}} < 0.5$, the respective state-space average models of $S_3S_4S_5 = [100, 001, 011]$ can be obtained as (5.50)–(5.52).

$$\begin{bmatrix} \frac{\mathrm{d}i_{L1}(t)}{\mathrm{d}t} \\ \frac{\mathrm{d}i_{L2}(t)}{\mathrm{d}t} \\ \frac{\mathrm{d}u_{C1}(t)}{\mathrm{d}t} \\ \frac{\mathrm{d}u_{C2}(t)}{\mathrm{d}t} \\ \frac{\mathrm{d}u_{C3}(t)}{\mathrm{d}t} \\ \frac{\mathrm{d}u_{low}(t)}{\mathrm{d}t} \end{bmatrix} = \begin{bmatrix} 0 & 0 & 0 & 0 & 0 & -\frac{1}{L_1} \\ 0 & 0 & \frac{1}{L_2} & 0 & 0 & -\frac{1}{L_2} \\ 0 & -\frac{1}{C_1} & 0 & 0 & 0 & 0 \\ 0 & 0 & 0 & -\frac{1}{C_2r_1} & -\frac{1}{C_2r_1} & 0 \\ 0 & 0 & 0 & -\frac{1}{C_3r_1} & -\frac{1}{C_3r_1} & 0 \\ \frac{1}{C_{\mathrm{low}}} & \frac{1}{C_{\mathrm{low}}} & 0 & 0 & 0 & -\frac{1}{R_{\mathrm{L-Buck}}C_{\mathrm{low}}} \end{bmatrix} \begin{bmatrix} i_{L1}(t) \\ i_{L2}(t) \\ u_{C1}(t) \\ u_{C2}(t) \\ u_{C3}(t) \\ u_{\mathrm{low}}(t) \end{bmatrix} + \begin{bmatrix} 0 \\ 0 \\ 0 \\ \frac{1}{C_2r_1} \\ \frac{1}{C_3r_1} \\ 0 \end{bmatrix} u_{\mathrm{high}}(t) \tag{5.50}$$

$$\begin{bmatrix} \frac{\mathrm{d}i_{L1}(t)}{\mathrm{d}t} \\ \frac{\mathrm{d}i_{L2}(t)}{\mathrm{d}t} \\ \frac{\mathrm{d}u_{C1}(t)}{\mathrm{d}t} \\ \frac{\mathrm{d}u_{C2}(t)}{\mathrm{d}t} \\ \frac{\mathrm{d}u_{C3}(t)}{\mathrm{d}t} \\ \frac{\mathrm{d}u_{\mathrm{low}}(t)}{\mathrm{d}t} \end{bmatrix} = \begin{bmatrix} 0 & 0 & 0 & \frac{1}{L_1} & 0 & -\frac{1}{L_1} \\ 0 & 0 & 0 & 0 & 0 & -\frac{1}{L_2} \\ 0 & 0 & -\frac{1}{C_1r} & 0 & \frac{1}{C_1r} & 0 \\ -\frac{1}{C_2} & 0 & 0 & -\frac{1}{C_2r_1} & -\frac{1}{C_2r_1} & 0 \\ 0 & 0 & \frac{1}{C_3r} & -\frac{1}{C_3r_1} & -\frac{1}{C_3r}-\frac{1}{C_3r_1} & 0 \\ \frac{1}{C_{\mathrm{low}}} & \frac{1}{C_{\mathrm{low}}} & 0 & 0 & 0 & -\frac{1}{R_{\mathrm{L-Buck}}C_{\mathrm{low}}} \end{bmatrix} \begin{bmatrix} i_{L1}(t) \\ i_{L2}(t) \\ u_{C1}(t) \\ u_{C2}(t) \\ u_{C3}(t) \\ u_{\mathrm{low}}(t) \end{bmatrix} + \begin{bmatrix} 0 \\ 0 \\ 0 \\ \frac{1}{C_2r_1} \\ \frac{1}{C_3r_1} \\ 0 \end{bmatrix} u_{\mathrm{high}}(t) \tag{5.51}$$

$$\begin{bmatrix} \frac{\mathrm{d}i_{L1}(t)}{\mathrm{d}t} \\ \frac{\mathrm{d}i_{L2}(t)}{\mathrm{d}t} \\ \frac{\mathrm{d}u_{C1}(t)}{\mathrm{d}t} \\ \frac{\mathrm{d}u_{C2}(t)}{\mathrm{d}t} \\ \frac{\mathrm{d}u_{C3}(t)}{\mathrm{d}t} \\ \frac{\mathrm{d}u_{\mathrm{low}}(t)}{\mathrm{d}t} \end{bmatrix} = \begin{bmatrix} 0 & 0 & 0 & 0 & 0 & -\frac{1}{L_1} \\ 0 & 0 & 0 & 0 & 0 & -\frac{1}{L_2} \\ 0 & 0 & -\frac{1}{C_1r} & 0 & \frac{1}{C_1r} & 0 \\ 0 & 0 & 0 & -\frac{1}{C_2r_1} & -\frac{1}{C_2r_1} & 0 \\ 0 & 0 & \frac{1}{C_3r} & -\frac{1}{C_3r_1} & -\frac{1}{C_3r}-\frac{1}{C_3r_1} & 0 \\ \frac{1}{C_{\mathrm{low}}} & \frac{1}{C_{\mathrm{low}}} & 0 & 0 & 0 & -\frac{1}{R_{\mathrm{L-Buck}}C_{\mathrm{low}}} \end{bmatrix} \begin{bmatrix} i_{L1}(t) \\ i_{L2}(t) \\ u_{C1}(t) \\ u_{C2}(t) \\ u_{C3}(t) \\ u_{\mathrm{low}}(t) \end{bmatrix} + \begin{bmatrix} 0 \\ 0 \\ 0 \\ \frac{1}{C_2r_1} \\ \frac{1}{C_3r_1} \\ 0 \end{bmatrix} u_{\mathrm{high}}(t) \tag{5.52}$$

Thus, the average model of the converter in step-down mode can be derived as below.

$$\begin{bmatrix} \frac{\mathrm{d}i_{L1}(t)}{\mathrm{d}t} \\ \frac{\mathrm{d}i_{L2}(t)}{\mathrm{d}t} \\ \frac{\mathrm{d}u_{C1}(t)}{\mathrm{d}t} \\ \frac{\mathrm{d}u_{C2}(t)}{\mathrm{d}t} \\ \frac{\mathrm{d}u_{C3}(t)}{\mathrm{d}t} \\ \frac{\mathrm{d}u_{\mathrm{low}}(t)}{\mathrm{d}t} \end{bmatrix} = \begin{bmatrix} 0 & 0 & 0 & \frac{d_4(t)}{L_1} & 0 & -\frac{1}{L_1} \\ 0 & 0 & \frac{d_3(t)}{L_2} & 0 & 0 & -\frac{1}{L_2} \\ 0 & -\frac{d_3(t)}{C_1} & -\frac{1-d_3(t)}{C_1r} & 0 & \frac{1-d_3(t)}{C_1r} & 0 \\ -\frac{d_4(t)}{C_2} & 0 & 0 & -\frac{1}{C_2r_1} & -\frac{1}{C_2r_1} & 0 \\ 0 & 0 & \frac{1-d_3(t)}{C_3r} & -\frac{1}{C_3r_1} & -\frac{1-d_3(t)}{C_3r}-\frac{1}{C_3r_1} & 0 \\ \frac{1}{C_{\mathrm{low}}} & \frac{1}{C_{\mathrm{low}}} & 0 & 0 & 0 & -\frac{1}{R_{\mathrm{L-Buck}}C_{\mathrm{low}}} \end{bmatrix} \begin{bmatrix} i_{L1}(t) \\ i_{L2}(t) \\ u_{C1}(t) \\ u_{C2}(t) \\ u_{C3}(t) \\ u_{\mathrm{low}}(t) \end{bmatrix} + \begin{bmatrix} 0 \\ 0 \\ 0 \\ \frac{1}{C_2r_1} \\ \frac{1}{C_3r_1} \\ 0 \end{bmatrix} u_{\mathrm{high}}(t) \tag{5.53}$$

Similarly, within the duty cycle range of $0.5 < d_{\mathrm{Buck}} < 1$, the small-signal model of the converter in step-down mode can be obtained as (5.54).

$$\begin{bmatrix} \frac{\mathrm{d}\hat{i}_{L1}(t)}{\mathrm{d}t} \\ \frac{\mathrm{d}\hat{i}_{L2}(t)}{\mathrm{d}t} \\ \frac{\mathrm{d}\hat{u}_{C1}(t)}{\mathrm{d}t} \\ \frac{\mathrm{d}\hat{u}_{C2}(t)}{\mathrm{d}t} \\ \frac{\mathrm{d}\hat{u}_{C3}(t)}{\mathrm{d}t} \\ \frac{\mathrm{d}\hat{u}_{\mathrm{low}}(t)}{\mathrm{d}t} \end{bmatrix} = \begin{bmatrix} 0 & 0 & 0 & \frac{D_{\mathrm{Buck}}}{L_1} & 0 & -\frac{1}{L_1} \\ 0 & 0 & \frac{D_{\mathrm{Buck}}}{L_2} & 0 & 0 & -\frac{1}{L_2} \\ 0 & -\frac{D_{\mathrm{Buck}}}{C_1} & -\frac{1-D_{\mathrm{Buck}}}{C_1 r} & 0 & \frac{1-D_{\mathrm{Buck}}}{C_1 r} & 0 \\ -\frac{D_{\mathrm{Buck}}}{C_2} & 0 & 0 & -\frac{1}{C_2 r_1} & -\frac{1}{C_2 r_1} & 0 \\ 0 & 0 & \frac{1-D_{\mathrm{Buck}}}{C_3 r} & -\frac{1}{C_3 r_1} & -\frac{1-D_{\mathrm{Buck}}}{C_3 r} - \frac{1}{C_3 r_1} & 0 \\ \frac{1}{C_{\mathrm{low}}} & \frac{1}{C_{\mathrm{low}}} & 0 & 0 & 0 & -\frac{1}{R_{\mathrm{L\text{-}Buck}} C_{\mathrm{low}}} \end{bmatrix} \begin{bmatrix} \hat{i}_{L1}(t) \\ \hat{i}_{L2}(t) \\ \hat{u}_{C1}(t) \\ \hat{u}_{C2}(t) \\ \hat{u}_{C3}(t) \\ \hat{u}_{\mathrm{low}}(t) \end{bmatrix}$$
$$+ \begin{bmatrix} 0 \\ 0 \\ 0 \\ \frac{1}{C_2 r_1} \\ \frac{1}{C_3 r_1} \\ 0 \end{bmatrix} \hat{u}_{\mathrm{high}}(t) + \begin{bmatrix} 0 & 0 & 0 & \frac{1}{L_1} & 0 & 0 \\ 0 & 0 & \frac{1}{L_2} & 0 & 0 & 0 \\ 0 & -\frac{1}{C_1} & \frac{1}{C_1 r} & 0 & -\frac{1}{C_1 r} & 0 \\ -\frac{1}{C_2} & 0 & 0 & 0 & 0 & 0 \\ 0 & 0 & -\frac{1}{C_3 r} & 0 & \frac{1}{C_3 r} & 0 \\ 0 & 0 & 0 & 0 & 0 & 0 \end{bmatrix} \begin{bmatrix} I_{L1}(t) \\ I_{L2}(t) \\ U_{C1}(t) \\ U_{C2}(t) \\ U_{C3}(t) \\ U_{\mathrm{low}}(t) \end{bmatrix} \hat{d}_{\mathrm{Buck}}(t) \quad (5.54)$$

5.5 Single-Switch Boost DC–DC Converter

As analyzed in the previous section, the input-parallel output-series Boost DC–DC converter is derived from the modified Cuk converter and the conventional Boost converter, which can achieve high voltage gain and low input current ripple. In order to reduce the component used in the topology, that further improves the power density of the converter, the inductors L_1 and L_2, as well as the power switches Q_1 and Q_2 can be consolidated to obtain the single-switch Boost DC–DC converter, as shown in Fig. 5.26.

The single-switch Boost DC–DC converter is analyzed in this section, which has the advantages of wide voltage gain range, low voltage stresses, and high conversion efficiency. Specifically, the voltage gain is twice that of the conventional Boost converter, and the voltage stresses are half of the high voltage side [131].

The topology of the single-switch Boost DC–DC converter is shown in Fig. 5.27, where the voltage difference between the negative terminals of the input and output is equal to the capacitor voltage. The converter is composed of power switch Q, diodes D_1–D_3, inductor L, and capacitors C_1–C_3.

5.5.1 *Operating Principle of the Single-Switch Boost DC–DC Converter*

In the single-switch Boost DC–DC converter, the charging branch of C_1 and C_2 is illustrated in Fig. 5.28a, in which C_1 and C_2 are charged by the input source U_{low}

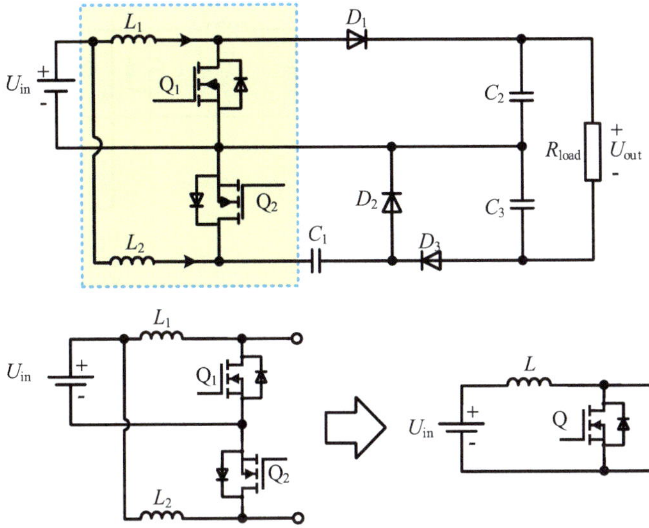

Fig. 5.26 Derivation of single-switch Boost DC–DC converter

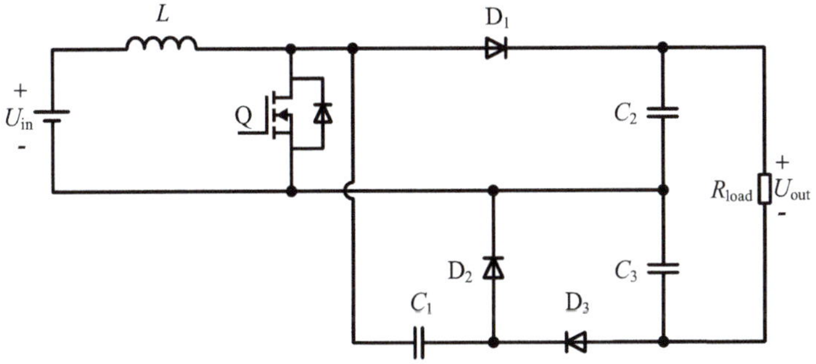

Fig. 5.27 Topology of the single-switch Boost DC–DC converter

and inductor L via D_2 and D_1, respectively. The capacitor voltage of C_3 is clamped by C_1, as shown in Fig. 5.28b. Since the output voltage is the sum of voltages across C_2 and C_3, the voltage boost can be achieved.

The key waveforms of the single-switch Boost DC–DC converter is depicted in Fig. 5.29, where S is the gate signal of power switch Q, U_Q and U_{D1}–U_{D3} are the respective voltage stresses across Q and D_1–D_3, and i_L is the inductor current of L. The operation of converter during each switching cycle contains two switching states, which are illustrated in Fig. 5.30.

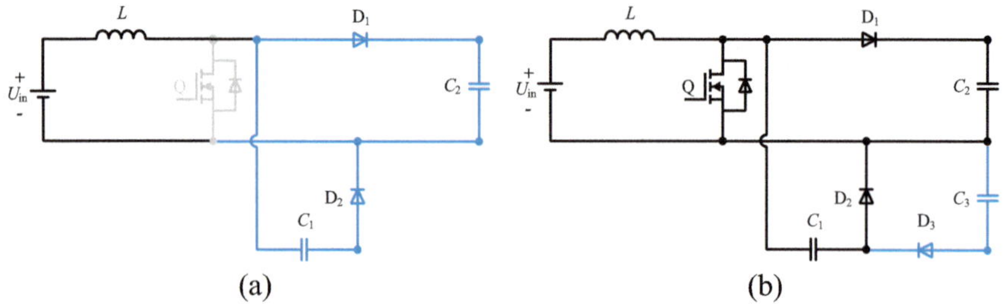

Fig. 5.28 Charging branches of single-switch Boost DC–DC converter. **a** Capacitors C_1 and C_2 branch. **b** Capacitor C_3 branch

Fig. 5.29 Key waveforms of the single-switch Boost DC–DC converter

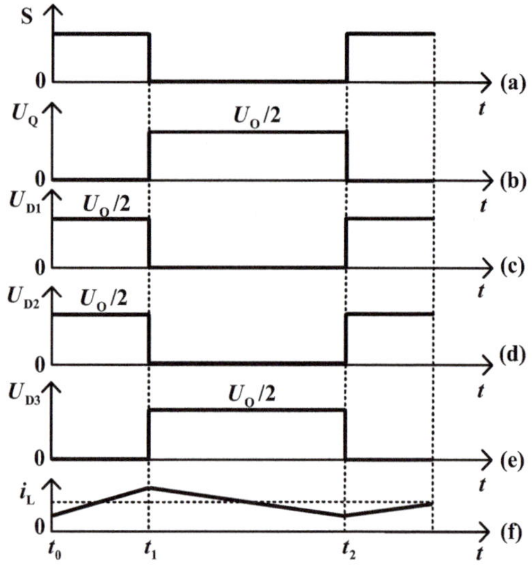

State 1) $S = 1$

When the power switch Q is turned ON, the inductor is charge by the input source U_{low} through Q. Meanwhile, the diode D_3 is conducted, and the capacitor C_3 is charged by C_1 through D_3. The output load on the high voltage side is powered by C_2.

State 2) $S = 0$

When the power switch Q is turned OFF, the diodes D_1 and D_2 are conducted. During this state, the capacitors C_1 and C_2 are charged by the input source U_{low} and inductor L. At the same time, the capacitor C_3 is discharged to power the load.

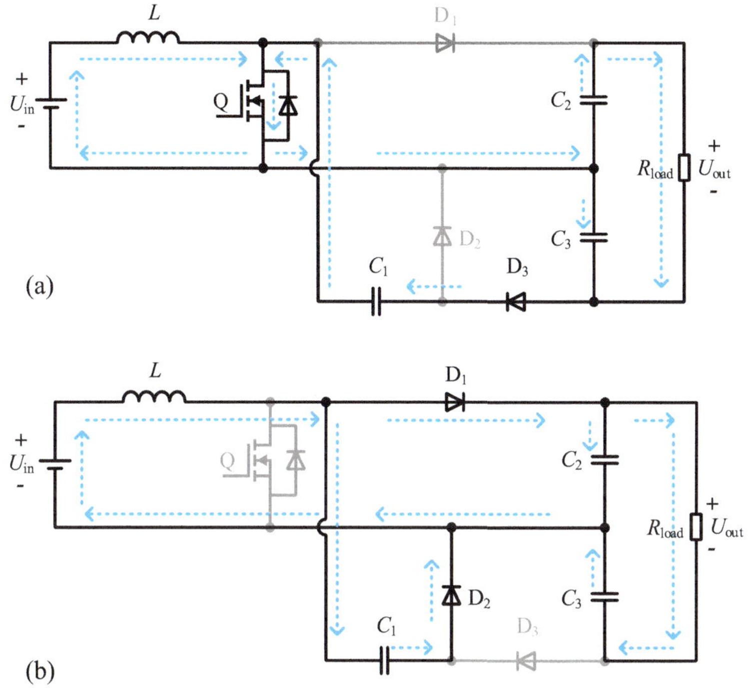

Fig. 5.30 Switching states of the single-switch Boost DC–DC converter during **a** State 1 and **b** State 2

5.5.2 *Small-Signal Modeling and Stability Analysis of the Single-Switch Boost DC–DC Converter*

In order to simplify the small-signal analysis, it is assumed that the capacitances of C_1–C_3 are equal to C, namely

$$C_1 = C_2 = C_3 = C \tag{5.55}$$

$u_{\text{in}}(t)$, $u_o(t)$, and $d(t)$ are taken as the input, output, and control variables, where d is the duty cycle of power switch Q. The state variables are $i_L(t)$, $u_{C1}(t)$–$u_{C3}(t)$. Since the capacitors C_1 and C_3 are connected in parallel during state 1, while C_1 and C_2 are connected in parallel during state 2, the voltages of C_1–C_3 are coupled. Hence, the series resistance r of C_1 is introduced for the voltage decoupling among $u_{C1}(t)$–$u_{C3}(t)$, so that the invalid state variables can be avoided. Thus, the state-space average models in state 1 and state 2 can be obtained as (5.56) and (5.57), respectively.

$$\begin{bmatrix} \frac{di_L(t)}{dt} \\ \frac{du_{C1}(t)}{dt} \\ \frac{du_{C2}(t)}{dt} \\ \frac{du_{C3}(t)}{dt} \end{bmatrix} = \begin{bmatrix} 0 & 0 & 0 & 0 \\ 0 & -\frac{1}{\mathrm{Cr}} & 0 & \frac{1}{\mathrm{Cr}} \\ 0 & 0 & -\frac{1}{\mathrm{CR}} & -\frac{1}{\mathrm{CR}} \\ 0 & \frac{1}{\mathrm{Cr}} & -\frac{1}{\mathrm{CR}} & -\frac{R+r}{\mathrm{CRr}} \end{bmatrix} \begin{bmatrix} i_L(t) \\ u_{C1}(t) \\ u_{C2}(t) \\ u_{C3}(t) \end{bmatrix} + \begin{bmatrix} \frac{1}{L} \\ 0 \\ 0 \\ 0 \end{bmatrix} u_{\mathrm{in}}(t)$$

$$u_{\mathrm{out}}(t) = \begin{bmatrix} 0 \; 0 \; 1 \; 1 \end{bmatrix} \begin{bmatrix} i_L(t) \; u_{C1}(t) \; u_{C2}(t) \; u_{C3}(t) \end{bmatrix}^T \tag{5.56}$$

$$\begin{bmatrix} \frac{di_L(t)}{dt} \\ \frac{du_{C1}(t)}{dt} \\ \frac{du_{C2}(t)}{dt} \\ \frac{du_{C3}(t)}{dt} \end{bmatrix} = \begin{bmatrix} 0 & 0 & -\frac{1}{L} & 0 \\ 0 & -\frac{1}{\mathrm{Cr}} & \frac{1}{\mathrm{Cr}} & 0 \\ \frac{1}{C} & \frac{1}{\mathrm{Cr}} & -\frac{R+r}{\mathrm{CRr}} & -\frac{1}{\mathrm{CR}} \\ 0 & 0 & -\frac{1}{\mathrm{CR}} & -\frac{1}{\mathrm{CR}} \end{bmatrix} \begin{bmatrix} i_L(t) \\ u_{C1}(t) \\ u_{C2}(t) \\ u_{C3}(t) \end{bmatrix} + \begin{bmatrix} \frac{1}{L} \\ 0 \\ 0 \\ 0 \end{bmatrix} u_{\mathrm{in}}(t)$$

$$u_{\mathrm{out}}(t) = \begin{bmatrix} 0 \; 0 \; 1 \; 1 \end{bmatrix} \begin{bmatrix} i_L(t) \; u_{C1}(t) \; u_{C2}(t) \; u_{C3}(t) \end{bmatrix}^T \tag{5.57}$$

Introducing the small-signal disturbances to variables, the small-signal mode of the converter can be written as (5.58).

$$\begin{bmatrix} \frac{d\hat{i}_L(t)}{dt} \\ \frac{d\hat{u}_{C1}(t)}{dt} \\ \frac{d\hat{u}_{C2}(t)}{dt} \\ \frac{d\hat{u}_{C3}(t)}{dt} \end{bmatrix} = \begin{bmatrix} 0 & 0 & -\frac{1-D}{L} & 0 \\ 0 & -\frac{1}{Cr} & \frac{1-D}{\mathrm{Cr}} & \frac{D}{\mathrm{Cr}} \\ \frac{1-D}{C} & \frac{1-D}{\mathrm{Cr}} & -\left(\frac{1-D}{\mathrm{Cr}} + \frac{1}{\mathrm{CR}}\right) & -\frac{1}{\mathrm{CR}} \\ 0 & \frac{D}{\mathrm{Cr}} & -\frac{1}{\mathrm{CR}} & -\left(\frac{1}{\mathrm{CR}} + \frac{D}{\mathrm{Cr}}\right) \end{bmatrix} \begin{bmatrix} \hat{i}_L(t) \\ \hat{u}_{C1}(t) \\ \hat{u}_{C2}(t) \\ \hat{u}_{C3}(t) \end{bmatrix} + \begin{bmatrix} \frac{1}{L} \\ 0 \\ 0 \\ 0 \end{bmatrix} \hat{u}_{\mathrm{in}}(t)$$

$$+ \begin{bmatrix} 0 & 0 & \frac{1}{L} & 0 \\ 0 & 0 & -\frac{1}{\mathrm{Cr}} & \frac{1}{\mathrm{Cr}} \\ -\frac{1}{C} & -\frac{1}{\mathrm{Cr}} & \frac{1}{\mathrm{Cr}} & 0 \\ 0 & \frac{1}{\mathrm{Cr}} & 0 & -\frac{1}{\mathrm{Cr}} \end{bmatrix} \begin{bmatrix} I_L \\ U_{C1} \\ U_{C2} \\ U_{C3} \end{bmatrix} \hat{d}(t)$$

$$\hat{u}_{\mathrm{out}}(t) = \begin{bmatrix} 0 \; 0 \; 1 \; 1 \end{bmatrix} \begin{bmatrix} \hat{i}_L(t) \; \hat{u}_{C1}(t) \; \hat{u}_{C2}(t) \; \hat{u}_{C3}(t) \end{bmatrix}^T \tag{5.58}$$

5.5.3 *Steady-State Characteristics of the Single-Switch Boost DC–DC Converter*

As the small-signal disturbances are zero during steady state, the following equations can be obtain from (5.56) and (5.57).

$$\begin{cases} I_L = \dfrac{2\text{DU}_{\text{high}}}{\text{RD}(1-D)+r} \\ U_{C1} = \dfrac{U_{\text{low}}[\text{RD}(1-D)+r-2\text{Dr}]}{(1-D)[\text{RD}(1-D)+r]} \\ U_{C2} = \dfrac{U_{\text{low}}}{1-D} \\ U_{C3} = \dfrac{U_{\text{low}}[\text{RD}(D-1)+r]}{(D-1)[\text{RD}(1-D)+r]} \\ U_{\text{high}} = \dfrac{2\text{RDU}_{\text{low}}}{\text{RD}(1-D)+r} \end{cases} \quad (5.59)$$

When the series resistance r is omitted, the average inductor current and capacitor voltages can be derived as follows.

$$\begin{cases} I_L = \dfrac{2}{1-D} I_{\text{high}} \\ U_{C1} = U_{C2} = U_{C3} = \dfrac{U_{\text{low}}}{1-D} \end{cases} \quad (5.60)$$

The voltage gain M of the single-switch Boost DC–DC converter can be written as below.

$$M = \frac{U_{\text{high}}}{U_{\text{low}}} = \frac{2}{1-D} \quad (5.61)$$

According to (5.60) and the operating principle analysis, the voltage stresses across components can be expressed by the output voltage as follows.

$$\begin{cases} U_Q = \dfrac{U_{\text{high}}}{2} \\ U_{D1} = U_{D2} = U_{D3} = \dfrac{U_{\text{high}}}{2} \\ U_{C1} = U_{C2} = U_{C3} = \dfrac{U_{\text{high}}}{2} \end{cases} \quad (5.62)$$

The current stresses of the power switch and diodes are be described as follows.

$$\begin{cases} I_Q = \left(\dfrac{2}{1-D} + \dfrac{1}{D}\right) I_{\text{high}} \\ I_{D1} = I_{D2} = \dfrac{1}{1-D} I_{\text{high}} \\ I_{D3} = \dfrac{1}{D} I_{\text{high}} \end{cases} \quad (5.63)$$

The single-switch Boost DC–DC converter is compared with the three-level Boost DC–DC convert and the relevant solutions in [128, 132]. The three-level Boost DC–DC converter also exhibits low voltage stresses across power switches and diodes, whereas the voltage gain is lower. Whereas, the converter in [128] has the same voltage gain as the single-switch Boost DC–DC converter and higher voltage stresses across diodes. In addition, the pulse voltage between the negative terminals of input and output would lead to severe EMI issues. However, the voltage stresses across the power switches and diodes are relatively higher. The same voltage gain and voltage stresses are obtained in [132], whereas one more inductor and capacitor are utilized in the converter. Therefore, compared with the counterparts, the single-switch Boost DC–DC converter has the advantages of high voltage gain, low voltage stresses, low cost and constant voltage difference between the input and output grounds.

5.5.4 Extended Single-Switch Boost DC–DC Converter

For the applications at high power and voltage level, the single-switch Boost DC–DC converter can be further extended with more stages. The extended topology of the single-switch Boost DC–DC converter with n stages is presented in Fig. 5.31, which contains $2n - 1$ capacitors and $2n - 1$ diodes. By means of the extended branches, higher voltage gain can be achieved. Figure 5.32 shows the switching states of the extended n-stage single-switch Boost DC–DC converter.

The equal capacitor voltages are denoted as U_C, so that the voltage gain of the converter with n extended stages can be written as follows.

$$M_{\text{Ext}} = \frac{U_{\text{high}}}{U_{\text{low}}} = \frac{nU_C}{U_{\text{low}}} = \frac{n}{1-D} \tag{5.64}$$

Moreover, the voltage stresses across power switch and diodes can be described as below.

$$U_Q = U_D = U_C = \frac{U_{\text{low}}}{1-D} = \frac{U_{\text{high}}}{n} \tag{5.65}$$

5.5.5 Fault-Tolerant Operation of the Single-Switch Boost DC–DC Converter

As illustrated in Fig. 5.33, the fault-tolerant circuit is utilized for the single-switch Boost DC–DC converter, where F, F_1–F_3 are fuses and Q_{oc}, Q_{oc1}–Q_{oc3} are auxiliary power switches. The fault-tolerant operation scheme is derived on the basis of [133], so as to avoid the failure of power semiconductors and improve the operation

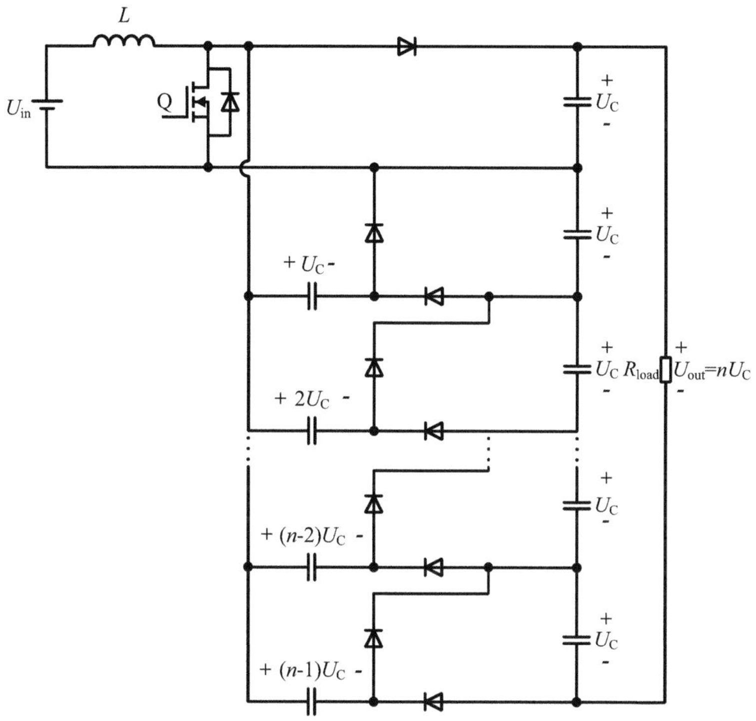

Fig. 5.31 Extended topology of the single-switch Boost DC–DC converter with n stages

reliability. When the converter is operating normally, the auxiliary power switches Q_{oc}, Q_{oc1}–Q_{oc3} are turned OFF. However, when a failure happens on power switch or diodes, the operation of converter will be switched to fault-tolerant condition, which is explained as below.

The voltage stress across Q is equal to $U_{high}/2$ when Q is turned OFF, and is clamped at zero when Q is turned ON. Thus, the operation failure of Q can be detected via the gate signal and voltage stress. The index sgn_Q is adopted to indicate the state of Q, which is obtained from by comparing the voltage across Q with the threshold δ as follows.

$$\text{sgn}_Q = \begin{cases} 0, & U_Q > \delta \\ 1, & U_Q < \delta \end{cases} \tag{5.66}$$

The value of δ is selected between 0 and $U_{high}/2$. In turn, the error signal is derived from the gate signal and sgn_Q as (5.67).

$$\text{err}_Q = S \oplus \text{sgn}_Q \tag{5.67}$$

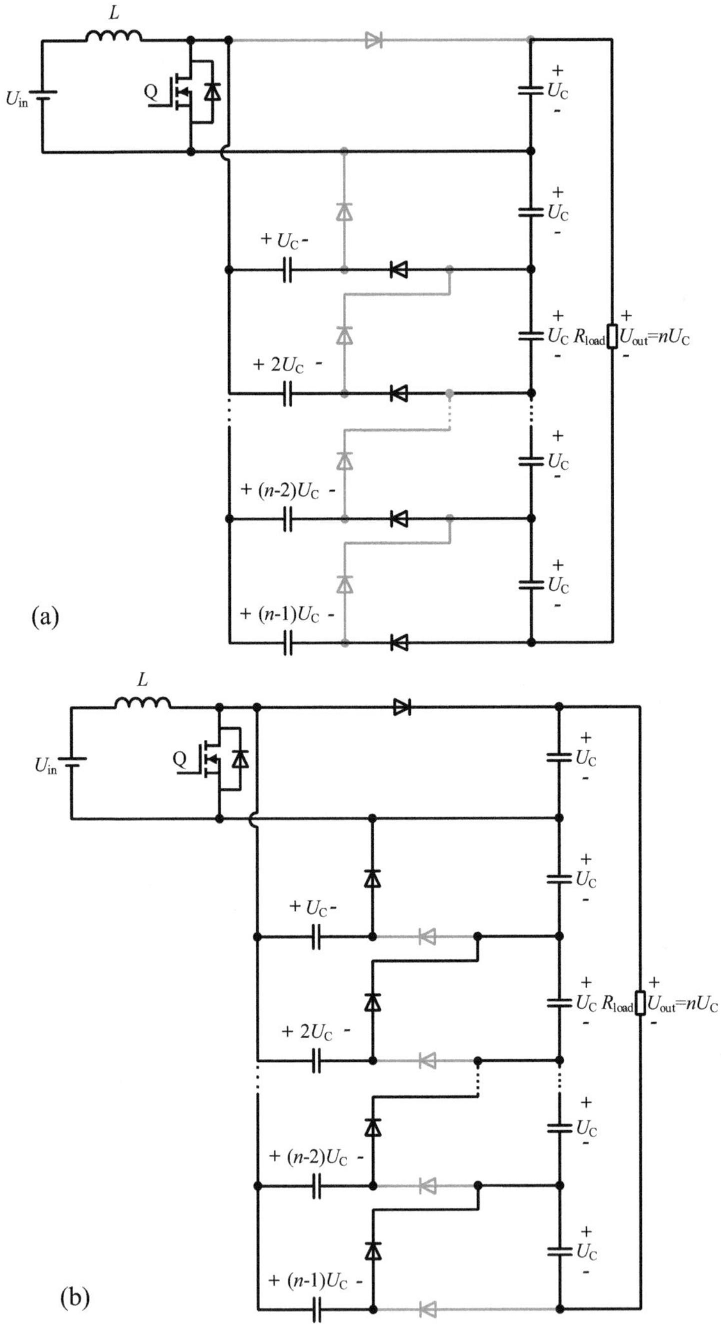

Fig. 5.32 Switching states of the extended single-switch Boost DC–DC converter with n stages **a** State 1, **b** State 2

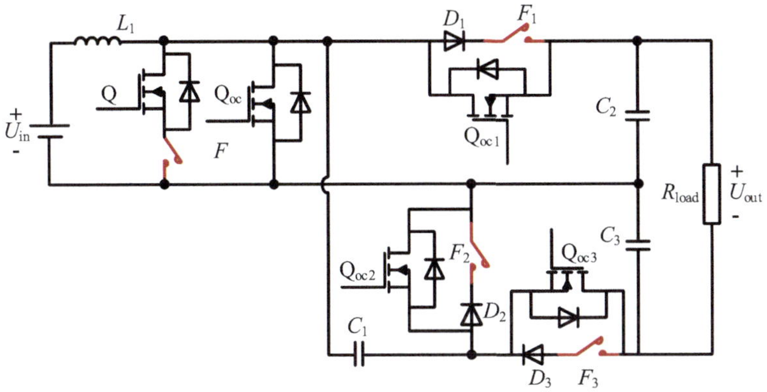

Fig. 5.33 Fault-tolerant circuit of the single-switch Boost DC–DC converter

As a result, when gate signal is 1 and sgn_Q is 1, the power switch Q is turned ON properly, so that err_Q is zero. On the contrary, when both gate signal and sgn_Q are equal to zero, erro_Q is zero, namely the power switch Q is turned OFF properly. Otherwise, when the values of gate signal and sgn_Q are different, the error signal err_Q becomes 1, which indicates the operation failure of Q. The fault detection and identification strategy are shown in Fig. 5.34. It can be seen that, when err_Q is equal to 1, the counting variable n_c is accumulated to count the fault duration. Once the counting variable n_c exceeds the threshold N, the program continues to identify the fault according to gate signal S. Specifically, the fault of Q is identified as open circuit if $S = 1$. Thereby, the open circuit fault (OCF) is set to 1 and short circuit fault (SCF) is set to 0. On the contrary, if $S = 0$, the fault of Q is identified as short circuit fault. Thus, OCF is set to 0 whereas SCF is set to 1.

The gate signal S_{Qoc} for the auxiliary power switch Q_{oc} is derived from the fault identification signals and gate signal S, as shown in Fig. 5.35. The fault signal is generated from OCF and SCF through OR gate. In turn, the gate signal S_{Qoc} is generated from S and the fault signal via AND gate. For example, in the case of open circuit fault, the fault signal is obtained as 1, so that S_{Qoc} will be synchronous with S. In this way, the faulted Q can be replaced by the auxiliary power switch Q_{oc}, so that the converter can maintain operation. Conversely, in the case of short circuit fault, the faulted Q will be isolated from the converter by the fuse F, and Q_{oc} works as the main power switch.

Furthermore, an additional power switch Q_c can be utilized for capacitor fault protection, as shown in Fig. 5.36. When the converter is operating normally under voltage control, the output voltage U_{high} is maintained at the reference voltage U_{ref}. However, when capacitor short circuit occurs, the output voltage will inevitably decrease. Hence, the fault indicator sgn_C is adopted for fault detection, which is obtained by comparing U_{high} with the threshold value as in (5.68), where k is a constant coefficient within (0.5, 1).

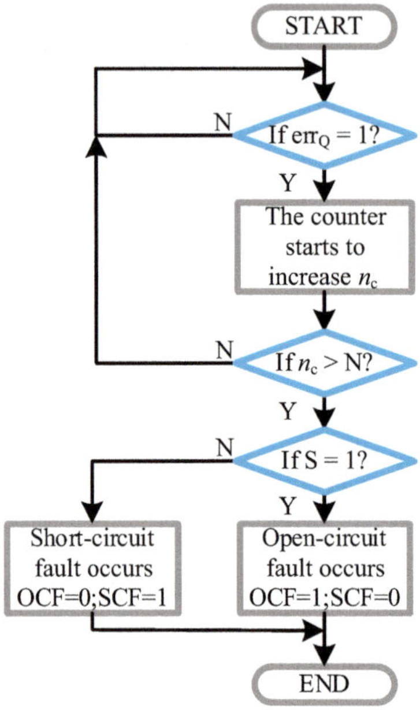

Fig. 5.34 Flowchart of fault detection and identification for the power switch Q

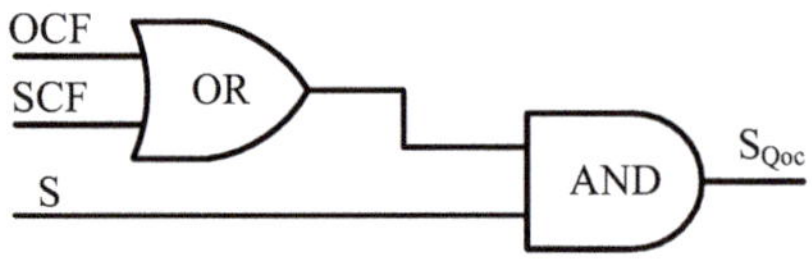

Fig. 5.35 Derivation of the gate signal for auxiliary power switch Q_{oc}

$$\mathrm{sgn}_C = \begin{cases} 0, & U_{\mathrm{high}} < kU_{\mathrm{ref}} \\ 1, & U_{\mathrm{high}} > kU_{\mathrm{ref}} \end{cases} \tag{5.68}$$

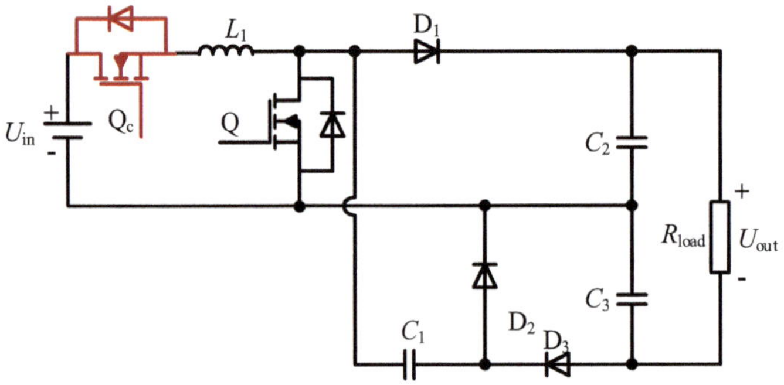

Fig. 5.36 Single-switch Boost DC–DC converter with capacitor fault protection

The capacitor short circuit fault is detected while sgn_C is set to 0. Meanwhile, the signal $m(t)$ is obtained according to the duration of transient process.

$$m(t) = \begin{cases} 0, & t < t_0 \\ 1, & t > t_0 \end{cases} \tag{5.69}$$

In turn, the fault indicator sgn_C and the transient time signal m(t) are utilized to generate the gate signal S_C of Q_C.

$$S_C = \mathrm{sgn}_C \| m(t) \tag{5.70}$$

Thereby, within the transient time (0, t_0), when the capacitor short circuit fault occurs, the gate signal S_C will be set to zero, so that the power switch Q_C will be turned OFF to prevent the converter from damage.

5.6 Quasi-Z-Source Boost DC–DC Converter

Based on the input-parallel output-series DC–DC converter in Sect. 5.3, the hybrid topology in Fig. 5.37a can be derived by adding two quasi-Z-source cells. Thereby, reduced voltage stresses across power semiconductors as well as higher voltage gain can be realized. Furthermore, the hybrid topology can be simplified as presented in Fig. 5.37b, where only one additional quasi-Z-source cell is required. The voltage difference between input and output grounds of the converter is equal to the capacitor voltage across C_3. Thereby, on the basis of the topology in Fig. 5.37b, the quasi-Z-source Boost DC–DC converter is deduced, which utilizes the same components whereas exhibits a common ground structure [134].

The topology of the quasi-Z-source Boost DC–DC converter is presented in Fig. 5.38, which consists of power switch Q and diodes D_1–D_5, inductors L_1 and L_2, and capacitors C_1–C_5. The converter can be regard as the series connection of a quasi-Z-source cell and a switched-capacitor cell.

5.6.1 *Operating Principle of the Quasi-Z-Source Boost DC–DC Converter*

The operation of the quasi-Z-source Boost DC–DC converter contains two states during each switching cycle, according to the switching state of the power switch Q. The key waveforms of the quasi-Z-source Boost DC–DC converter is depicted in Fig. 5.39, where S is the gate signal of Q, u_{C1}–u_{C5} and i_{L1}, i_{L2} are the respective capacitor voltages and inductor currents of C_1–C_5 and i_{L1}, i_{L2}. And u_{D1}–u_{D5} are the

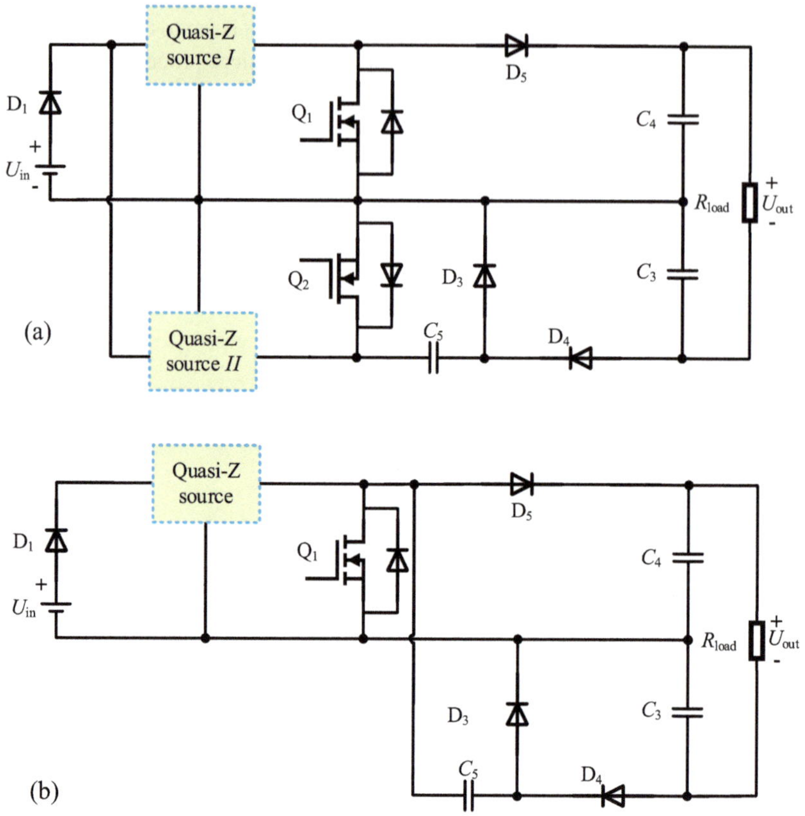

Fig. 5.37 **a** Hybrid topology of input-parallel output-series DC–DC converter and quasi-*Z*-source. **b** Simplified circuit of the hybrid topology

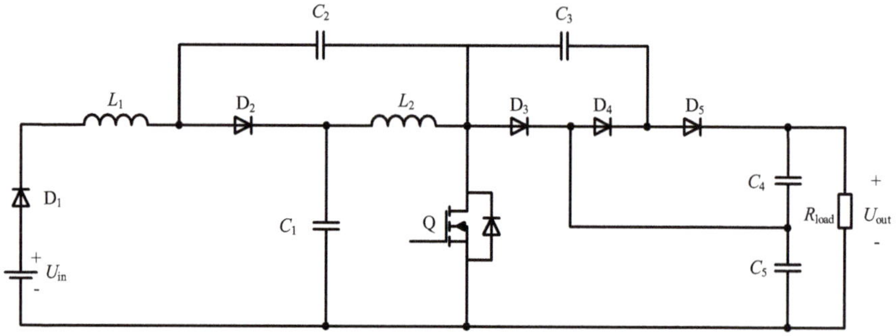

Fig. 5.38 Topology of the quasi-*Z*-source Boost DC–DC converter

voltage stresses across diodes D_1–D_5. As can be observed from the illustration, the switching states consist of $S = 1$ and $S = 0$, which are explained in detail as follows.

State 1) $S = 1$

When the power switch Q is turned ON, the inductors L_1 and L_2 are both charged via Q, as shown in Fig. 5.40a. The output load is powered by the capacitor C_4. Meanwhile, the capacitor C_3 is charged by C_5 via diode D_4, while C_1 and C_2 are discharged.

State 2) $S = 0$

When the power switch Q is turned OFF, the inductors L_1 and L_2 are discharged while the diodes D_2, D_3, and D_5 are conducted, as shown in Fig. 5.40b. During this state, the capacitor C_1 is charged by L_1 and the input source, and C_2 is charged by L_2. The capacitors C_4 and C_5 are both charged, whereas C_3 is discharged to power the load.

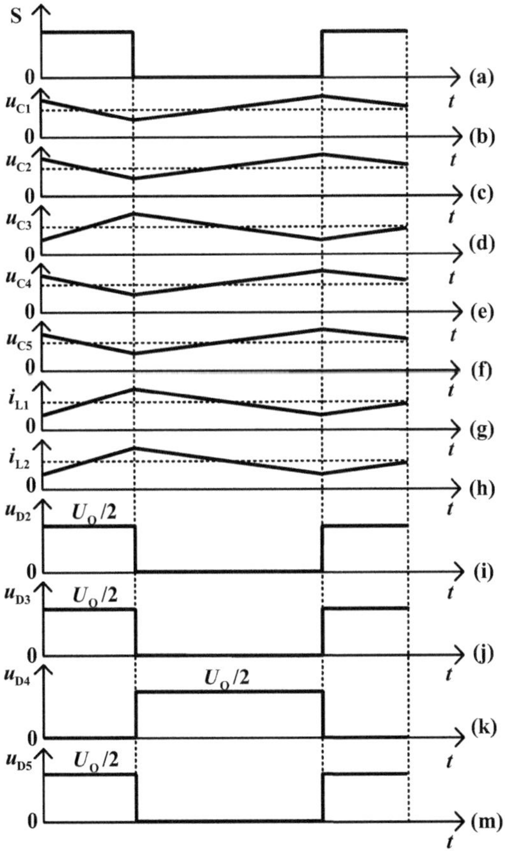

Fig. 5.39 Key waveforms of the quasi-Z-source Boost DC–DC converter

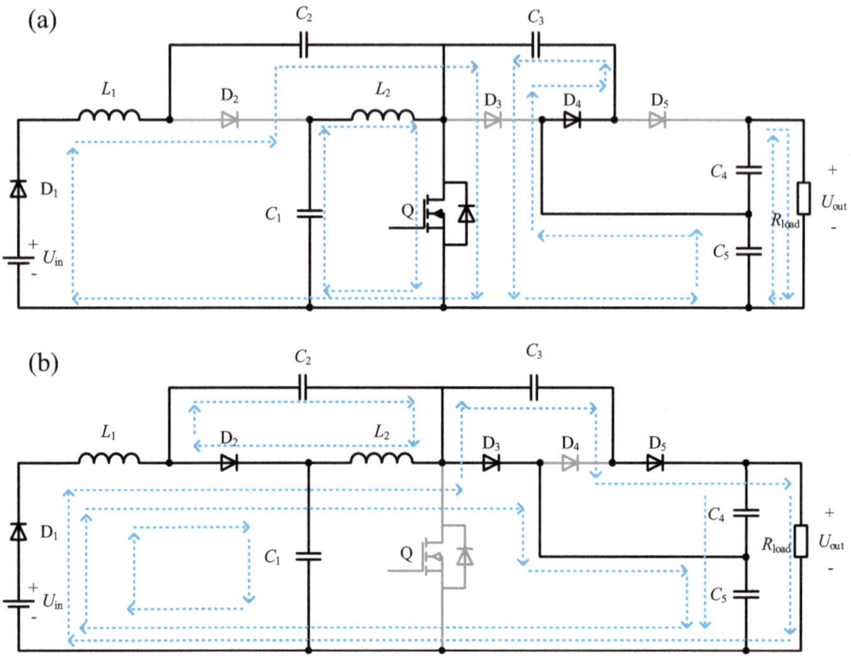

Fig. 5.40 Switching states of the quasi-*Z*-source Boost DC–DC converter during **a** State 1 and **b** State 2

5.6.2 Steady-State Characteristics of the Quasi-Z-Source Boost DC–DC Converter

Based on the above operating analysis, the voltages across components during each switching states can be obtained by applying Kirchhoff's voltage laws as follows.

For State 1,

$$\begin{cases} u_{L1} = U_{\text{low}} + U_{C2} \\ u_{L2} = U_{C1} \\ U_{C3} = U_{C5} \\ U_{\text{high}} = U_{C4} + U_{C5} \end{cases} \tag{5.71}$$

And for State 2,

$$\begin{cases} u_{L1} = U_{\text{low}} - U_{C11} \\ u_{L1} = U_{\text{low}} - U_{\text{high}} + U_{C2} + U_{C3} \\ u_{L2} = -U_{C2} \\ U_{C3} = U_{C4} \end{cases} \tag{5.72}$$

According to volt-second balance principle, the following equations can be derived by combing (5.71) and (5.72).

$$\begin{cases} U_{C1} = \dfrac{1-d}{1-2d} U_{\text{low}} \\ U_{C2} = \dfrac{d}{1-2d} U_{\text{low}} \\ U_{C3} = U_{C4} = U_{C5} = U_{\text{high}}/2 \\ U_{\text{high}} = \dfrac{2}{1-2d} U_{\text{low}} \end{cases} \tag{5.73}$$

In turn, the voltage gain M of the quasi-Z-source Boost DC–DC converter can be written as below.

$$M = \frac{U_{\text{high}}}{U_{\text{low}}} = \frac{2}{1-2d} \tag{5.74}$$

Thereby, the duty cycle should be limited within the range of $0 < d < 0.5$. The capacitor voltages can also be expressed by the output voltage as follows.

$$\begin{cases} U_{C1} = \dfrac{1-d}{2} U_{\text{high}} \\ U_{C2} = \dfrac{d}{2} U_{\text{high}} \\ U_{C3} = U_{C4} = U_{C5} = U_{\text{high}}/2 \end{cases} \tag{5.75}$$

It can be observed from Fig. 5.40a that, the diode D_2 is connected with C_1 and C_2 in parallel, while D_3 and D_5 are connected with C_5 and C_4 in parallel, respectively. Thus, the voltage stresses U_{D2}, U_{D3}, and U_{D5} across diodes D_2, D_3 and D_5 can be obtained, as shown below.

$$U_{D2} = U_{D3} = U_{D5} = \frac{U_{\text{high}}}{2} \tag{5.76}$$

Moreover, it can be seen in Fig. 5.40b that, the voltages cross Q and D_4 is equal to the capacitor voltages of C_5 and C_3.

$$U_Q = U_{D4} = \frac{U_{\text{high}}}{2} \tag{5.77}$$

Therefore, the voltage stresses for power semiconductors are all half of the output voltage on the high voltage side. In terms of the current stresses, (5.78) and (5.79) can be obtained in State 1 and State 2, respectively, where i_{C1}–i_{C5} are respective the capacitor currents of C_1–C_5.

$$\begin{cases} i_{C1} = -I_{L2} \\ i_{C2} = -I_{L1} \\ i_{C4} = -I_{\text{high}} \\ i_{C5} = -(-i_{C4} + i_{C3}) \end{cases} \tag{5.78}$$

$$\begin{cases} i_{C1} = i_{C2} + I_{L1} - I_{L2} \\ i_{C3} = -i_{C4} - I_{\text{high}} \\ i_{C5} = I_{L1} - I_{\text{high}} - i_{C1} \end{cases} \tag{5.79}$$

According to amp-second balance principle, the average input current I_{low} as well as the average inductor currents I_{L1} and I_{L2} of L_1 and L_2 can be derived as follows.

$$I_{\text{low}} = I_{L1} = I_{L2} = \frac{2}{1-2d} I_{\text{high}} \tag{5.80}$$

During State 1, the power switch Q is turned ON and D_4 is conducted. Thus, the respective current stresses I_Q and I_{D4} of Q and D_4 can be obtained from Fig. 5.40a and (5.78) as below.

$$\begin{cases} I_Q = \dfrac{1+2d}{d(1-2d)} I_{\text{high}} \\ I_{D4} = \dfrac{d+1}{d} I_{\text{high}} \end{cases} \tag{5.81}$$

During State 2, the diodes D_2, D_3, and D_5 are conducted, and the corresponding current stresses I_{D2}, I_{D3}, and I_{D5} can be derived from Fig. 5.40b and (5.79) as follows.

$$\begin{cases} I_{D2} = \dfrac{2}{(1-2d)(1-d)} I_{\text{high}} \\ I_{D3} = I_{D5} = \dfrac{1}{(1-d)} I_{\text{high}} \end{cases} \tag{5.82}$$

The quasi-Z-source Boost DC–DC converter can achieve higher voltage gain and lower voltage stresses across semiconductors than the conventional quasi-Z-source converter. The voltage gain curves of the comparative Boost DC–DC converters are presented in Fig. 5.41.

On the basis of the conventional converter, the Boost DC–DC converter in [135] exhibits reduced voltage stresses but fails to achieve wide voltage gain range. High voltage gain of the converter in [136] can be realized by the cascade structure with a Z-source cell and a switched-capacitor cell. In addition, while this converter does have a wider voltage gain range, particularly at lower duty cycles, it still presents several drawbacks when compared to the quasi-Z-source Boost DC–DC converter. These include a higher number of inductors, increased voltage stress on semiconductors, and non-shared ground connections between the input and output sides. The use of

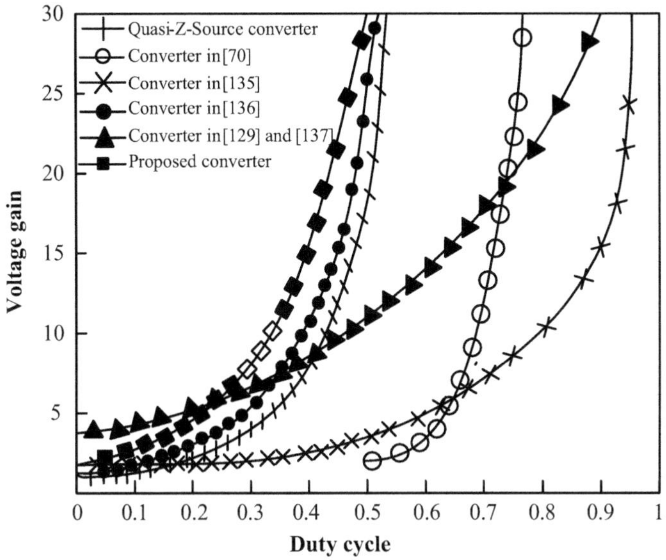

Fig. 5.41 Voltage gain curves of the comparative Boost DC–DC converters

dual-coupled inductors in the converter, as outlined in [129], can lead to a reduction in the number of required inductors. However, the voltage gain is conditional on the turns ratio N, which presents a limitation. Furthermore, in the non-common ground structure, the potential difference between the input and output sides manifests as a PWM voltage, which could result in the occurrence of electromagnetic interference issues. By utilizing an interleaved approach with a diode-capacitor multiplier [137], the step-up DC–DC converter can attain a common ground, and its voltage gain is equivalent to that of the converter presented in [129]. Furthermore, the interleaved structure that incorporates an additional power switch enables a reduction in the ripple of the input current, particularly when the duty cycles of the two power switches are roughly at 0.5. The quasi-Z-source Boost DC–DC converter boasts a wider voltage gain range compared to the previous design [137] in the lower duty cycle range (i.e., $d = 0 - 1/3$). In the higher duty cycle range, the voltage gain of the converter in [137] is lower than that of the quasi-Z-source Boost DC–DC converter. It is noteworthy that the reduction in ripple of the input current may be influenced in both the lower and higher duty cycle ranges, which are situated far from 0.5. The authors of [70] presented a three-level boost DC–DC converter with a quasi-Z-source for fuel cell vehicles. This converter allows for power switches with duty cycles in the range of 0.5–0.75, which is near 0.5, and the voltage stress is halved due to the three-level structure. However, its basic voltage gain is still lower than that of the quasi-Z-source Boost DC–DC converter.

Thus, as demonstrated through the comparisons made above, it becomes evident that the quasi-Z-source Boost DC–DC converter offers a range of benefits, including a high voltage gain that does not require extreme duty cycles, low voltage stress across semiconductors, and a common ground shared between the input and output sides.

5.7 Quadratic Boost DC–DC Converter

In order to achieve higher voltage gain, the quasi-Z-source Boost DC–DC converter can be modified into quadratic Boost DC–DC converter by replacing the quasi-Z-source cell with quadratic voltage lifting network, as shown in Fig. 5.42 [138].

In the voltage lifting cell, the input inductor L_1 and power switch Q_1 comprise a Boost topology with capacitor C_1 and diode D_3, and another Boost topology with C_2 and D_4. The energy storage inductor is charged by these two capacitors C_1 and C_2 through Q_2.

Hence, through the integration of the voltage lifting cell, the resulting quadratic Boost DC–DC converter attains an expanded range of higher voltage gain. The converter's second stage preserves the configuration of the quasi-Z-source Boost DC–DC converter, requiring only the incorporation of one extra MOSFET and an additional diode.

The quadratic Boost DC–DC converter with high voltage gain is shown in Fig. 5.43. It includes two active power switches Q_1 and Q_2, two inductors L_1 and

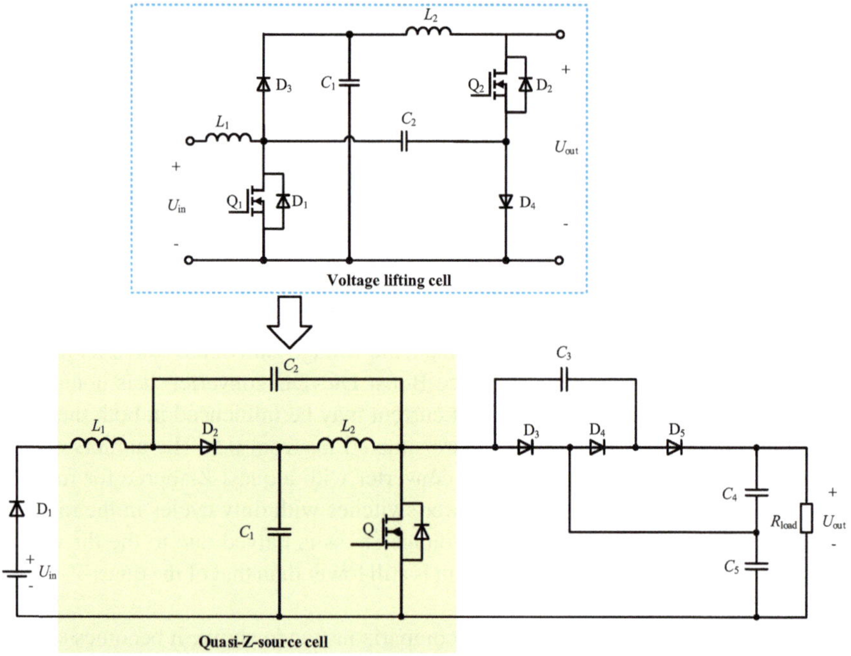

Fig. 5.42 Derivation of quadratic Boost DC–DC converter

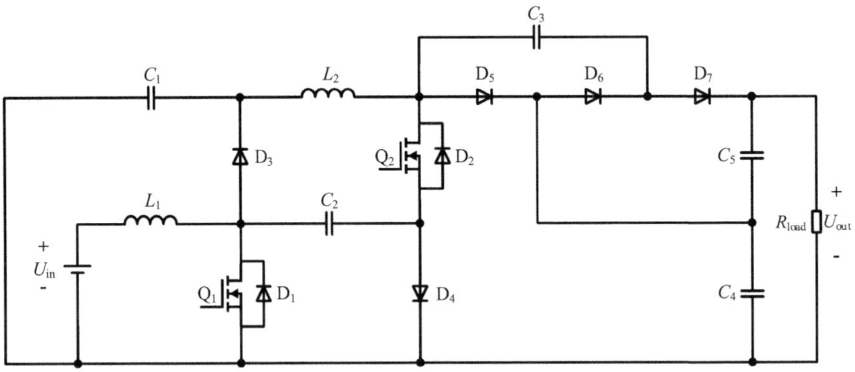

Fig. 5.43 Topology of the quadratic Boost DC–DC converter

L_2, and five capacitors C_1–C_5 in addition to five power diodes D_3–D_7. The fuel cell source U_{in} and inductor L_1 are connected in series to charge parallel capacitors C_1 and C_2. Inductor L_2 is used as another energy storage component to achieve high voltage gain. The converter also features a ladder-type voltage multiplier made up of capacitors C_3–C_5 and diodes D_5–D_7 to enhance the voltage gain and reduce the voltage stress on the high voltage side's power semiconductors.

5.7.1 Operating Principle of the Quadratic Boost DC–DC Converter

This simultaneous switching of Q_1 and Q_2 in the quadratic Boost DC–DC converter results in two distinct switching states during each switching period, as illustrated in Fig. 5.44.

State 1) $S_1S_2 = 11$

During this state, Q_1 and Q_2 are turned ON, and both inductors L_1 and L_2 are charged. The inductor L_1 is charged by the DC source on the low voltage side, whereas the inductor L_2 is charged by capacitors C_1 and C_2. At the same time, the capacitor C_3 is charged by C_2 and C_4 via diode D_6. The output load R is powered by the capacitors C_4 and C_5.

State 2) $S_1S_2 = 00$

While Q_1 and Q_2 are both turned OFF, the inductors L_1 and L_2 are discharged. The capacitors C_1, C_2, C_4, and C_5 are all charged. Meanwhile, the capacitor C_3 is discharged.

In the quadratic Boost DC–DC converter, the inductor currents i_{L1} and i_{L2} follow a similar energy transfer process, as per the operating waveforms depicted in Fig. 5.45. Under the switching state that $S_1S_2 = 11$, power switches Q_1 and Q_2, along with

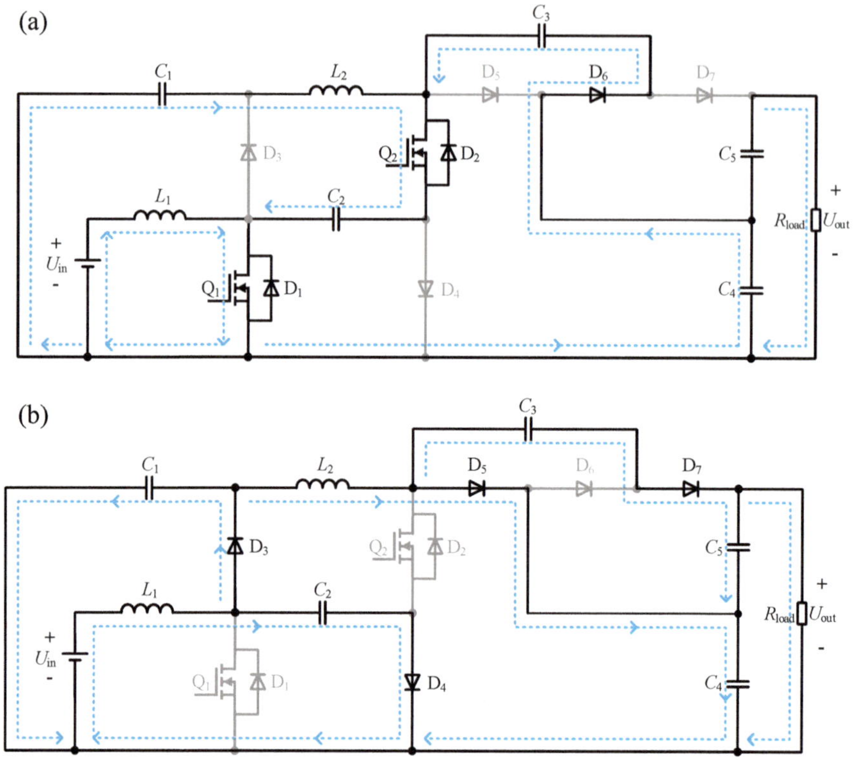

Fig. 5.44 Switching states of the quadratic Boost DC–DC converter during **a** State 1 and **b** State 2

diode D_6, are turned on. This causes the current i_{Q1} to increase linearly, while i_{Q2} and i_{D6} decrease linearly. The negative output capacitor current i_{C5} indicates that capacitor C_5 is discharging. On the other hand, during another switching state (S_1S_2 = 00), power switches Q_1 and Q_2, along with diode D_6, are turned off. This causes the currents i_{D5} and i_{C5} to decrease linearly, and the capacitor voltage fluctuations reflect the charging and discharging processes. Capacitors C_2 and C_3 exhibit opposite charging and discharging states, which can be observed from their respective voltage fluctuations U_{C2} and U_{C3}.

5.7.2 *Steady-State Characteristics of the Quadratic Boost DC–DC Converter*

The on-time and off-time of the power switches in the converter can be calculated based on the duty cycle d, and the switching period T_s. Specifically, the on-time is equal to dT_s, while the off-time is equal to $(1 - d)T_s$. In the analysis of the circuit,

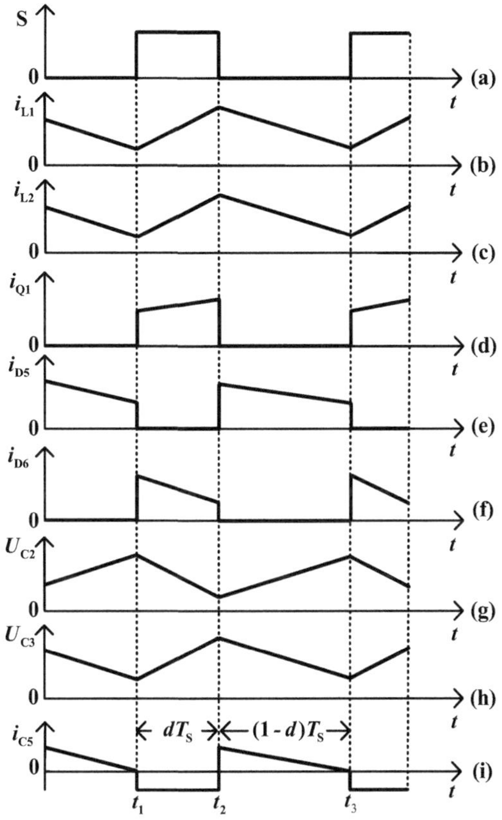

Fig. 5.45 Key waveforms of the quadratic Boost DC–DC converter

it is assumed that the inductor current and capacitor voltage remain constant during each switching period, and the forward voltage drop and on-state resistance of the power semiconductors are negligible.

Applying the volt-second balance principle to the analysis of switching states, along with considering the duty cycle and switching frequency, can yield the relationship between capacitor voltages.

$$\begin{cases} \mathrm{d}U_{\mathrm{in}}T_s + (U_{\mathrm{in}} - U_{C2})(1-d)T_s = 0 \\ (U_{C1} - U_{C2})\mathrm{d}T_s + (U_{C2} - U_{C4})(1-d)T_s = 0 \end{cases} \tag{5.83}$$

$$\begin{cases} U_{C1} = U_{C2} \\ U_{C3} = U_{C5} = U_{C2} + U_{C4} \\ U_{\mathrm{out}} = U_{C4} + U_{C5} \end{cases} \tag{5.84}$$

From the two switching states illustrated in Fig. 5.44, the voltage relationship between the output and capacitor voltages can be determined. Consequently, the voltage gain M can be calculated using Eqs. (5.83) and (5.84) as below, where U_{out} and U_{in} are the input and output voltage of the quadratic Boost DC–DC converter, respectively.

$$M = \frac{U_{out}}{U_{in}} = \frac{3 + d}{(1 - d)^2} \tag{5.85}$$

Hence, the quadratic Boost DC–DC converter can achieve a wide and high voltage gain range in theory. To evaluate the potential advantages and limitations of the converter and relative works, a comparison with similar converters was carried out to assess their performance. The comparison was made based on the voltage gain versus duty cycle, and the results were presented graphically in Fig. 5.46.

Based on the comparison of voltage gain curves versus duty cycle, it can be deduced that the quadratic Boost DC–DC converter outperforms the converters discussed in [135, 136, 139, 140] in terms of voltage gain, particularly when the duty cycle is > 0.2. The converter described in [136] exhibits a better voltage gain curve, but it suffers from low conversion efficiency and non-common ground, which can lead to higher power losses and increased d*u*/d*t* issues. Therefore, considering the voltage gain, conversion efficiency, and common ground, the quadratic Boost DC–DC converter presents significant advantages, including a high and wide voltage gain range, acceptable conversion efficiency, and a common ground. These benefits highlight the superiority of the quadratic Boost DC–DC converter over its counterparts.

Based on the analysis of the operation states shown in Fig. 5.44 and the voltage gain, it can be concluded that the voltage stresses across the different power devices

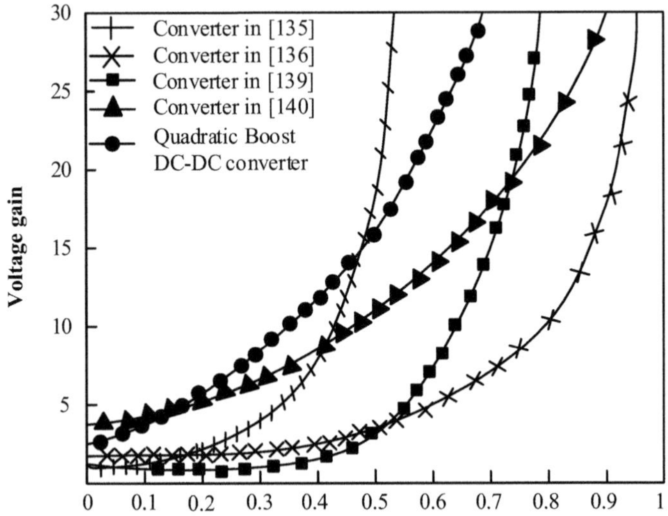

Fig. 5.46 Comparison of converter voltage gain versus duty cycle with relative works

in the converter can be determined as follows.

$$
\begin{cases}
U_{Q1} = U_{D3} = U_{D4} = U_{C1} = U_{C2} = \dfrac{1-d}{3+d}U_{\text{out}} \\
U_{Q2} = U_{C4} = \dfrac{1+d}{3+d}U_{\text{out}} \\
U_{D5} = U_{D6} = U_{D7} = U_{C3} = U_{C5} = \dfrac{2}{3+d}U_{\text{out}}
\end{cases}
\tag{5.86}
$$

The voltage stresses across the active power switches Q_1 and Q_2 are less than half of the output voltage U_{out}. The diodes D_3 and D_4 have voltage stresses that are less than one third of U_o, whereas the voltage stresses across D_5–D_7 and capacitors C_1–C_5 are less than two-thirds of U_{out}. Specifically, the voltage stresses across capacitors C_1 and C_2 are less than one third of U_{out}, whereas the voltage stress across C_4 is less than half of the output high voltage U_{out}. Additionally, the voltage stresses across C_3 and C_5 are less than two-thirds of U_{out}. It is important to note that the voltage stresses across the power devices are a crucial factor to consider in the design of power converters, as they can impact the reliability and efficiency of the converter. The analysis of voltage stresses in the converter reveals that the stresses are within acceptable limits, indicating that the converter is a viable and reliable option for power conversion applications. This analysis can provide useful insights for researchers and engineers in designing efficient and reliable power converters for a wide range of applications.

$$
\begin{cases}
I_{Q1} = \dfrac{1+3d-d^2-d^3}{d(1-d)^2}I_{\text{out}} \\
I_{Q2} = \dfrac{1+d}{d(1-d)}I_{\text{out}} \\
I_{D3} = \dfrac{2}{(1-d)^2}I_{\text{out}} \\
I_{D4} = \dfrac{1+d}{(1-d)^2}I_{\text{out}} \\
I_{D5} = \dfrac{1}{(1-d)}I_{\text{out}} \\
I_{D6} = \dfrac{1}{d}I_{\text{out}} \\
I_{D7} = \dfrac{1}{(1-d)}I_{\text{out}}
\end{cases}
\tag{5.87}
$$

The results of current stress analysis across the power devices are displayed in (5.87), which can be obtained through the application of Kirchhoff's current laws and referring to the current analysis depicted in Fig. 5.2. These current stresses are not fixed and depend on the operating duty cycle, which is typically set between 0.2

and 0.4. As an example, active power switch Q_2 has the highest current stress, which is 7.5 times the output current Io, making it a useful benchmark for component parameter design. It is important to note that the current stresses across Q_1–D_7 represent average values, while those across capacitors C_1–C_5 are root mean square values. These values provide important information for evaluating the current ratings and selecting suitable components to ensure stable operation of the converter.

By adding only small number of diodes, the quadratic Boost DC–DC converter is able to achieve a high and wide voltage gain range. On the other hand, Converter [136] can also achieve high voltage gain, but only when the duty cycle approaches 0.5. However, this converter suffers from a high voltage stress, which is almost equal to the output voltage when the duty cycle is close to zero, and has a lower efficiency compared to the other converters. In contrast, the quadratic Boost DC–DC converter is more suitable for applications that require a large step-up ratio when compared to Converters [135, 139]. The converter in [27] experiences a maximum device voltage stress, which is higher than half of the output voltage, when d is not equal to 0.5, whereas the maximum voltage stress across the power switches in the quadratic Boost DC–DC converter is less than half of the output voltage. When it comes to diode selection, the quadratic Boost DC–DC converter has a lower maximum voltage stress across the diodes than Converters [135, 139]. Converter [140] has the advantage of lower voltage stress, but it does not have a common ground between the input and output sides. This can cause additional EMI issues, which can impact its performance.

5.7.3 Small-Signal Modeling of the Quadratic Boost DC–DC Converter

For small-signal modeling of the quadratic Boost DC–DC converter, the power semiconductors, inductors, and capacitors are analyzed under ideal conditions. The state-space averaging method is used to obtain the average model and the small-signal model. To simplify the analysis, the capacitances are set such that C_1–C_5 are considered as equal. The inductances are defined as L_1 and L_2, and the output load resistance is R. The input variable, output variable, and control variable are $u_{\text{in}}(t)$, $u_o(t)$, and d, respectively, while $i_{L1}(t)$, $i_{L2}(t)$, $u_{C1}(t)$, $u_{C2}(t)$, $u_{C3}(t)$, $u_{C4}(t)$, and $u_{C5}(t)$ are the state variables. In the loop circuit when Q_1 and Q_2 are turned ON (as shown in Fig. 5.44a), C_2, C_3, and C_4 are connected in series. Therefore, the sum of voltages across C_2, C_3, and C_4 is zero, resulting in an invalid state variable. To avoid this, an equivalent series resistance is included in the same loop circuit, which removes the coupling among C_2, C_3, and C_4. Similarly, when Q_1 and Q_2 turn off (as shown in Fig. 5.44b), C_1 and C_2 are connected in parallel, resulting in the voltages across C_1 and C_2 being equal, which is another invalid state variable. To avoid this, an equivalent series resistance is included in the loop circuits, removing the coupling relationship between C_1 and C_2.

The state-space average model of the converter can be obtained using (5.88) and (5.89) for the State 1 and State 2, respectively.

$$\begin{bmatrix} \frac{di_{L1}(t)}{dt} \\ \frac{di_{L2}(t)}{dt} \\ \frac{du_{C1}(t)}{dt} \\ \frac{du_{C2}(t)}{dt} \\ \frac{du_{C3}(t)}{dt} \\ \frac{du_{C4}(t)}{dt} \\ \frac{du_{C5}(t)}{dt} \end{bmatrix} = \begin{bmatrix} 0 & 0 & 0 & 0 & 0 & 0 & 0 \\ 0 & -\frac{r}{L_2} & \frac{1}{L_2} & \frac{1}{L_2} & 0 & 0 & 0 \\ 0 & -\frac{1}{C} & 0 & 0 & 0 & 0 & 0 \\ 0 & -\frac{1}{C} & 0 & -\frac{1}{\mathrm{Cr}} & \frac{1}{\mathrm{Cr}} & -\frac{1}{\mathrm{Cr}} & 0 \\ 0 & 0 & 0 & \frac{1}{\mathrm{Cr}} & -\frac{1}{\mathrm{Cr}} & \frac{1}{\mathrm{Cr}} & 0 \\ 0 & 0 & 0 & -\frac{1}{\mathrm{Cr}} & \frac{1}{\mathrm{Cr}} & -\frac{R+r}{\mathrm{CRr}} & -\frac{1}{\mathrm{CR}} \\ 0 & 0 & 0 & 0 & 0 & -\frac{1}{\mathrm{CR}} & -\frac{1}{\mathrm{CR}} \end{bmatrix} \begin{bmatrix} i_{L1}(t) \\ i_{L2}(t) \\ u_{C1}(t) \\ u_{C2}(t) \\ u_{C3}(t) \\ u_{C4}(t) \\ u_{C5}(t) \end{bmatrix} + \begin{bmatrix} \frac{1}{L_1} \\ 0 \\ 0 \\ 0 \\ 0 \\ 0 \\ 0 \end{bmatrix} u_{\mathrm{in}}(t)$$

$$u_{\mathrm{out}}(t) = \begin{bmatrix} 0\ 0\ 0\ 0\ 0\ 1\ 1 \end{bmatrix} \begin{bmatrix} i_{L1}(t)\ i_{L2}(t)\ u_{C1}(t)\ u_{C2}(t)\ u_{C3}(t)\ u_{C4}(t)\ u_{C5}(t) \end{bmatrix}^{\mathrm{T}} \tag{5.88}$$

$$\begin{bmatrix} \frac{di_{L1}(t)}{dt} \\ \frac{di_{L2}(t)}{dt} \\ \frac{du_{C1}(t)}{dt} \\ \frac{du_{C2}(t)}{dt} \\ \frac{du_{C3}(t)}{dt} \\ \frac{du_{C4}(t)}{dt} \\ \frac{du_{C5}(t)}{dt} \end{bmatrix} = \begin{bmatrix} 0 & 0 & 0 & -\frac{1}{L_1} & 0 & 0 & 0 \\ 0 & 0 & 0 & \frac{1}{L_2} & 0 & -\frac{1}{L_2} & 0 \\ 0 & 0 & -\frac{1}{\mathrm{Cr}} & \frac{1}{\mathrm{Cr}} & 0 & 0 & 0 \\ \frac{1}{C} & -\frac{1}{C} & \frac{1}{\mathrm{Cr}} & -\frac{1}{\mathrm{Cr}} & 0 & 0 & 0 \\ 0 & 0 & 0 & 0 & -\frac{1}{\mathrm{Cr}} & 0 & \frac{1}{\mathrm{Cr}} \\ 0 & \frac{1}{C} & 0 & 0 & 0 & -\frac{1}{\mathrm{CR}} & -\frac{1}{\mathrm{CR}} \\ 0 & 0 & 0 & 0 & \frac{1}{\mathrm{Cr}} & -\frac{1}{\mathrm{CR}} & -\frac{R+r}{\mathrm{CRr}} \end{bmatrix} \begin{bmatrix} i_{L1}(t) \\ i_{L2}(t) \\ u_{C1}(t) \\ u_{C2}(t) \\ u_{C3}(t) \\ u_{C4}(t) \\ u_{C5}(t) \end{bmatrix} + \begin{bmatrix} \frac{1}{L_1} \\ 0 \\ 0 \\ 0 \\ 0 \\ 0 \\ 0 \end{bmatrix} u_{\mathrm{in}}(t)$$

$$u_{\mathrm{out}}(t) = \begin{bmatrix} 0\ 0\ 0\ 0\ 0\ 1\ 1 \end{bmatrix} \begin{bmatrix} i_{L1}(t)\ i_{L2}(t)\ u_{C1}(t)\ u_{C2}(t)\ u_{C3}(t)\ u_{C4}(t)\ u_{C5}(t) \end{bmatrix}^{\mathrm{T}} \tag{5.89}$$

By combining these two equations, the average model of the converter can be expressed as (5.90).

$$\begin{bmatrix} \frac{di_{L1}(t)}{dt} \\ \frac{di_{L2}(t)}{dt} \\ \frac{du_{C1}(t)}{dt} \\ \frac{du_{C2}(t)}{dt} \\ \frac{du_{C3}(t)}{dt} \\ \frac{du_{C4}(t)}{dt} \\ \frac{du_{C5}(t)}{dt} \end{bmatrix} = \begin{bmatrix} 0 & 0 & 0 & -\frac{d-1}{L_1} & 0 & 0 & 0 \\ 0 & -\frac{rd}{L_2} & \frac{d}{L_2} & \frac{1}{L_2} & 0 & \frac{d-1}{L_2} & 0 \\ 0 & -\frac{d}{C} & -\frac{d-1}{\mathrm{Cr}} & \frac{1-d}{\mathrm{Cr}} & 0 & 0 & 0 \\ \frac{1-d}{C} & -\frac{1}{C} & \frac{1-d}{\mathrm{Cr}} & -\frac{1}{\mathrm{Cr}} & \frac{d}{\mathrm{Cr}} & -\frac{d}{\mathrm{Cr}} & 0 \\ 0 & 0 & 0 & \frac{d}{\mathrm{Cr}} & -\frac{1}{\mathrm{Cr}} & \frac{d}{\mathrm{Cr}} & \frac{1-d}{\mathrm{Cr}} \\ 0 & \frac{1-d}{C} & 0 & -\frac{d}{\mathrm{Cr}} & \frac{d}{\mathrm{Cr}} & -\frac{1}{\mathrm{CR}} - \frac{d}{\mathrm{Cr}} & -\frac{1}{\mathrm{CR}} \\ 0 & 0 & 0 & 0 & \frac{1-d}{\mathrm{Cr}} & -\frac{1}{\mathrm{CR}} & \frac{d}{\mathrm{Cr}} - \frac{R+r}{\mathrm{CRr}} \end{bmatrix} \begin{bmatrix} i_{L1}(t) \\ i_{L2}(t) \\ u_{C1}(t) \\ u_{C2}(t) \\ u_{C3}(t) \\ u_{C4}(t) \\ u_{C5}(t) \end{bmatrix} + \begin{bmatrix} \frac{1}{L_1} \\ 0 \\ 0 \\ 0 \\ 0 \\ 0 \\ 0 \end{bmatrix} u_{\mathrm{in}}(t)$$

$$u_{\mathrm{out}}(t) = \begin{bmatrix} 0\ 0\ 0\ 0\ 0\ 1\ 1 \end{bmatrix} \begin{bmatrix} i_{L1}(t)\ i_{L2}(t)\ u_{C1}(t)\ u_{C2}(t)\ u_{C3}(t)\ u_{C4}(t)\ u_{C5}(t) \end{bmatrix}^{\mathrm{T}} \tag{5.90}$$

The small-signal disturbance variables describe the state variables, input variable, output variable, and control variable, as given in Eq. (5.91). Here, I_{L1}, I_{L2}, U_{C1}, U_{C2}, U_{C3}, U_{C4}, U_{C5}, U_{in}, U_{out}, and D represent the steady-state components, while $\hat{i}_{L1}(t)$, $\hat{i}_{L2}(t)$, $\hat{u}_{C1}(t)$, $\hat{u}_{C2}(t)$, $\hat{u}_{C3}(t)$, $\hat{u}_{C4}(t)$, $\hat{u}_{C5}(t)$, $\hat{u}_{\text{in}}(t)$, $\hat{u}_{\text{out}}(t)$, and $\hat{d}(t)$ are the corresponding small-signal disturbance variables.

$$\begin{cases} i_{L1}(t) = I_{L1} + \hat{i}_{L1}(t) \\ i_{L2}(t) = I_{L2} + \hat{i}_{L2}(t) \\ u_{C1}(t) = U_{C1} + \hat{u}_{C1}(t) \\ u_{C2}(t) = U_{C2} + \hat{u}_{C2}(t) \\ u_{C3}(t) = U_{C3} + \hat{u}_{C3}(t) \\ u_{C4}(t) = U_{C4} + \hat{u}_{C4}(t) \\ u_{C5}(t) = U_{C5} + \hat{u}_{C5}(t) \\ u_{\text{in}}(t) = U_{\text{in}} + \hat{u}_{\text{in}}(t) \\ u_{\text{out}}(t) = U_{\text{out}} + \hat{u}_{\text{out}}(t) \\ d = D + \hat{d} \end{cases} \tag{5.91}$$

The small-signal model of the converter is given by (5.92), which can be derived from (5.90) and (5.91).

$$\begin{aligned} \begin{bmatrix} \frac{d\hat{i}_{L1}(t)}{dt} \\ \frac{d\hat{i}_{L2}(t)}{dt} \\ \frac{d\hat{u}_{C1}(t)}{dt} \\ \frac{d\hat{u}_{C2}(t)}{dt} \\ \frac{d\hat{u}_{C3}(t)}{dt} \\ \frac{d\hat{u}_{C4}(t)}{dt} \\ \frac{d\hat{u}_{C5}(t)}{dt} \end{bmatrix} &= \begin{bmatrix} 0 & 0 & 0 & \frac{d-1}{L_1} & 0 & 0 & 0 \\ 0 & -\frac{rd}{L_2} & \frac{d}{L_2} & \frac{1}{L_2} & 0 & \frac{d-1}{L_2} & 0 \\ 0 & -\frac{d}{C} & \frac{d-1}{\text{Cr}} & \frac{1-d}{\text{Cr}} & 0 & 0 & 0 \\ \frac{1-d}{C} & -\frac{1}{C} & \frac{1-d}{\text{Cr}} & -\frac{1}{\text{Cr}} & \frac{d}{\text{Cr}} & -\frac{d}{\text{Cr}} & 0 \\ 0 & 0 & 0 & \frac{d}{\text{Cr}} & -\frac{1}{\text{Cr}} & \frac{d}{\text{Cr}} & \frac{1-d}{\text{Cr}} \\ 0 & \frac{1-d}{C} & 0 & -\frac{d}{\text{Cr}} & \frac{d}{\text{Cr}} & -\frac{1}{\text{CR}} - \frac{d}{\text{Cr}} & -\frac{1}{\text{CR}} \\ 0 & 0 & 0 & 0 & \frac{1-d}{\text{Cr}} & -\frac{1}{\text{CR}} & \frac{d}{\text{Cr}} - \frac{R+r}{\text{CRr}} \end{bmatrix} \begin{bmatrix} \hat{i}_{L1}(t) \\ \hat{i}_{L2}(t) \\ \hat{u}_{C1}(t) \\ \hat{u}_{C2}(t) \\ \hat{u}_{C3}(t) \\ \hat{u}_{C4}(t) \\ \hat{u}_{C5}(t) \end{bmatrix} \\ &+ \begin{bmatrix} \frac{1}{L_1} \\ 0 \\ 0 \\ 0 \\ 0 \\ 0 \\ 0 \end{bmatrix} \hat{u}_{\text{in}}(t) + \begin{bmatrix} 0 & 0 & 0 & \frac{1}{L_1} & 0 & 0 & 0 \\ 0 & -\frac{r}{L_2} & \frac{1}{L_2} & 0 & 0 & \frac{1}{L_2} & 0 \\ 0 & -\frac{1}{C} & \frac{1}{\text{Cr}} & -\frac{1}{\text{Cr}} & 0 & 0 & 0 \\ -\frac{1}{C} & 0 & -\frac{1}{\text{Cr}} & 0 & \frac{1}{\text{Cr}} & -\frac{1}{\text{Cr}} & 0 \\ 0 & 0 & 0 & \frac{1}{\text{Cr}} & 0 & \frac{1}{\text{Cr}} & -\frac{1}{\text{Cr}} \\ 0 & -\frac{1}{C} & 0 & -\frac{1}{\text{Cr}} & \frac{1}{\text{Cr}} & -\frac{1}{\text{Cr}} & 0 \\ 0 & 0 & 0 & 0 & -\frac{1}{\text{Cr}} & 0 & \frac{1}{\text{Cr}} \end{bmatrix} \begin{bmatrix} I_{L1} \\ I_{L2} \\ U_{C1} \\ U_{C2} \\ U_{C3} \\ U_{C4} \\ U_{C5} \end{bmatrix} \\ \hat{u}_{\text{out}}(t) &= \begin{bmatrix} 0\,0\,0\,0\,0\,1\,1 \end{bmatrix} \begin{bmatrix} \hat{i}_{L1}(t) \; \hat{i}_{L2}(t) \; \hat{u}_{C1}(t) \; \hat{u}_{C2}(t) \; \hat{u}_{C3}(t) \; \hat{u}_{C4}(t) \; \hat{u}_{C5}(t) \end{bmatrix}^{\text{T}} \end{aligned} \tag{5.92}$$

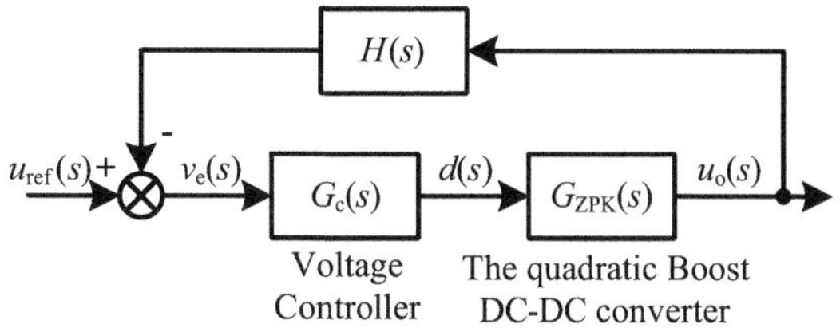

Fig. 5.47 Control scheme for the quadratic Boost DC–DC converter

The voltage loop control scheme for the quadratic Boost DC–DC converter is illustrated in Fig. 5.47. The transfer function of the converter is denoted as $G_{ZPK}'(s)$, which represents the behavior of the converter in response to changes in input and output voltage. The voltage controller transfer function is represented as $G_c(s)$, which is a proportional-integral controller. The feedback transfer function is represented as $H(s)$. The voltage controller can be designed for the converter to achieve suitable static and dynamic performances. The static performance refers to the steady-state behavior of the system, whereas dynamic performance refers to the behavior of the system over time. The voltage controller can ensure that the output voltage of the converter is regulated to a desired value, while also minimizing any overshoot or undershoot in the output voltage in response to changes in the input voltage or load current.

5.8 Coupled-Inductor Bidirectional DC–DC Converter

The coupled-inductor bidirectional DC–DC converter topology is developed based on the quasi-Z-source Boost DC–DC converter [141]. The derivation process is outlined in Fig. 5.48. Initially, a step-up DC–DC converter is derived from the second stage of the quasi-Z-source Boost DC–DC converter. Utilizing inductor L_1 as the primary side of the coupled inductor, the secondary side L_2 can be connected in series with C_3 to further elevate the output voltage. Additionally, an LC network is incorporated on the input side to mitigate current ripple. To enable bidirectional operation, the diodes D_3–D_5 are substituted with MOSFETs Q_2–Q_4.

Due to the adoption of coupled-inductor, the derived converter is capable of achieving high voltage gain by adjusting the turns ratio of the coupled-inductor. This provides flexibility in meeting various voltage requirements in different applications. The converter also achieves high conversion ratios in both step-up and step-down modes, while reducing voltage stresses across power switches. In addition, the converter design enables almost zero-current ripple on both the input voltage source and load sides, which can further improve efficiency and reduce electromagnetic interference. Another advantage of the converter is its common ground structure between the input voltage source and load sides. This can simplify the overall system design and reduce complexity, potentially leading to cost savings. Overall, the coupled-inductor bidirectional DC–DC converter introduced in this section offers

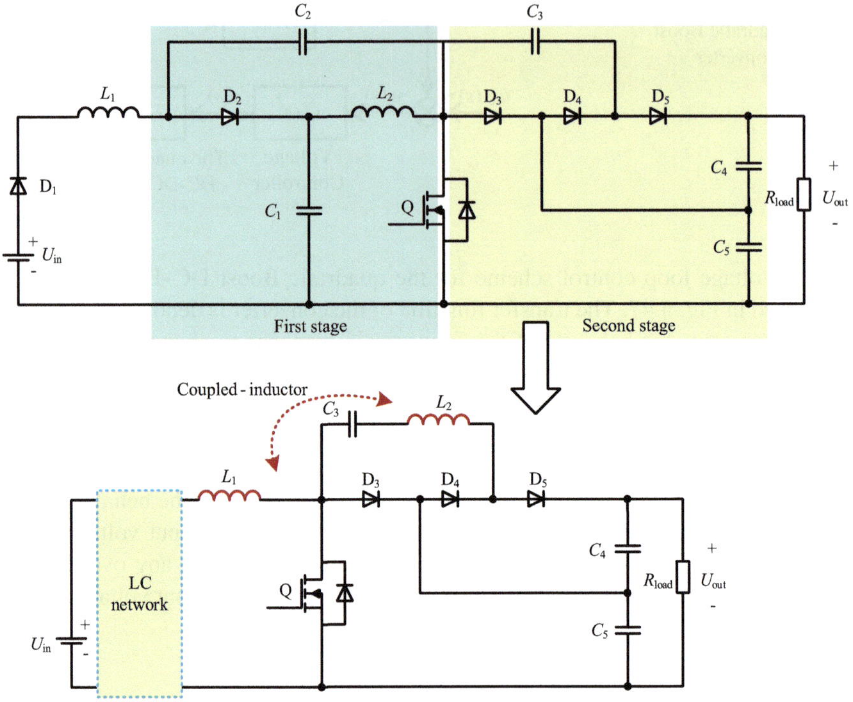

Fig. 5.48 Derivation of coupled-inductor bidirectional DC–DC converter

several advantages over conventional converters and could be a promising solution for various power conversion applications.

As shown in Fig. 5.49, the derived coupled-inductor bidirectional DC–DC converter is comprised of Q_1–Q_4, labeled as four active power switches, and their inherent parasitic diodes D_1–D_4. It also includes a coupled inductor with leakage inductance L_k and magnetic inductance L_m, as well as primary and secondary sides

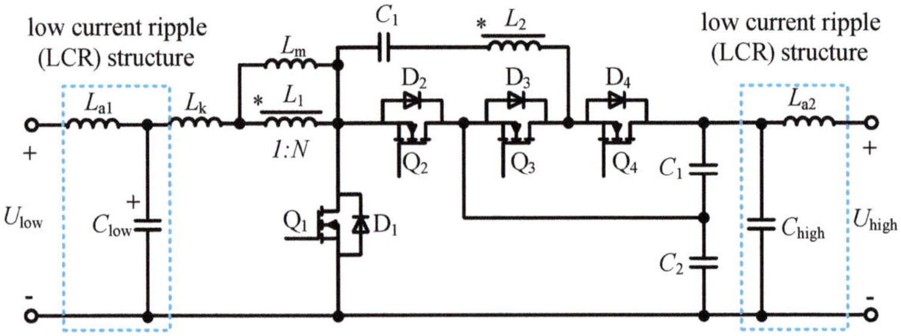

Fig. 5.49 Topology of the coupled-inductor bidirectional DC–DC converter

labeled L_1 and L_2, respectively. Additionally, the converter features two small auxiliary inductors, L_{a1} and L_{a2}, and five capacitors labeled as C_1–C_3, C_{low}, and C_{high}. The converter can achieve a high conversion ratio using its turns ratio, while the LCR structures significantly decrease the current ripple on the U_{low} and U_{high} sides.

By examining the relationship between the current and voltage of the auxiliary inductor L_a, the following equation can be derived, where $\mathrm{d}i_{\text{La}}$ represents the change in inductor current and is proportional to the current ripple rate, and $\mathrm{d}t$ is the elapsed time for the variation and is related to the switching frequency.

$$u_{\text{La}} = L\frac{\mathrm{d}i_{\text{La}}}{\mathrm{d}t} \tag{5.93}$$

To obtain the current ripple rate of L_a, the following equation is deduced, where d is the duty cycle, u_{La} is the voltage across the auxiliary inductor L_a,f_s is the switching frequency, L is the inductance, and i_{La} is the mean value of the inductor current. The values of f_s, L, and i_{La} are fixed and depend on the design parameters and operating state of the converter. As such, the current ripple rate α is directly proportional to the voltage u_{La} across the auxiliary inductor.

The current ripple rate of the auxiliary inductor L_a, plays an essential role in the design and operation of the converter. The ripple rate, denoted as α, can be derived using the following formula:

$$\alpha = \frac{u_{\text{La}}\mathrm{d}T_s}{\mathrm{L}i_{\text{La}}} = \frac{\mathrm{d}u_{\text{La}}}{f_s\mathrm{L}i_{\text{La}}} \tag{5.94}$$

Here, d represents the duty cycle, uLa is the voltage across the auxiliary inductor, L is the inductance of L_a, i_{La} is the mean value of the inductor current, and f_s is the switching frequency. It is important to note that f_s, L, and i_{La} are fixed values that are dependent on the converter's design parameters and operating state. Therefore, α is directly proportional to the voltage u_{La} across the auxiliary inductor, which can be controlled to regulate the current ripple rate.

5.8.1 Operating Principle of the Coupled-Inductor Bidirectional DC–DC Converter

Figure 5.50 displays the waveforms of u_L with respect to time t for the two distinct structures. Without the LCR structure, the voltage across inductor L_k varies periodically with a cycle of $1/f_s$, resulting in the generation of oscillatory current ripple rate α with a large amplitude of u_{Lk}. This oscillation can cause large inductor current ripple, which is undesired for proper converter operation.

To mitigate this effect, LCR structures were implemented, resulting in significant improvements in the converter's performance. During steady state, the voltages across the auxiliary inductors, L_{a1} and L_{a2}, can be evaluated using the following calculations.

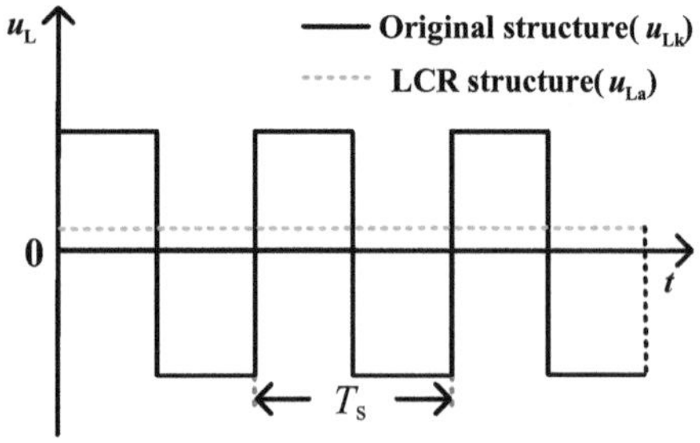

Fig. 5.50 Voltage waveforms across the auxiliary inductor with and without the LCR structure

$$\begin{cases} u_{La1} = u_{Clow} - u_{low} \approx 0 \\ u_{La2} = u_{Chigh} - u_{high} \approx 0 \end{cases} \tag{5.95}$$

It can be observed in (5.95) that the voltage across the auxiliary inductors (L_{a1} and L_{a2}) is almost negligible when LCR structures are used. This is a significant improvement as compared to the no LCR structure, where the voltage across the inductors would oscillate periodically, leading to a high current ripple rate and a large amplitude of the inductor voltages. By reducing the amplitude of the inductor voltages through the use of LCR structures, the ripple of the inductor currents can be effectively reduced, resulting in a longer lifespan of the voltage sources and improved power conversion efficiency. In other words, the LCR structures help to minimize the current ripple rate, which can be particularly useful in high-frequency applications where a large current ripple rate can cause electromagnetic interference and reduce the overall efficiency of the power conversion process. By reducing the amplitude of the inductor voltages, the LCR structures help to minimize the energy losses in the system and improve the stability of the power supply.

When operating in boost mode, the coupled-inductor bidirectional DC–DC converter utilizes power switch Q_1 as the main switch, while Q_2–Q_4 serve as the freewheel diodes D_2–D_4. As a result, there are two switching states during each switching period in boost mode, which are illustrated in Fig. 5.51.

State 1) $S_1S_2S_3S_4 = 1000$

When Q_1 is turned ON, L_1 is charged by the DC source U_{low} on the low voltage side, followed by C_3 being charged by L_2 and C_2 in series. Moreover, the load R is charged by the series branch C_1 and C_2.

State 2) $S_1S_2S_3S_4 = 0000$

When Q_1 is turned OFF, C_2 is charged by the DC source and L_1 in series, followed by the series branch C_1 and C_2 being charged by the DC source, L_1, L_2, and C_3 in series, along with the load R. The output voltage U_{high} is nearly equivalent to the total voltage across C_1 and C_2.

In step-down mode, the main switches are Q_2–Q_4, and Q_1 acts as the freewheel diode D_1. The switching period in buck mode consists of two switching states, which are illustrated in Fig. 5.52.

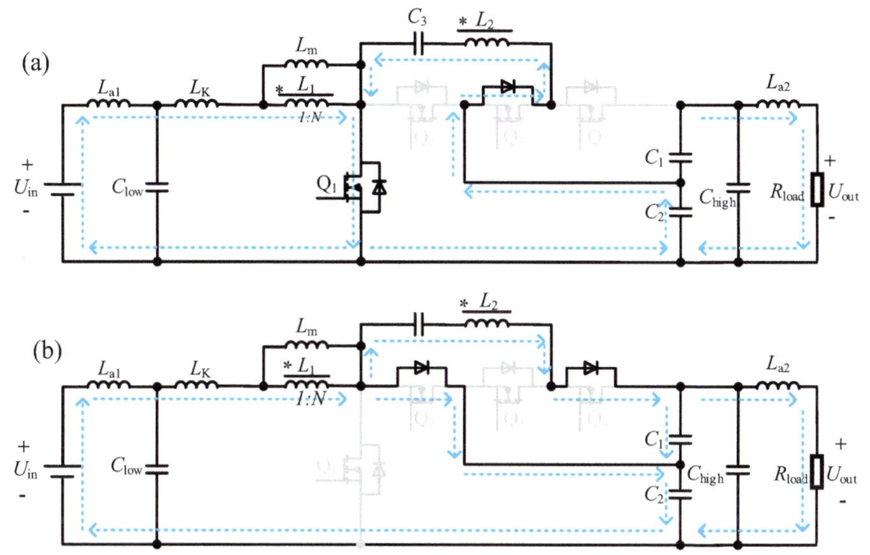

Fig. 5.51 Switching states of the coupled-inductor bidirectional DC–DC converter in step-up mode during **a** State 1 and **b** State 2

State 1) $S_1S_2S_3S_4 = 0111$

During this state, Q_3 is turned ON and Q_2 and Q_4 are turned OFF. The load R is charged by L_1, and C_3 is charged by L_2 and C_2 in series. The charging sequence can be represented as R–L_{a1}–L_k–L_1–D_1 for the load, and L_1–L_k–L_{a1}–R–C_2–Q_3–L_2–C_3 for the capacitor.

State 2) $S_1S_2S_3S_4 = 0111$

This state is similar to state 1, except C_2 is charged by L_2 and C_3 in series.

State 3) $S_1S_2S_3S_4 = 0000$

When Q_2–Q_4 are all turned OFF, L_1 charges the load R, while C_3 and L_2 charge the series branch of C_1 and C_2.

State 4) $S_1S_2S_3S_4 = 0000$

In this state, L_1 alone charges the load R, similar to state III.

State 5) $S_1S_2S_3S_4 = 0101$

When Q_2 and Q_4 are turned ON, Q_3 is turned OFF, the load R is charged by the paralleled branch of C_2 and L_1, as well as the branch of C_2, C_1, L_2, and C_3.

State 6) $S_1S_2S_3S_4 = 0000$

Finally, state 6 is similar to state 1, but Q_3 is turned OFF during the dead time.

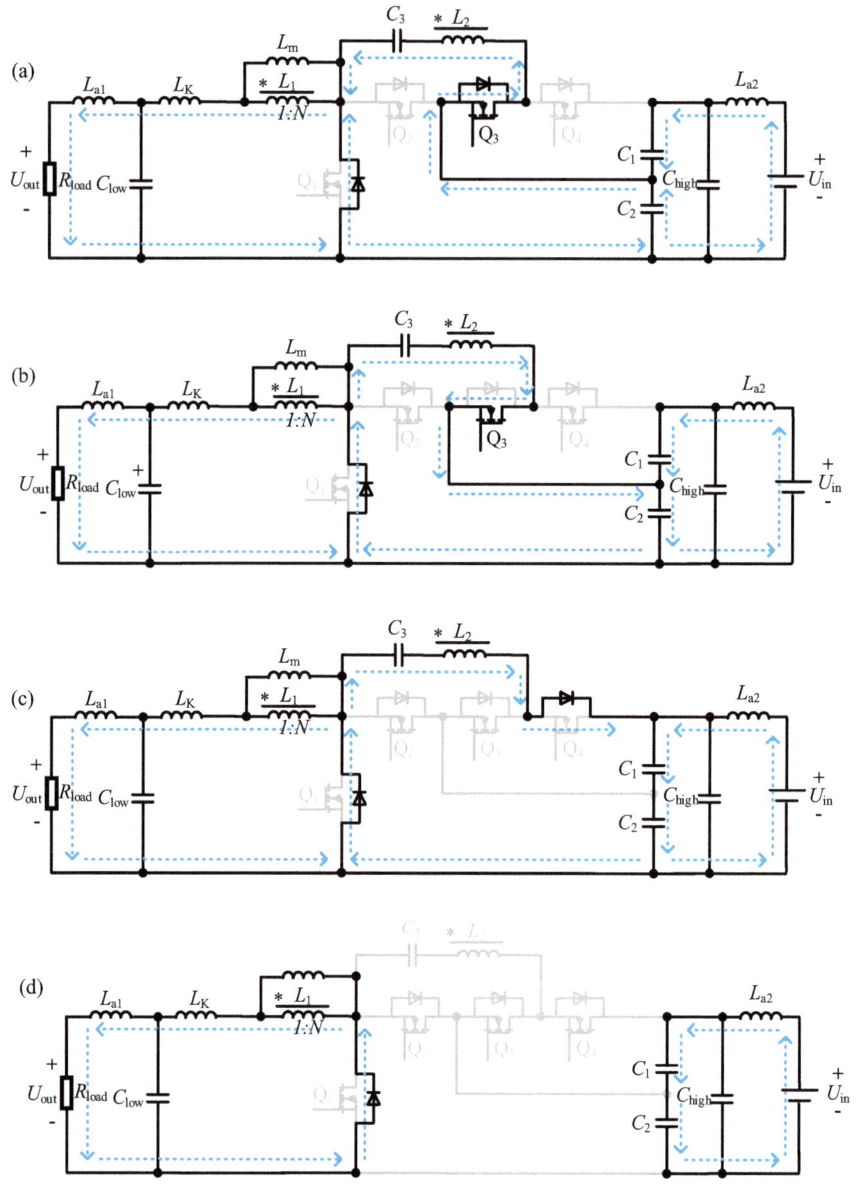

Fig. 5.52 Switching states of the coupled-inductor bidirectional DC–DC converter in step-down mode during **a** State 1, **b** State 2, **c** State 3, **d** State 4, **e** State 5, and **f** State 6

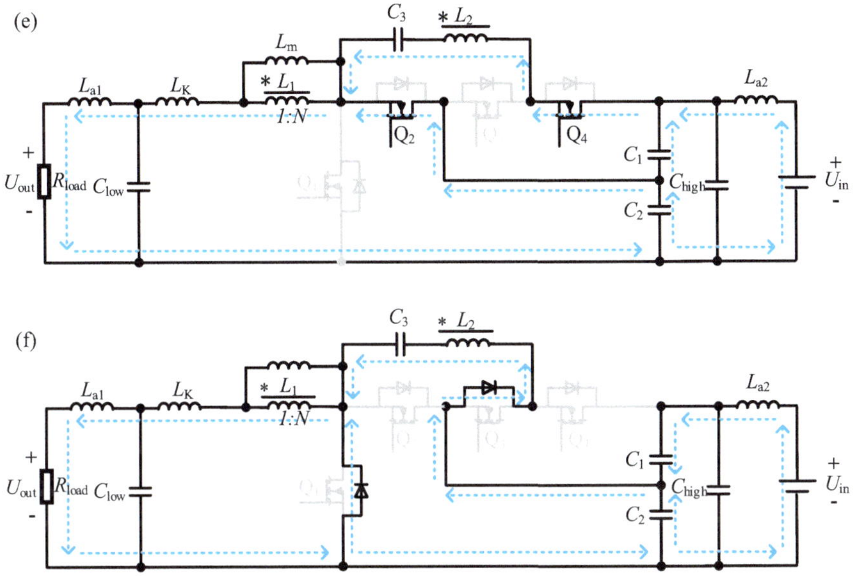

Fig. 5.52 (continued)

The operating waveforms of the coupled-inductor bidirectional DC–DC converter are depicted in Fig. 5.53, which provides an insight into the current stress experienced by each power switch. In Boost mode, the current and voltage waveforms of the key power devices are presented in Fig. 5.53a, where the currents flowing through the switches gradually increase from zero. Conversely, in Buck mode operation, Fig. 5.53b illustrates the current and voltage waveforms of power switches, where ZVS operation for Q_3 is achieved and the ZCS ON for Q_4 is attained. These soft-switching techniques significantly reduce switching losses, thereby enhancing the overall conversion efficiency of the converter. By achieving ZVS and ZCS, the power switches can operate at higher frequencies, which results in a reduction of the size and weight of the magnetic components in the converter.

5.8.2 *Steady-State Characteristics of the Coupled-Inductor Bidirectional DC–DC Converter*

To simplify the analysis of the coupled-inductor bidirectional DC–DC converter's voltage gain and component stresses in Boost mode, certain components such as the leakage inductor L_k, the auxiliary inductors (L_{a1} and L_{a2}), and the capacitors (C_{low} and C_{high}) can be disregarded. As a result, the switching states in boost mode can be represented in a simplified form as demonstrated in Fig. 5.54. By applying the volt-sec balance equations of L_1 and L_2 and taking into account the primary

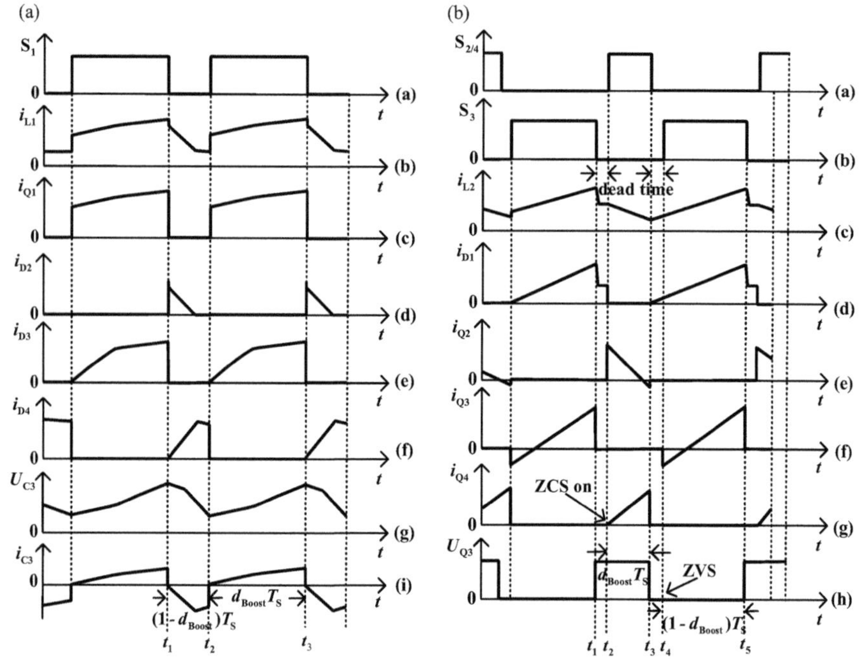

Fig. 5.53 Key waveforms of the coupled-inductor bidirectional DC–DC converter **a** in step-up mode. **b** In step-down mode

and secondary voltage relationships of the coupled-inductor, (5.96) can be deduced, which provides insights into the behavior of the system.

$$\begin{cases} d_{\text{Boost}} U_{\text{low}} = (1 - d_{\text{Boost}})(U_{C2} - U_{\text{low}}) \\ d_{\text{Boost}}(U_{C2} - U_{C3}) = (1 - d_{\text{Boost}})(U_{C1} - U_{C3}) \\ kNU_{\text{low}} = U_{C3} - U_{C2} \\ kN(U_{C2} - U_{\text{low}}) = U_{C1} - U_{C3} \end{cases} \tag{5.96}$$

The following equation presents the voltage gain of the converter, as well as the voltage stresses across capacitors C_1–C_3.

$$\begin{cases} U_{C1} = \dfrac{1 + kN}{1 - d_{\text{Boost}}} U_{\text{low}} \\ U_{C2} = \dfrac{1}{1 - d_{\text{Boost}}} U_{\text{low}} \\ U_{C3} = \dfrac{1 + kN - kNd_{\text{Boost}}}{1 - d_{\text{Boost}}} U_{\text{low}} \\ U_{\text{high}} = U_{C1} + U_{C2} = \dfrac{2 + kN}{1 - d_{\text{Boost}}} U_{\text{low}} \end{cases} \tag{5.97}$$

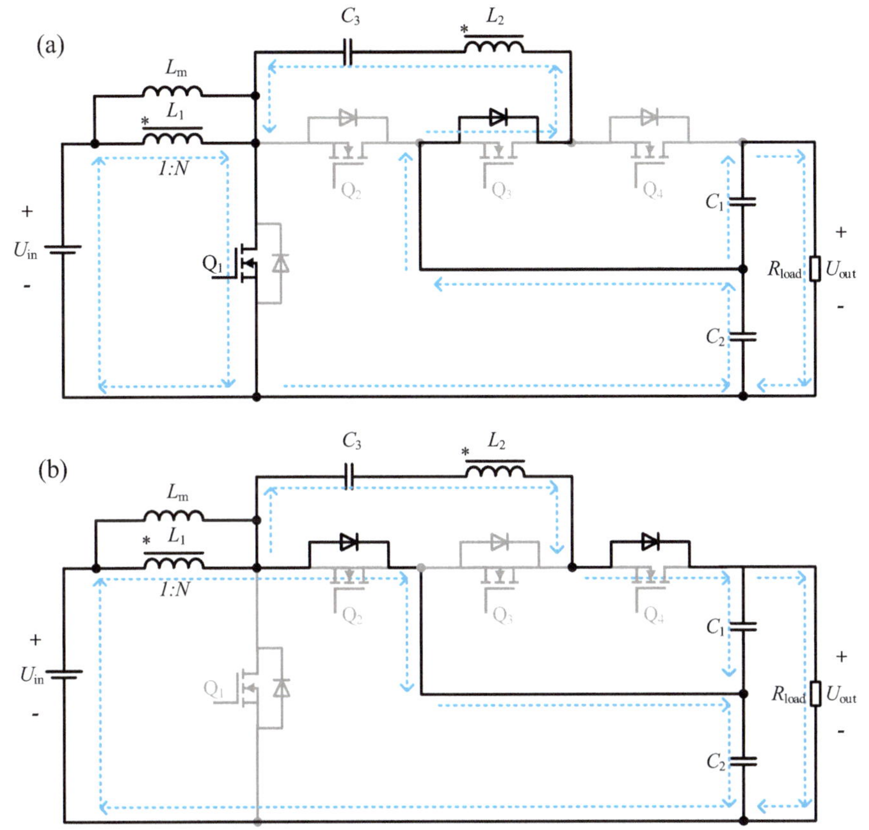

Fig. 5.54 Simplified switching states of the coupled-inductor bidirectional DC–DC converter in step-up mode during **a** State 1 and **b** State 2

The voltage gain of the coupled-inductor bidirectional DC–DC converter can be obtained from (5.97), where k denotes the coupling coefficient of the coupled-inductor, N represents the ideal turns ratio of the coupled-inductor, and d_{Boost} is the duty cycle in Boost mode. If k is assumed to be 1, the voltage gain of the converter can be simplified as follows:

$$M_{\text{Boost}} = \frac{U_{\text{high}}}{U_{\text{low}}} = \frac{2+N}{1-d_{\text{Boost}}} \tag{5.98}$$

The normalized voltage gain of the converter, which is a key parameter for evaluating the performance of any DC–DC converter, is plotted in Fig. 5.55. In this plot, the turns ratio of the coupled inductor, denoted by N, is used as a variable to investigate the behavior of the converter. It can be observed that the converter can achieve a step-up voltage gain M in the range of 5–20, which is a significant improvement over conventional converters, especially during the duty cycle range of 0.2–0.8.

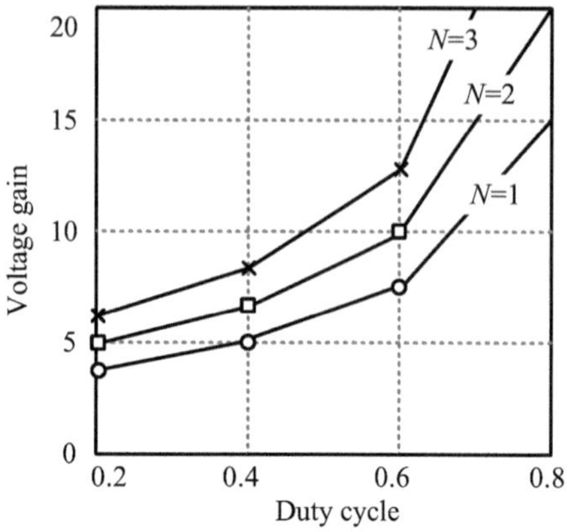

Fig. 5.55 Normalized voltage gain of the coupled-inductor bidirectional DC–DC converter

In addition to the high step-up voltage gain, the coupled-inductor bidirectional DC–DC converter can also achieve high step-down voltage gain as well, which is useful in Buck mode. Furthermore, the converter eliminates the need for extremely large or small duty cycles that are typically associated with conventional bidirectional DC–DC converters.

To investigate the voltage gain and the device stresses of the coupled-inductor bidirectional DC–DC converter in Buck mode, the leakage inductor L_k, the auxiliary inductors L_{a1} and L_{a2}, and the capacitors C_{low} and C_{high} can be disregarded, similar to the above Boost mode analysis. This enables the switching states in Buck mode to be simplified, as presented in Fig. 5.56. Hence, the (5.99) can be derived based on the volt-sec balance equations of L_1 and L_2, along with the primary and secondary voltage relationships of the coupled-inductor.

$$\begin{cases} (1-d_{\text{Buck}})U_{\text{low}} = d_{\text{Buck}}(U_{C2}-U_{\text{low}}) \\ (1-d_{\text{Buck}})(U_{C2}-U_{C3}) = d_{\text{Buck}}(U_{C1}-U_{C3}) \\ kNU_{\text{low}} = U_{C3}-U_{C2} \\ kN(U_{C2}-U_{\text{low}}) = U_{C1}-U_{C3} \end{cases} \tag{5.99}$$

The voltage gain and the voltage stresses across capacitors C_1–C_3 can be determined by (5.100), where the coupling coefficient of the coupled inductor is represented by k, the ideal turns ratio of the coupled inductor by N, and the duty cycle in buck mode by d_{Buck}.

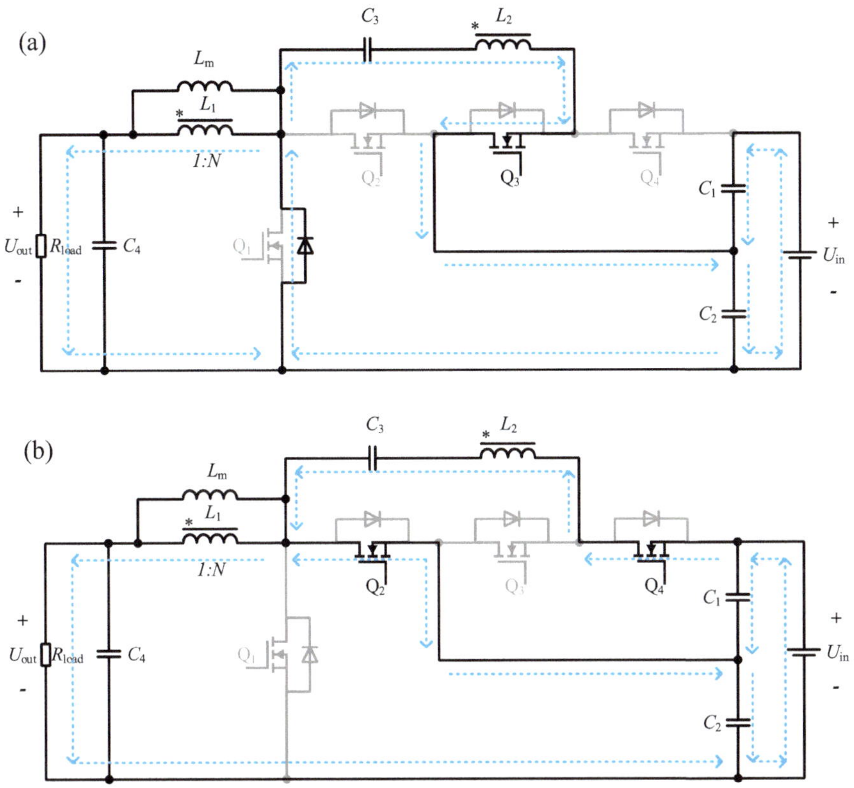

Fig. 5.56 Simplified switching states of the coupled-inductor bidirectional DC–DC converter in step-down mode during **a** State 1 **b** State 2

$$\begin{cases} U_{C1} = \dfrac{1 + kN}{2 + kN} U_{\text{high}} \\ U_{C2} = \dfrac{1}{2 + kN} U_{\text{high}} \\ U_{C3} = \dfrac{1 + kN d_{\text{Buck}}}{2 + kN} U_{\text{high}} \\ U_{\text{low}} = \dfrac{d_{\text{Buck}}}{2 + kN} U_{\text{high}} \end{cases} \tag{5.100}$$

The voltage gain of the converter in step-down mode can be obtained based on (5.100). If the coupling coefficient of the coupled inductor k is set to 1, the voltage gain can be simplified as follows:

$$M_{\text{Buck}} = \frac{U_{\text{low}}}{U_{\text{high}}} = \frac{d_{\text{Buck}}}{2 + N} \tag{5.101}$$

Based on the switching states shown in Figs. 5.54 and 5.56, it is determined that the voltage stresses across power switches Q_1 and Q_2 are equivalent to that of C_2, and the voltage stresses across power switches Q_3 and Q_4 are equivalent to that of C_1. When k is taken as 1, the voltage stresses across the power switches can be obtained using the following method.

$$\begin{cases} U_{Q1} = U_{Q2} = U_{C2} = \dfrac{1}{2+N} U_{\text{high}} \\ U_{Q3} = U_{Q4} = U_{C1} = \dfrac{1+N}{2+N} U_{\text{high}} \end{cases} \tag{5.102}$$

By utilizing the ampere-sec balance equations of capacitors C_1–C_3, the current stresses on the power switches are obtained as below.

$$\begin{cases} I_{Q1} = \dfrac{1+(1+N)d_{\text{Boost}}}{(2+N)d_{\text{Boost}}} I_{\text{low}} = \dfrac{1+(1+N)(1-d_{\text{Buck}})}{(2+N)(1-d_{\text{Buck}})} I_{\text{low}} \\ I_{Q2} = \dfrac{1}{2+N} I_{\text{low}} \\ I_{Q3} = \dfrac{1-d_{\text{Boost}}}{(2+N)d_{\text{Boost}}} I_{\text{low}} = \dfrac{d_{\text{Buck}}}{(2+N)(1-d_{\text{Buck}})} I_{\text{low}} \\ I_{Q4} = \dfrac{1}{2+N} I_{\text{low}} \end{cases} \tag{5.103}$$

The coupled-inductor bidirectional DC–DC converter is competitive with other counterparts based on various characteristics such as the number of power switches, diodes, coupled inductors, inductors, and capacitors. The conventional buck/boost converter employs the fewest number of power switches but suffers from high voltage stress equal to U_{high}. Although another converter has lower voltage stress, it lacks a common ground structure between the input and output sides. In contrast, the coupled-inductor bidirectional DC–DC converter introduced in this section employs only four low-rated voltage power switches and achieves a high and wide voltage gain range while producing negligible current ripple.

Compared with other converters, the converter in [111] employs fewer power switches and passive/active elements and has a common ground structure. The converter in [113] has the same voltage gain and power switch number as that of the coupled-inductor bidirectional DC–DC converter, but it has issues such as large current ripple and non-common ground structure, which are not present in the converter. The converter in [142] uses two coupled inductors and a switched-capacitor circuit to achieve high voltage gain with fewer capacitors. The converter in [143] has a higher voltage gain but similar current ripple as the converter in [111]. The converter in [112] has the same voltage gain and power switch number as that of the converter in [143], but the voltage stresses across the power switches are limited to the turns ratio N of the coupled inductor. As a result, the acceptable voltage stress can only be obtained when N is sufficiently large.

5.9 Efficiency and Application Analysis

To validate the practical performance of the switched-capacitor DC–DC converters introduced in this chapter, experimental prototypes are developed to implement efficiency measurements. And the application potential of the family of switched-capacitor DC–DC converters is analyzed in detail.

5.9.1 *Efficiency Analysis*

The prototype of switched-capacitor bidirectional DC–DC converter is designed with rated power at 300 W, while the high voltage side is regulated at 300 V and the low voltage side range is from 40 to 100 V. The switching frequency is chosen as 20 kHz. The measured efficiency is plotted in Fig. 5.57. It can be observed that the operation efficiency increases as the low voltage side increases. In step-down mode, the maximum efficiency of the converter is 94.45% and the minimum efficiency is 90.86%. In step-up mode, the maximum and minimum efficiency of the converter is 94.39 and 90.08%, respectively. It is obvious that the switched-capacitor bidirectional DC–DC converter can achieve high efficiency within a wide operation range, and the efficiency in step-down mode is relatively higher than that in step-up mode.

Due to the multiphase structure, the interleaved switched-capacitor bidirectional DC–DC converter can operate at higher power level. Thus, the prototype of the converter is designed with rated power at 800 W, and the low voltage side and high voltage side are 30–100 V and 400 V, respectively. The switching frequency of the prototype is also 20 kHz.

Figure 5.58a and b present the efficiency of the three-phase interleaved switched-capacitor bidirectional DC–DC converter in step-down mode and step-up mode, respectively. The conversion efficiency varies with the output power and low voltage side. In step-down mode, the converter reaches its maximum efficiency at 95.9% when the output power is 70 W and the low voltage side is 100 V. And the minimum efficiency is 89% with output power at 200 W and low voltage side is 100 V. And

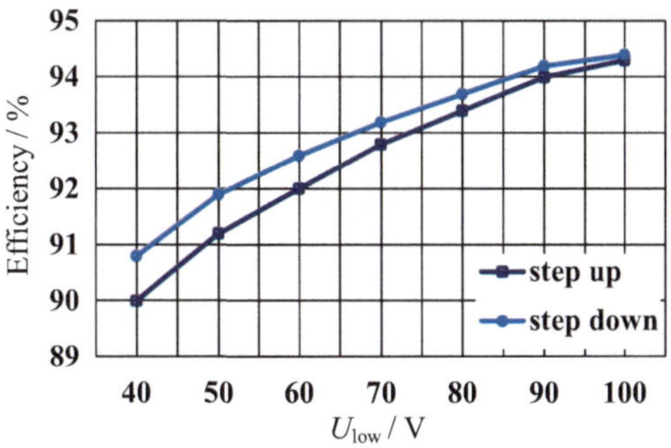

Fig. 5.57 Efficiency of the switched-capacitor bidirectional DC–DC converter

the minimum efficiency is 89% with output power at 200 W and low voltage side at 30 V. In step-up mode, the maximum efficiency is 95.8% with output power at 700 W and low voltage side at 90 V. The minimum efficiency is 89.2% when the output power is 200 W and the low voltage side is 30 V. In conclusion, the efficiency of the converter is lower at high voltage gain and light load.

A 1 kW prototype of the interleaved switched-capacitor bidirectional DC–DC converter is built for experimental demonstration. The efficiency curves of the prototype in step-down mode and step-up mode are displayed in Fig. 5.59, where the range of the low voltage side is from 50 to 120 V, and the high voltage side is constant at 400 V. It can be seen from the figure that the efficiency increases as the low side voltage increases. Thus, the converter reaches its maximum efficiency when the low voltage side is 120 V, which is 95.3% in step-down mode and 95.21% in step-up

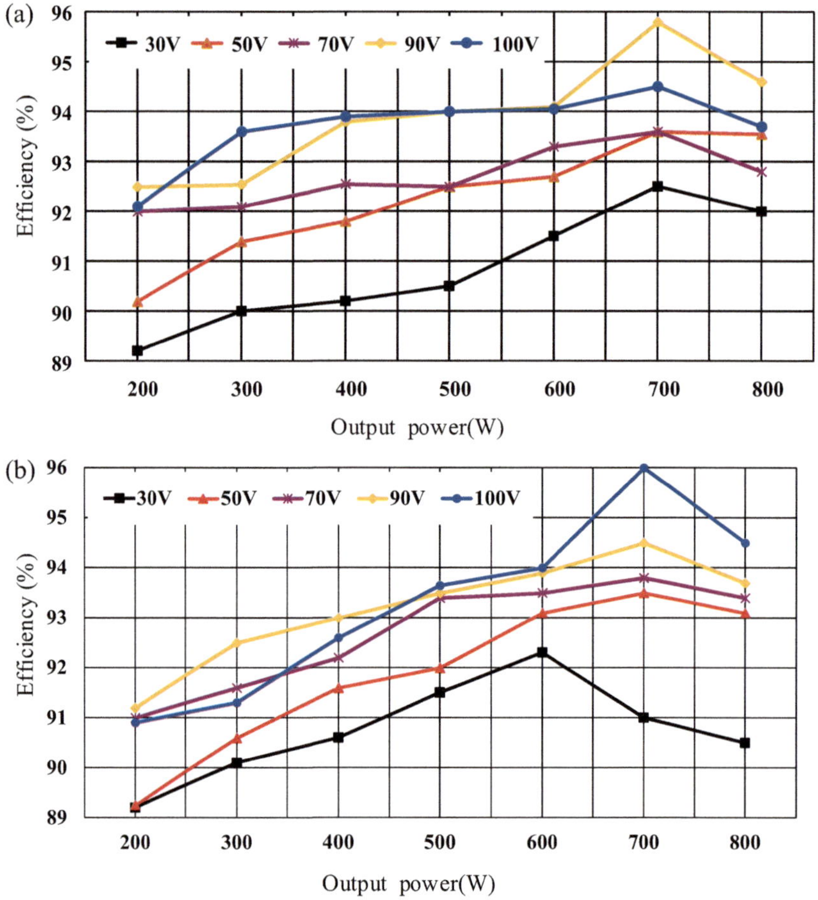

Fig. 5.58 Efficiency of the three-phase interleaved switched-capacitor bidirectional DC–DC converter in **a** step-down and **b** step-up modes

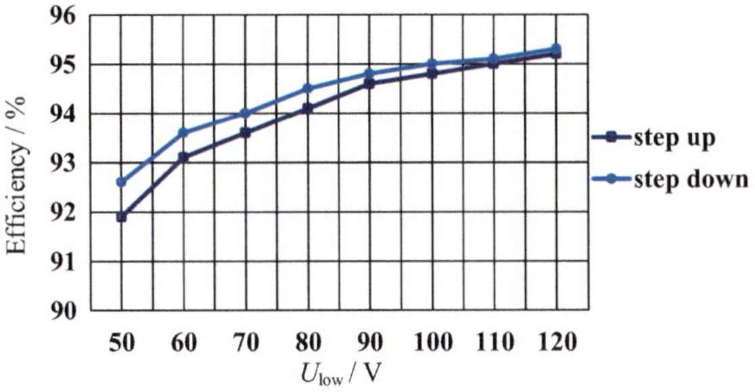

Fig. 5.59 Efficiency of the interleaved switched-capacitor bidirectional DC–DC converter

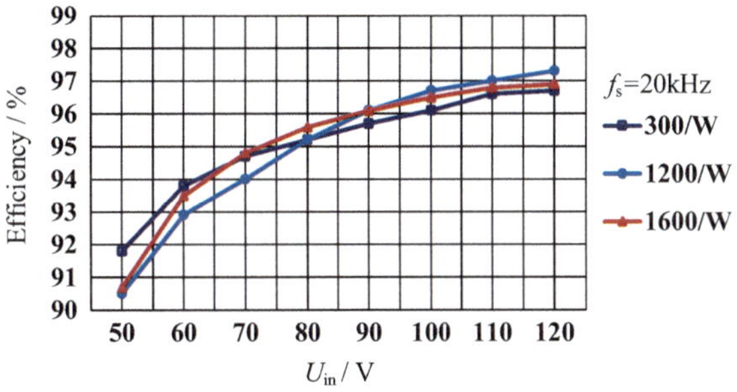

Fig. 5.60 Efficiency of the single-switch Boost DC–DC converter

mode. The minimum efficiency is observed with the low voltage side at 50 V, which is 92.6% in step-down mode and 91.88% in step-up mode.

The operation efficiency of the single-switch Boost DC–DC converter is obtained based on the 1.6 kW prototype, which is displayed in Fig. 5.60. The output voltage on the high voltage side is constant at 400 V. Meanwhile, the input voltage on the low voltage side varies within 50–120 V.

As can be observed from Fig. 5.60, the conversion efficiency of the single-switch Boost prototype is positively associated with the low voltage side and shows little difference at different output power levels. To be specific, at the rate power level, namely 1.2 kW, the maximum operation efficiency is measured when the low voltage side is at 120 V. Correspondingly, the minimum efficiency of the converter within the operation range is 90.53% when the low voltage side is at 50 V.

Based on the quasi-Z-source Boost DC–DC converter, a prototype is developed with rated power level at 400 W. Same as the single-switch Boost prototype, the output voltage of the quasi-Z-source Boost DC–DC converter is also maintained at 400 V, while the input voltage on the low voltage side varies within 50–150 V. When the converter operates at rated power level with different voltage gains, the conversion efficiency is measured ant plotted in Fig. 5.61. It can be seen from the figure that, as the input voltage on the low voltage side increases from 50 to 150 V, the operation efficiency of the quasi-Z-source Boost DC–DC converter increases from 84.5 to 95.13%. In other words, the conversion efficiency of the prototype is positively related to the low voltage side.

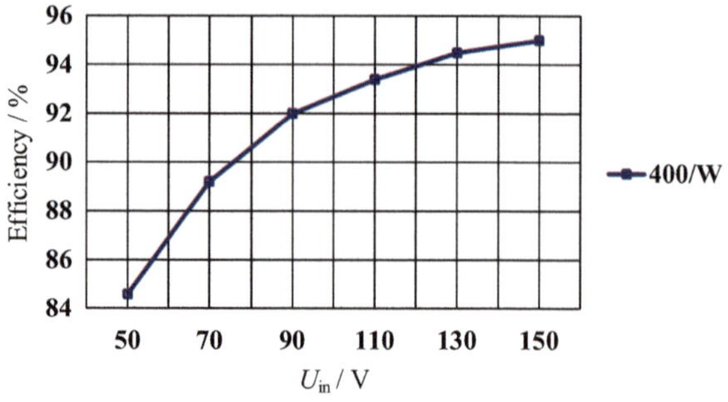

Fig. 5.61 Efficiency of the quasi-Z-source Boost DC–DC converter

In terms of the quadratic Boost DC–DC converter, the experimental test is conducted based on a 300 W prototype. By means of a closed voltage loop, the output voltage on the high voltage side is maintained at 400 V. Whereas the input voltage on the low voltage side varies within 40–80 V. According to the efficiency curves presented in Fig. 5.62, the converter exhibits higher efficiency across a wide operation range than most of the DC–DC converters. The overall efficiency of the converter with output power at 300 W is higher than conditions at 200 and 400 W. When the converter operates at the rated power level, namely 400 W, the conversion efficiency varies from 90.06 to 93.8% as the input voltage varies from 40 to 80 V. When the output power is at 300 W, the converter reaches its maximum efficiency at 95.01% when the input voltage is at 80 V.

The prototype of the coupled-inductor bidirectional DC–DC converter is designed with the rated power level at 400 W. The range of low voltage side is from 30 to 70 V, while the high voltage side is maintained constant at 300 V. The switching frequency is chosen as 20 kHz. Figure 5.63 shows the efficiency curves of the coupled-inductor bidirectional DC–DC converter in step-down mode and step-up mode at different voltage gain. It is noticed that when the low voltage side is relatively low, the conversion efficiency of the prototype in step-down mode is ~ 0.5% higher than that in step-up mode. On the contrary, when the low voltage side is relatively high, the conversion efficiency of the prototype in step-up mode is higher than that in step-down mode. To the specific, the converter reaches its maximum efficiency with

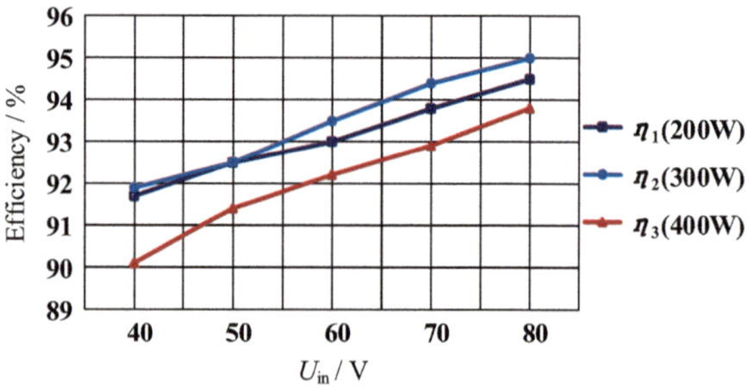

Fig. 5.62 Efficiency of the quadratic Boost DC–DC converter

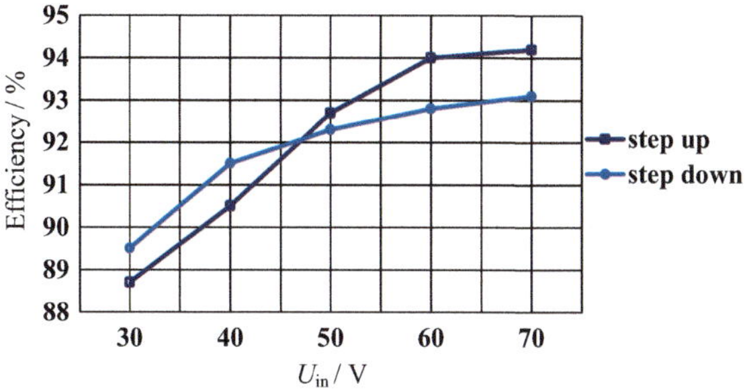

Fig. 5.63 Efficiency of the coupled-inductor bidirectional DC–DC converter

the low voltage side at 70 V, which is 93.1% in step-down mode and 94.2% in step-up mode. The minimum efficiency of the converter in step-down mode and step-up mode are 89.4 and 88.7%, respectively.

5.9.2 *Application Analysis*

As analyzed in the previous section, the interleaved switched-capacitor bidirectional DC–DC converter is derived by expand the basic switched-capacitor bidirectional topology to three-phase. Therefore, these two converter exhibit similar operation characteristics including common ground and bidirectional power flow, etc. However, due to the three-phase cascaded structure, the interleaved switched-capacitor bidirectional DC–DC converter has a higher voltage gain range, as well as higher power level. Moreover, the input current ripple can be suppressed by the phase-shifted charging/discharging states of the inductors.

According to the efficiency analysis, both the converters can achieve high efficiency within a wide operation range. Because of the additional inductors and power switches, the three-phase interleaved bidirectional DC–DC converter shows relatively lower efficiency than the basic converter at the same operation condition. Therefore, the switched-capacitor bidirectional DC–DC converters can be applied in energy storage systems to perform efficient charging and discharging. To be specific, the three-phase interleaved topology is more suitable for higher power level applications, and battery applications which requiring low current ripple.

Compared with the input-parallel output-series structure, the interleaved switched-capacitor bidirectional DC–DC converter introduced in 5.4 is capable of bidirectional power conversion, thus have extensive application prospects. Similar to other interleaved DC–DC converters, the topology can achieve low current ripple on the low voltage side. Thereby, it can be applied as battery charging/discharging regulator in energy storage systems. However, the lack of common ground between the input and output makes it less competitive in high-power-level applications.

With reduced amount of power switches and inductors, the single-switch Boost DC–DC converter can achieve higher power density in comparison with the interleaved structure. The disadvantage is that larger inductance is require to maintain the same low current ripple on the input side. Furthermore, the issue of non-common grounding between inputs and outputs persists in this topology. Consequently, it can only be utilized in low-power applications.

The topology of quasi-Z-source Boost DC–DC converter enhances the output voltage gain and resolves the issue of non-common grounding between inputs and outputs. Additionally, the voltage stress on the components in the converter is relatively low. However, this converter can only operate in a boost mode. Therefore, it is more suitable for applications with high voltage levels, such as fuel cells, where unidirectional power conversion is required.

Similarly, the quadratic Boost topology is derived from the quasi-Z-source Boost topology, inheriting the common grounding characteristic between inputs and outputs and having relatively low component voltage stress. However, this quadratic Boost topology can achieve a higher voltage gain with a relatively reduced range. As a Boot converter, it can only operate in a unidirectional step-up mode. Considering the aforementioned features, this converter can be applied in photovoltaic systems and fuel cell vehicles. It is particularly suitable for scenarios where a higher voltage gain is needed, and the energy source voltage fluctuation range is not significant.

The coupled-inductor bidirectional DC–DC converter possesses advantages such as common grounding between inputs and outputs, low component voltage stress, small required magnetic component volume, and high voltage gain. Additionally, this converter can achieve bidirectional energy flow for both step-up and step-down operations. Moreover, in both operation modes, the coupled-inductor bidirectional DC–DC converter can achieve high energy conversion efficiency. Therefore, this topology is suitable for energy storage systems, providing efficient charge and discharge control for energy storage devices such as batteries and super-capacitors.

5.10 Conclusion

DC–DC converters that utilize switched-capacitor cells have the potential to significantly improve the performance and efficiency of power electronics in various applications, including fuel cell vehicles. The multiple converter topologies discussed in this chapter offer numerous advantages, such as high voltage gains, low voltage stresses, and common ground points between input and output sides. These benefits make them suitable for various power interface applications in fuel cell vehicles.

Moreover, some of these converters have the potential to reduce switching losses and offer the possibility of soft-switching, which enhances the efficiency and reliability of power electronics. The converters' capability to operate over wide voltage gain ranges and their ability to handle large input and output voltage ranges make them suitable for different types of applications.

Overall, switched-capacitor cell-based DC–DC converters offer potential solutions for improving power electronics performance and efficiency in renewable energy applications. And the specific converter topology to be used depends on the specific requirements of the system. In conclusion, the use of switched-capacitor cell-based DC–DC converters provides an opportunity to improve the efficiency and reliability of power electronics, which is essential for the development and adoption of renewable energies.

Chapter 6
Design and Application of DC–DC Converters

6.1 Comparison and Application Analysis

6.1.1 Comparison of DC–DC Converters

To evaluate the practicality of the various non-isolated DC–DC converters discussed in this book for use in renewable energy power conversion systems, a thorough analysis is necessary. This assessment should cover key aspects such as the voltage gain range, circuit component count, common ground utilization, and bidirectional power conversion capability.

The voltage gain range is crucial as it influences how well the converter can match the energy source's voltage fluctuations with the DC bus voltage. This alignment significantly affects energy source utilization and overall system efficiency. A wider voltage gain range allows for smaller low-side currents, reducing component conduction losses and enhancing overall energy transfer efficiency.

In addition, the number of components in the converter significantly impacts the system's power density, a critical consideration in applications where strict volume and weight requirements are essential. A simplified circuit structure is vital for optimizing overall performance. Conversely, a higher component count leads to increased switching and conduction losses, hindering efforts to improve the converter's operational efficiency. The grounding configuration is also crucial. A shared ground between input and output reduces electromagnetic noise, especially in applications with high power levels and frequencies. Conversely, a potential difference between input and output grounds can result in significant electromagnetic interference, posing a risk to the system's reliability.

The bidirectional power conversion capability of the converter is crucial for its effectiveness in renewable energy power conversion systems. Energy storage devices, including batteries and super-capacitors, must operate in step-up mode during discharge to supply energy to the bus-side load. Conversely, during charging,

Y. Zhang and S. Gao, *High Efficiency Non-isolated DC-DC Converters with Wide Voltage Gain Range for Renewable Energies*, CPSS Power Electronics Series,
https://doi.org/10.1007/978-981-97-0648-8_6

these devices need to operate in step-down mode to efficiently convert bus energy to the required lower voltage. Table 6.1 provides a comprehensive comparison of characteristic parameters for the non-isolated DC converters featured in this book.

6.1.2 Application Analysis of DC–DC Converters for Energy Storage Devices

In renewable energy power conversion systems, batteries and super-capacitors serve as crucial energy storage devices, capturing excess energy from sources like photovoltaic panels and fuel cells through the charging process. When renewable sources alone cannot meet the bus load demands, these stored energies are discharged to power the loads, thereby enhancing the system's overall energy utilization efficiency and ensuring consistent and reliable operation.

Batteries, known for their high energy density and stable port voltage, are typically employed as primary energy storage solutions. On the other hand, super-capacitors, while having lower energy density, boast higher power density, allowing them to deliver rapid and substantial power compensation or feedback. Nevertheless, super-capacitors exhibit significant port voltage fluctuations during both charge and discharge cycles.

Considering these traits, DC–DC converters connecting batteries or super-capacitors to the DC bus must possess bidirectional operation capabilities, accommodating both step-up and step-down functions. This flexibility is crucial for effectively controlling the charging and discharging processes of these energy storage devices. In battery converters, it's imperative to minimize low-side current ripple, especially given the high charging and discharging currents, which can accelerate battery aging. Conversely, for super-capacitor converters, a wide voltage gain range is essential to fully leverage the capacity of the super-capacitors. Consequently, the DC converter needs a substantial voltage gain range to align the super-capacitor voltage with the DC bus voltage.

Hence, the bidirectional DC–DC converters discussed in Sects. 2.1, 2.2, 3.4, 4.2, 5.2, 5.3, and 5.8 can all serve as regulators for charging and discharging energy storage devices. However, the three-phase bidirectional DC–DC converter in Sect. 5.3 features an interleaved structure, making it well-suited for batteries due to its small current ripple on the low voltage side. On the other hand, the asymmetric H-bridge bidirectional DC–DC converter in Sect. 2.3 offers a wider voltage gain range and a common ground, making it ideal for super-capacitor applications. In terms of the conversion efficiency, the switched-quasi-Z-source bidirectional DC–DC converter in Sect. 4.2 shows better performance on achieving high operation efficiency within wide operation range.

Table 6.1 Comparison of the introduced non-isolated DC–DC converters

Topology	Voltage gain	Number of semiconductors	Common ground	Power flow
H-bridge DC–DC converter in Sect. 2.1	$M_{\text{Boost}} = \frac{1}{1-(d_2+d_3)}$	4	No	Bidirectional
Asymmetric H-bridge bidirectional DC–DC converter in Sect. 2.2	$M_{\text{Boost}} = \frac{1}{1-(d_2+d_4)}$	4	Yes	Bidirectional
Transformerless Buck three-level DC–DC converter in Sect. 3.2	$M_{\text{Buck}} = d_1 + d_2 - 1$	12	No	Step-down
Hybrid Boost three-level DC–DC converter in Sect. 3.3	$M_{\text{Boost}} = \frac{1}{1-(d_1+d_2)}$	12	No	Step-up
Bidirectional three-level DC–DC converter in Sect. 3.4	$M_{\text{Boost}} = \frac{1}{1-(d_1+d_2)}$	12	No	Bidirectional
Quasi-Z-source Boost three-level DC–DC converter in Sect. 3.5	$M_{\text{Boost}} = \frac{2}{3-4d}$	5	Yes	Step-up
Switched-quasi-Z-source bidirectional DC–DC converter in Sect. 4.2	$M_{\text{Boost}} = \frac{1+d}{1-d}$	3	Yes	Bidirectional
Hybrid switched-capacitors/ switched-quasi-Z-source bidirectional DC–DC converter in Sect. 4.3	$M_{\text{Boost}} = \frac{2+d}{1-d}$	5	No	Bidirectional
Switched-capacitor bidirectional DC–DC converter in Sect. 5.2	$M_{\text{Boost}} = \frac{2}{1-d}$	4	Yes	Bidirectional
Three-phase interleaved switched-capacitor bidirectional DC–DC converter in Sect. 5.3	$M_{\text{Boost}} = \frac{3}{1-d}$	8	Yes	Bidirectional
Input-parallel output-series Boost DC–DC converter in Sect. 5.4	$M_{\text{Boost}} = \frac{2}{1-d}$	5	No	Step-up
Single-switch Boost DC–DC converter in Sect. 5.5	$M_{\text{Boost}} = \frac{2}{1-d}$	4	No	Step-up
Quasi-Z-source Boost DC–DC converter in Sect. 5.6	$M_{\text{Boost}} = \frac{2}{1-2d}$	5	Yes	Step-up
Quadratic Boost DC–DC converter in Sect. 5.7	$M_{\text{Boost}} = \frac{3+d}{(1-d)^2}$	7	Yes	Step-up
Coupled-inductor bidirectional DC–DC converter in Sect. 5.8	$M_{\text{Boost}} = \frac{2+N}{1-d}$	4	Yes	Bidirectional

6.1.3 Application Analysis of DC–DC Converters for Fuel Cells

Fuel cells operate based on the principles of electrochemical reactions between oxidants and reducers. The output characteristics of fuel cells are nonlinear. Therefore, to ensure the high-efficiency and stable operation of fuel cells, DC–DC converters used in fuel cell systems must exhibit excellent performance to meet the system's operational requirements. Typically, a single fuel cell unit operates at a loaded voltage between 0.5 and 1.2 V. To meet both the system load and DC–DC bus voltage requirements, fuel cell stacks are connected in series and parallel, thereby increasing their power and voltage levels. Fuel cell stack electrical output characteristics are relatively gentle, necessitating the use of DC–DC converters to connect them to the system's DC bus. Because the output voltage levels of fuel cells in series are relatively high, it is essential to employ DC–DC converter topologies with reduced component stress, simplifying component selection and enhancing operational reliability. Furthermore, since fuel cells are frequently utilized in vehicles, aircraft, and other transportation means, the system's size and weight are crucial considerations for performance evaluation. High-power-density DC–DC converters can effectively reduce the system's volume and enhance its overall integration.

Hence, the H-bridge and asymmetric H-bridge configurations outlined in Chap. 2 are unsuitable for fuel cell systems due to their limited power density caused by the extensive use of semiconductors. In contrast, the quasi-Z-source bidirectional DC–DC converter presented in Sect. 3.5 demonstrates superior performance in achieving both high power density and exceptional conversion efficiency. Additionally, the switched quasi-Z-source bidirectional DC–DC converter discussed in Sect. 4.2 and the input-parallel output-series Boost DC–DC converters in Sect. 5.4 are viable options for fuel cell systems due to their ability to maintain low voltage stresses across components.

6.1.4 Application Analysis of DC–DC Converters for Photovoltaic Systems

Similar to fuel cells, photovoltaic arrays are composed of multiple PV cells arranged in series and parallel configurations, exhibiting nonlinear characteristics in their output. In PV systems, DC converters connect the PV cells to the DC bus, efficiently transforming solar energy to power the loads on the bus and charge the 0batteries. PV DC converters with wide gain ranges offer greater flexibility in PV panel design and provide a broader power point matching range. For PV systems applied in vehicles and aircraft, PV DC converters need to have high power density and low electromagnetic noise, ensuring reduced system weight and volume.

The DC–DC converter discussed earlier, suitable for fuel cell systems, can also find effective application in photovoltaic systems due to the analogous traits shared by photovoltaic panels and fuel cells. Both H-bridge and asymmetric H-bridge topologies can be employed in ground-mounted photovoltaic systems. Additionally, the switched-capacitors introduced in Chap. 5 can serve as photovoltaic regulators, tailored to specific voltage gain requirements.

6.2 Example of the Asymmetric H-Bridge Bidirectional DC–DC Converter

6.2.1 Parameter Design of the Asymmetric H-Bridge Bidirectional DC–DC Converter

To achieve the best performance of the asymmetric H-bridge bidirectional DC–DC converter, it is crucial to consider the parameter design for the passive components, which includes the filtering inductor L and the capacitors C_{low} and C_{high} on the low and high voltage sides, respectively. To ensure optimal design and functionality of the converter for specific application conditions, it is imperative to design the components in accordance with the operating principle during switching states. Referring to the operating principle of the bidirectional DC–DC converter in step-down mode, it can be observed that Q_1 and Q_3 are both ON while Q_2 and Q_4 are both OFF (i.e., $S_1S_3 = 11$) initiates the charging of the inductor. The charging phase within each carrier period lasts $(d_1 + d_3 - 1) \times T_s$, where T_s denotes the carrier period. Furthermore, the high-frequency component current of the inductor flows through the capacitor on the low voltage side during this phase, leading to the derivation of (6.1), where the voltage ripple of the low voltage side capacitor is represented by ΔU_{low}, while the inductor current ripple in the step-down mode is denoted as $\Delta I_{\mathrm{L\text{-}Buck}}$.

$$\begin{cases} \Delta U_{\mathrm{low}} \times C_{\mathrm{low}} = \frac{T}{4} \times \frac{\Delta I_{\mathrm{L\text{-}Buck}}}{2} \times \frac{1}{2} \\ L_{\mathrm{Buck}} \frac{\Delta I_{\mathrm{L\text{-}Buck}}}{(d_1+d_3-1)T_s/2} = U_{\mathrm{high}} - U_{\mathrm{low}} \end{cases} \tag{6.1}$$

Furthermore, based on the relationship of U_{low} and U_{high} in (6.2), the capacitance of the low voltage side capacitor and the inductance L_{Buck} in the step-down mode can be derived using (6.3).

$$U_{\mathrm{low}} = (d_1 + d_3 - 1) \times U_{\mathrm{high}} \tag{6.2}$$

$$\begin{cases} C_{\mathrm{low}} = \frac{\Delta I_{\mathrm{L\text{-}Buck}} \times T_s}{16 \Delta U_{\mathrm{low}}} \\ L_{\mathrm{Buck}} = \frac{U_{\mathrm{low}}(2-d_1-d_3)T_s}{2\Delta I_{\mathrm{L\text{-}Buck}}} \end{cases} \tag{6.3}$$

In the step-up mode, the inductor is in a state of discharge when $S_2S_4 = 00$. The discharging time in each carrier period can be calculated as $(1 - d_2 - d_4) \times T_s$, and this leads to the derivation of the following equation, where the voltage ripple of the high voltage side capacitor is represented by ΔU_{high}, and the inductor current ripple in the step-up mode is denoted as $\Delta I_{\text{L-Boost}}$.

$$\begin{cases} C_{\text{high}} \frac{\Delta U_{\text{high}}}{(1-d_2-d_4)T_s/2} = I_{\text{L - Boost}} - I_{\text{out}} \\ L_{\text{Boost}} \frac{\Delta I_{\text{L - Boost}}}{(1-d_2-d_4)T_s/2} = U_{\text{high}} - U_{\text{low}} \end{cases} \tag{6.4}$$

It is noted that the load current in the high voltage side and the voltage ripple of the low voltage side capacitor can be obtained from (6.5).

$$\begin{cases} I_{\text{out}} = (1 - d_2 - d_4) I_{\text{L - Boost}} \\ U_{\text{low}} = (1 - d_2 - d_4) U_{\text{high}} \end{cases} \tag{6.5}$$

Thus, (6.6) can be utilized to determine the capacitance of the high voltage side capacitor and the inductance L_{Boost} in the step-up mode, as presented below.

$$\begin{cases} C_{\text{high}} = \frac{I_{\text{out}}(d_2+d_4)T_s}{2\Delta U_{\text{high}}} \\ L_{\text{Boost}} = \frac{U_{\text{low}}(d_2+d_4)T_s}{2\Delta I_{\text{L - Boost}}} \end{cases} \tag{6.6}$$

By combining (6.3) and (6.6), the capacitance of both the high and low voltage-side capacitors and the inductance of the inductor can be designed accordingly.

6.2.2 *Experimental Results and Analysis of the Asymmetric H-Bridge Bidirectional DC–DC Converter in Fuel Cell Vehicle*

To test the practicality and efficiency of the asymmetric H-bridge bidirectional DC–DC converter in fuel cell vehicles, a 300 W prototype was created. To control the bidirectional voltage and current loops, a TMS320F28335 DSP is used, while MOSFETs of the IXYS-IXFK64N50P model were selected as the active power switches. The prototype consists of a variable low voltage side with a range of 24–48 V and a constant high voltage side at 200 V. The switching frequency for the prototype is set at $f_s = 10$ kHz, and the dead time is $t_d = 1$ μs. The above analysis allows for the calculation of the range of capacitances and inductances required for the converter. To ensure optimal performance, certain margins are reserved during the parameter design process. The resulting capacitor and inductor parameters are presented in Table 6.2.

Figure 6.1 displays the gate signals and voltage stress of the slave active power switches during the synchronous rectification operation. Specifically, in the step-down mode, Q_2 and Q_4 achieve both ZVS ON and OFF, as illustrated by the driven signal u_{GS2} and voltage stress u_{Q2} in Fig. 6.1a. Conversely, in the step-up mode, Q_1 and Q_3 accomplish ZVS ON and OFF, represented by the driven signal u_{GS3} and

Table 6.2 Experimental parameters

Parameter	Value
Rated power P	300 W
Storage/filter capacitor C_{low}	200 μF
Storage/filter capacitor C_{high}	330 μF
Storage/filter inductor L	306 μH
High voltage side U_{high}	200 V
Low voltage side U_{low}	24–48 V
Switching frequency f_s	10 kHz
Power semiconductors Q_1–Q_4	IXYS-IXFK64N50P

voltage stress U_{Q3} in Fig. 6.1b. Consequently, the slave active power switches of the asymmetric H-bridge bidirectional DC–DC converter can utilize ZVS operation without requiring any additional hardware. This approach enhances conversion efficiency, minimizing power loss in the switches and improving overall performance of the system.

The asymmetric H-bridge bidirectional DC–DC converter is capable of both step-down and step-up voltage conversion over a wide range. Figure 6.2 presents the experimental waveforms of voltages on the high side and low side of the converter. In step-down mode, the input voltage is fixed at 200 V, while the output voltage can be continuously varied from 24 to 48 V over 8 s using a voltage control loop. In step-up mode, the input voltage is continuously varied from 48 to 24 V over 8 s, while the output voltage is held constant at 200 V by the voltage control loop. The converter can operate in a wide range of voltage conversion ratios, from 0.12 to 0.24 in step-down mode and from 4.2 to 8.4 in step-up mode. The voltages on the high voltage side are kept constant at 200 V throughout the operation.

In step-down mode, the prototype converts 200–24 V using a narrow pulse output PWM voltage $U_{bg} = U_{ag}$ - U_{ab}. The experimental results are shown in Fig. 6.3, which depicts the pulse voltages and inductor current i_L. Figure 6.3a shows the pulse voltages of U_{ag} and U_{ab} from half-bridges, while the narrow pulse voltage U_{bg}, obtained from U_{ag} - U_{ab}, is illustrated in Fig. 6.3b. Therefore, the narrow pulse voltage U_{bg} occurs twice in each switching period with $T_s = 100$ μs. The inductor is charged when U_{bg} is at 200 V and discharged when it's at zero, as shown in Fig. 6.3b.

Even though the converter achieves high step-down ratio, all active power switches maintain proper duty cycles. For instance, the active power switch Q_1 operates with the duty cycle d_1 of approximately 0.58, which is close to 0.5, as shown in Fig. 6.4. It's worth noting that during each switching period, the converter charges and discharges the inductor twice. The converter has two additional power semiconductors compared to the Buck/Boost converter. However, its equivalent switching frequency is twice as high. Therefore, all capacitor and inductor volumes in the converter can be reduced by nearly half compared to those in the Buck/Boost converter.

In step-up mode, where the input voltage of 24 V is converted to 200 V, a narrow pulse output voltage is also generated. Figure 6.5 illustrates the pulse voltages and

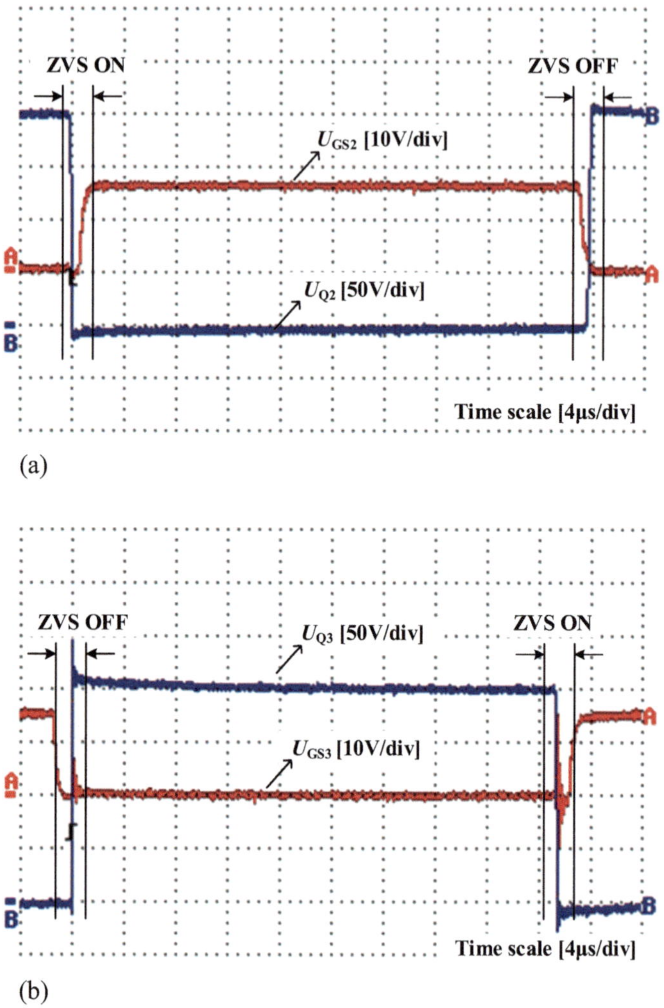

Fig. 6.1 **a** Driven signal U_{GS2} and voltage stress U_{Q2} across Q_2 and **b** driven signal U_{GS3} and voltage stress U_{Q3} across Q_3 under synchronous rectification operation

inductor current i_L, as well as corresponding driven signals and voltage stresses. The pulse voltages U_{ag} and U_{ab} from the half-bridges are depicted in Fig. 6.5a, while the narrow pulse voltage U_{bg} is obtained by subtracting U_{ab} from U_{ag} as shown in Fig. 6.5b. Each switching period $T_s = 100\ \mu s$ results in the narrow pulse voltage U_{bg} at twice the switching frequency, with the inductor being charged when U_{bg} is zero and discharged when U_{bg} is at 200 V (the narrow pulse), as indicated in Fig. 6.5b.

Despite achieving a high step-down ratio, the active power switches still operate with appropriate duty cycles. For example, taking active power switch Q_4 shown in Fig. 6.6 with the duty cycle d_4 of approximately 0.44, which is close to 0.5.

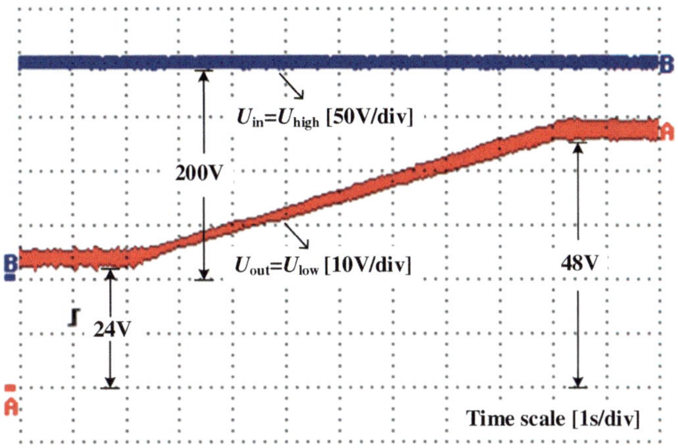

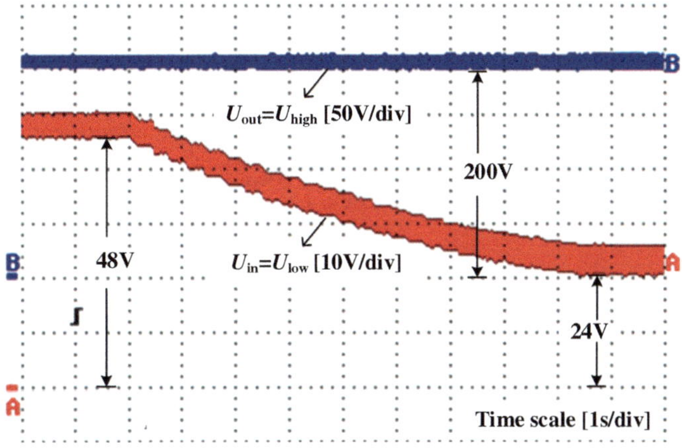

Fig. 6.2 Voltages on the high side and low side of the prototype **a** in step-down mode and **b** in step-up mode

Moreover, like the step-down mode, the inductor of the converter is charged and discharged twice during each switching period, with the charging and discharging frequency of the inductor being twice the switching frequency.

Figure 6.7 shows the experimental results of the converter's bidirectional operation between the step-down and step-up modes. The converter acts as a current-source converter, interfacing between the DC-link bus (at a constant 200 V) and the battery stacks (at a nominal 48 V), allowing easy control of power flow. In Fig. 6.7a, the converter steps down with a reference inductor current of – 4 A, resulting in power flow from the DC-link bus to the battery stacks ($U_{\text{low}} = 54$ V). When the converter is

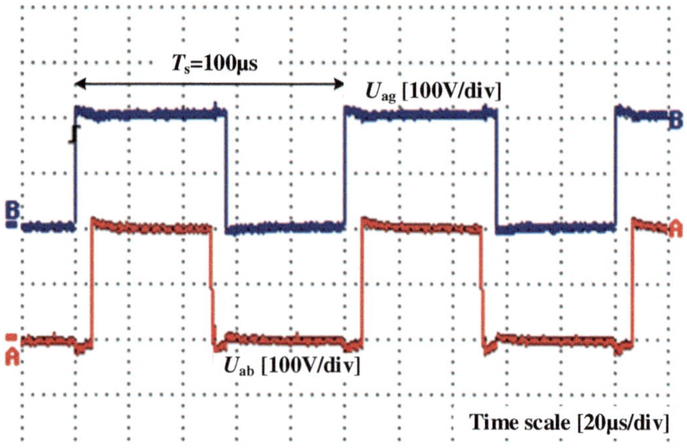

(a)

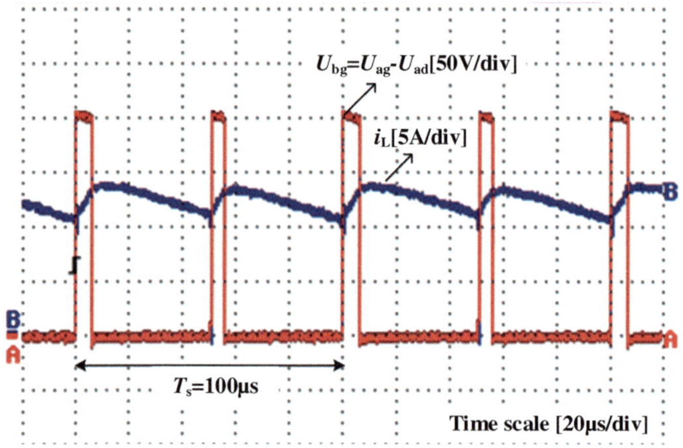

(b)

Fig. 6.3 **a** Pulse voltages U_{ab} and U_{ag} **b** Pulse voltage U_{bg} and inductor current i_L of the prototype in step-down mode

controlled to operate continuously from the step-down to the step-up mode, the power flow is reversed, with the converter operating in the step-up mode and a reference inductor current of 4A (U_{low} = 52 V). Figure 6.7b shows the detailed transient process of the step-down to the step-up mode with I_L steps from − 4A to 4A, where the battery stacks are charged with a constant current of 4 A before the controlled power flow command is given, and then discharged with the same current for 3 ms. In contrast, Fig. 6.7c shows the opposite operating condition with I_L steps from 4 A to − 4 A, where the charged current of the battery stacks stabilizes at a constant 4 A for 8 ms after the battery current rapidly falls to zero following the new power flow

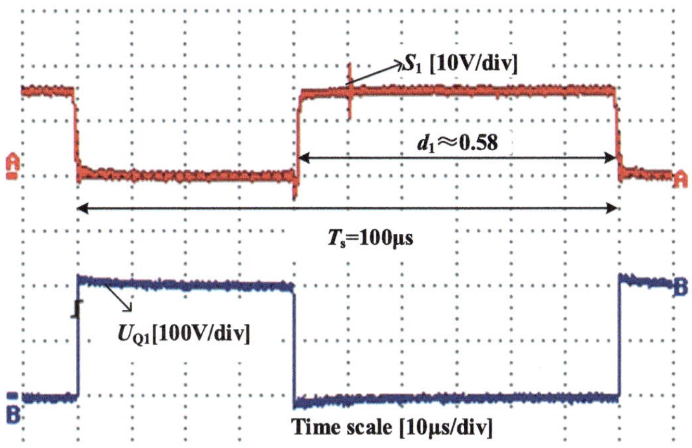

Fig. 6.4 Voltage stress U_{Q1} and driven signal U_{gs1} of Q_1 of the prototype in step-down mode

command. The converter responds quickly to the control signal due to the effective modulation method and the reduced inductance, as demonstrated in Fig. 6.7.

Figure 6.8 shows the measured conversion efficiency of the asymmetric H-bridge bidirectional DC–DC converter at different load powers from 100 to 300 W and different output/input voltages from 24 to 48 V. A Yokogawa-WT3000 Power Analyzer was used for the measurements. In the step-down mode, with an input voltage of 200 V, the maximum efficiency was achieved when the output voltage was 48 V and the load power was 300 W, with an efficiency of 94.7%. Conversely, the minimum efficiency was 90.7% when the output voltage was 24 V and the load power was 300W. In the step-up mode, with an output voltage of 200 V, the maximum efficiency was 93.5% when the input voltage was 48 V and the load power was 200 W, whereas the minimum efficiency was 83% when the input voltage was 24 V and the load power was 300 W. These results demonstrate that the conversion efficiency increases as the voltage on the low voltage side increases.

Figure 6.9 shows the calculated power loss distributions for the experiment conducted at $U_{low} = 48$ V, $U_{high} = 200$ V, and $P = 300$ W. In the step-down mode, the converter exhibits a total loss of 12.09 W, as depicted in Fig. 6.9a. By analyzing the power loss distribution, it can be concluded that the major losses are contributed by power switches Q_1 and Q_3. Specifically, the turn-on and turn-off losses, along with the conduction losses of Q_1 and Q_3, account for 38.627% of the total losses. The conduction losses of power switches Q_1-Q_4 constitute 25.475% of the total losses. The conduction losses and the core loss of the inductor contribute 20.595% and 15.303%, respectively.

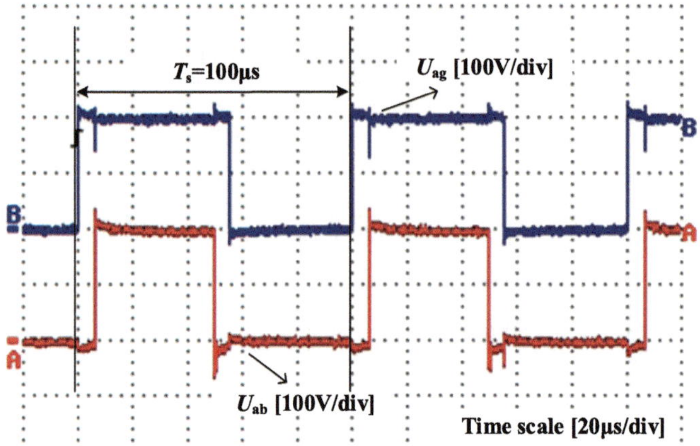

(a)

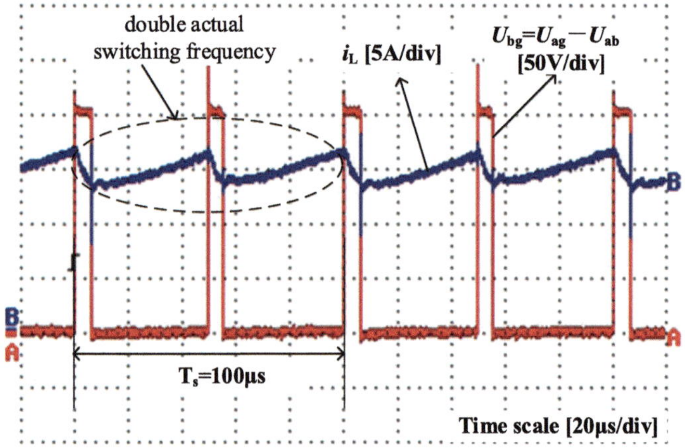

(b)

Fig. 6.5 **a** Pulse voltages U_{ag} and U_{ab} and **b** pulse voltage U_{bg} and inductor current i_L of the prototype in step-up mode

In the step-up mode, the converter shows a total loss of 13.3 W, as shown in Fig. 6.9b. The largest power losses are also attributed to the turn-on and turn-off losses of Q_2 and Q_4, along with the conduction losses of Q_1–Q_4, accounting for 58.271% of the total losses. The losses incurred by the capacitors and the copper loss of the inductor make up 27.819% of the total losses, while the remaining 13.91% corresponds to the core loss of the inductor.

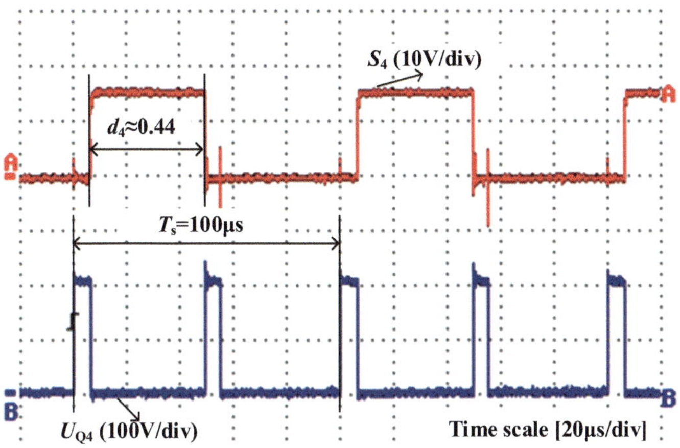

Fig. 6.6 Voltages and driven signals of Q_1 of the prototype in step-up mode

6.3 Example of the Bidirectional Three-Level DC–DC Converter

6.3.1 Parameter Design of the Bidirectional Three-Level DC–DC Converter

When it comes to designing the parameters for the bidirectional three-level DC–DC converter introduced in Chapter III, careful attention must be given to the selection of power switches and diodes, taking into consideration their current and voltage stresses. This is crucial to ensure the components can withstand the anticipated operating conditions without compromising their performance or reliability. The voltage stresses for all of the semiconductors in the bidirectional three-level DC–DC converter can be obtained by analyzing the energy flow paths among the voltage source, inductor, and capacitors during their effective switching states. In this converter, the voltages U_{C1} and U_{C2} are half the high voltage U_{high}.

$$U_{C1} = U_{C2} = \frac{U_{high}}{2} \tag{6.7}$$

By considering these factors, the voltage stress of the power switches and diodes employed in the topology can be deduced.

$$\begin{cases} U_{Q1} = U_{Q2} = U_{Q3} = U_{Q4} = U_{Q5} = U_{Q6} = U_{Q7} = U_{Q8} = \dfrac{U_{high}}{2} \\ U_{Dc1} = U_{Dc2} = \dfrac{U_{high}}{2} \end{cases} \tag{6.8}$$

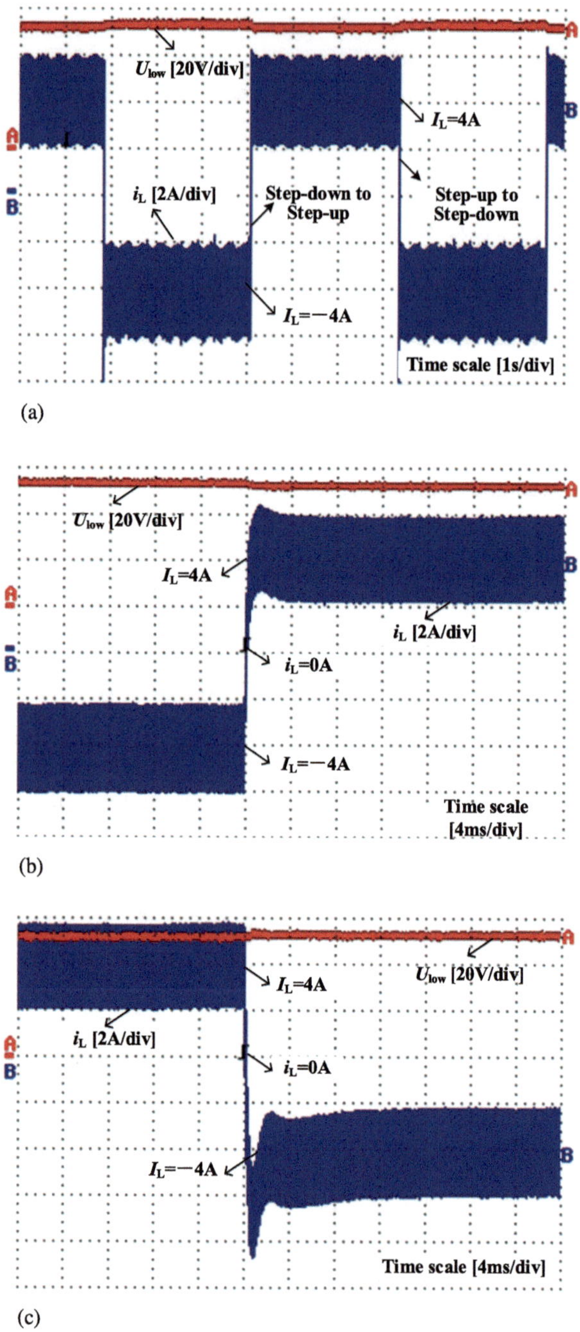

Fig. 6.7 Transient process during **a** bidirectional power flow, **b** step-down to step-up mode, and **c** step-up to step-down mode

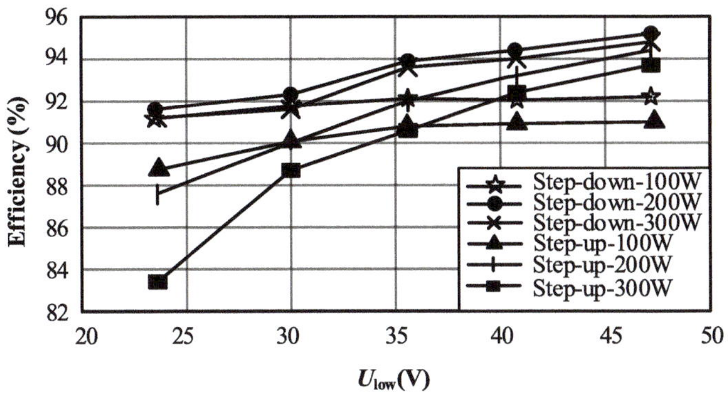

Fig. 6.8 Conversion efficiency curve of the prototype

Fig. 6.9 Power loss distributions for the prototype during **a** step-down and **b** step-up modes when $U_{\text{low}} = 48$ V, $U_{\text{high}} = 200$ V, $P = 300$ W

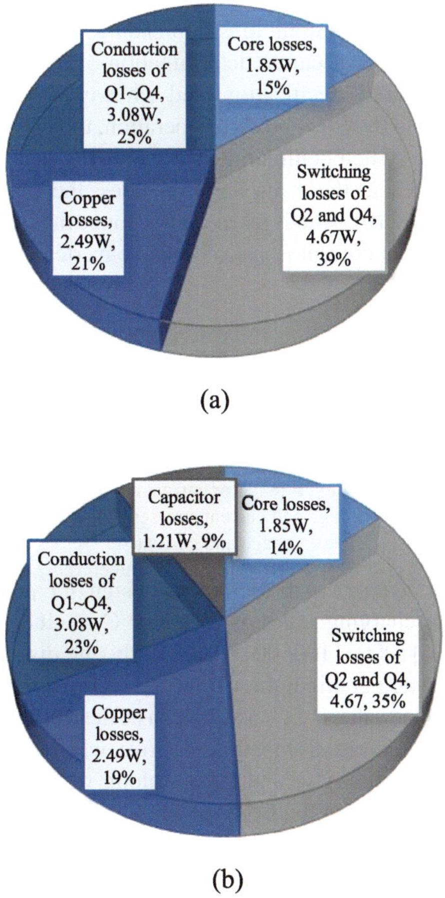

Equation (6.14) can be used to obtain the average currents of the semiconductors Q_1–Q_8 and D_{c1}–D_{c4}, which represent the current stresses when they are in the ON state. It takes into account the energy flow paths among the voltage source, inductor, and capacitors during effective switching states. Specifically, I_{Q1}-I_{Q8} and I_{Dc1}-I_{Dc4} denote the average currents of Q_1-Q_8 and D_{c1}-D_{c4}, respectively. Moreover, I_{Lf} is the current passing through the inductor L_f, which is identical to the current flowing through the low voltage side.

$$\begin{cases} I_{Q1} = I_{Q2} = I_{Q3} = I_{Q4} = I_{Q5} = I_{Q6} = I_{Q7} = I_{Q8} = I_{Lf} \\ I_{Dc1} = I_{Dc2} = I_{Dc3} = I_{Dc4} = I_{Lf} \end{cases} \tag{6.9}$$

In addition to the power switches and diodes, the capacitances of C_1, C_2, and C_f, as well as the filtering inductance L_f, play vital roles in the converter's overall performance. The capacitances contribute to smoothing out voltage ripples and stabilizing the output, thus influencing the converter's efficiency and transient response. As such, accurate determination of the capacitance values is essential to meet the required specifications and achieve optimal converter performance.

In the Buck mode of operation, the capacitor C_f enters the discharging state when the switch configuration is $S_1S_2S_7S_8 = 1100$ and 0011. By observing the behavior of C_f in these discharging states, the capacitance value of C_f can be determined as (6.10), which takes into account the capacitor voltage fluctuation ΔU_{Cf} and the output current $I_{\text{out-Buck}}$ of the converter operating in step-down mode.

$$C_f = \frac{(I_{Lf} - I_{\text{out - Buck}})(1 - m_a)}{\Delta U_{Cf} f_s} \tag{6.10}$$

The capacitance of C_f can also be represented by the current fluctuation Δi_{Lf} of L_f as below.

$$C_f = \frac{\Delta i_{Lf}(1 - m_a)}{\Delta U_{Cf} f_s} \tag{6.11}$$

In terms of the parameter design of C_1 and C_2, in the Boost mode of operation, the discharging states of capacitor C_1 include the switch configurations that $S_3S_4S_5S_6$ = [1000, 1100, 0011]. Similarly, the discharging states of capacitor C_2 include the switch configurations that $S_3S_4S_5S_6$ = [0001, 1100, 0011]. Therefore, the capacitance values of C_1 and C_2 can be determined using (6.12) based on the specific switch configurations mentioned above, which takes into account the capacitor voltage fluctuations ΔU_{C1} and ΔU_{C2}, along with the output current $I_{\text{out-Boost}}$ of the converter operating in the Boost mode.

$$\begin{cases} C_1 = \dfrac{I_{\text{out - Boost}}(1 - m_b)}{\Delta U_{C1} f_s} \\ C_2 = \dfrac{I_{\text{out - Boost}}(1 - m_b)}{\Delta U_{C2} f_s} \end{cases} \tag{6.12}$$

Furthermore, the filtering inductance L_f plays a crucial role in controlling output current fluctuations and maintaining the stability of the converter. Its value needs to be carefully selected to balance the current ripple requirements and system efficiency. The inductor L_f exhibits similar charging and discharging behavior in both the Buck mode and the Boost mode, demonstrating symmetry between bidirectional power flows. Consequently, the design of the inductance is solely based on the considerations for the Buck mode. By observing the charging and discharging states of the inductor L_f, it can be noted that L_f enters the discharging state when the switch configuration is $S_1S_2S_7S_8 = 1100$. Based on this observation, the inductance value of L_f can be deduced as (6.13), where the current fluctuation is denoted as Δi_{Lf}.

$$L_f = \frac{(1 - m_a)U_{high}}{\Delta i_{Lf} f_s} \tag{6.13}$$

The inclusion of Eqs. (6.15)–(6.17) facilitates the design process for the inductance of L_f and the capacitances of C_f, C_1, and C_2. These design considerations are vital in achieving optimal performance and functionality of the bidirectional three-level DC–DC converter. By addressing these design aspects, the converter can operate reliably and deliver the desired performance.

6.3.2 Experimental Results of the Bidirectional Three-Level DC–DC Converter in Energy Storage System

To validate the effectiveness of the bidirectional three-level DC–DC converter and control strategies, an experimental prototype with voltage and current control loops is established in the laboratory. The experimental setup aims to investigate the converter's performance and behavior under different operation modes, namely Buck, Boost, and bidirectional. The experiment parameters, outlining the key specifications of the prototype, are provided in Table 6.1. These parameters serve as the basis for configuring and conducting the experiments.

Under rated conditions, the selection of power switches and diodes is based on their voltage and current stress considerations. The chosen components should be able to withstand the anticipated electrical stresses to ensure reliable operation. This careful selection ensures that the power switches and diodes are suitable for the intended application and can handle the expected power levels. Furthermore, the capacitance values and inductance are determined using the parameter design analysis conducted earlier. By considering the specific requirements and design considerations, the appropriate capacitance values for capacitors and inductance for inductors are determined. The selected power components, along with the values of capacitors and inductors, are also listed in Table 6.3.

Due to the inherent rise time delays and disparities among the series-connected capacitors in practical scenarios, it is common for the capacitor voltages to become imbalanced. This imbalance can potentially impact the performance and stability

Table 6.3 Experimental parameters of the bidirectional three-level DC–DC converter

Parameter	Value
Rated power P_n	1.2 kW
Series-connected capacitor C_1	940 μF
Series-connected capacitor C_2	940 μF
Filtering capacitor C_f	940 μF
Filtering inductor L_f	270 μH
High voltage side U_{high}	400 V
Low voltage side U_{low}	60–160 V
Switching frequency	10 kHz
Power switches Q_1–Q_8	IXTK 102N30P
Diodes D_{c1}–D_{c4}	DSEC 60-03A
Non-extreme duty cycle range	0.2–0.8
Restriction factor $k_{Buckmin}$	0.1
Restriction factor $k_{Buckmax}$	1.5
Restriction factor $k_{Boostmin}$	0.1
Restriction factor $k_{Boostmax}$	1.5

of the converter. To address these challenges and ensure the healthy and efficient operation of the converter, a duty cycle disturbance control method is implemented to balance the voltages of the series-connected capacitors. By dynamically adjusting the duty cycle of the converter's switching signals, active compensation for the voltage imbalances among the capacitors can be realized. Thereby, the operation of converter can be maintained within safe and optimal voltage limits.

An experimental setup is conducted with a step-down voltage gain M_{Buck} of 0.15, where the low voltage is set at 60 V and the high voltage is set at 400 V. Figures 6.10 and 6.11 present the experimental results, showcasing the blocking voltages across Q_1 and Q_2, the pulse voltage u_{ab}, and inductor currents i_L under different values of k_{Buck}. The experimental findings reveal that the voltage stresses across Q_1 and Q_2 are identical, indicating symmetrical behavior. Similarly, the voltages across the series-connected capacitors exhibit equal values. Notably, these voltages are measured at 200 V, which is half the value of U_{high}.

A suitable restriction function of $k_{Buck} = 0.1$ is applicable when the duty cycles approach 0.5, which is corroborated by the experimental results presented in Fig. 6.10a, where the duty cycles d_1 and d_2 are roughly 0.56 and 0.59, respectively. However, it is important to note that the inductor current experiences significant fluctuations, ranging from approximately 17–30 A, as demonstrated in Fig. 6.10b.

Considering the permissible non-extreme duty cycle range of 0.2–0.8, a more appropriate restriction function that $k_{Buck} = 1.5$, can be determined. This is supported by the experimental observations in Fig. 6.11a, where the duty cycles d_1 ad d_2 are roughly 0.35 and 0.8, respectively. In this scenario, the inductor current fluctuates between 20 and 27 A, as displayed in Fig. 6.11b.

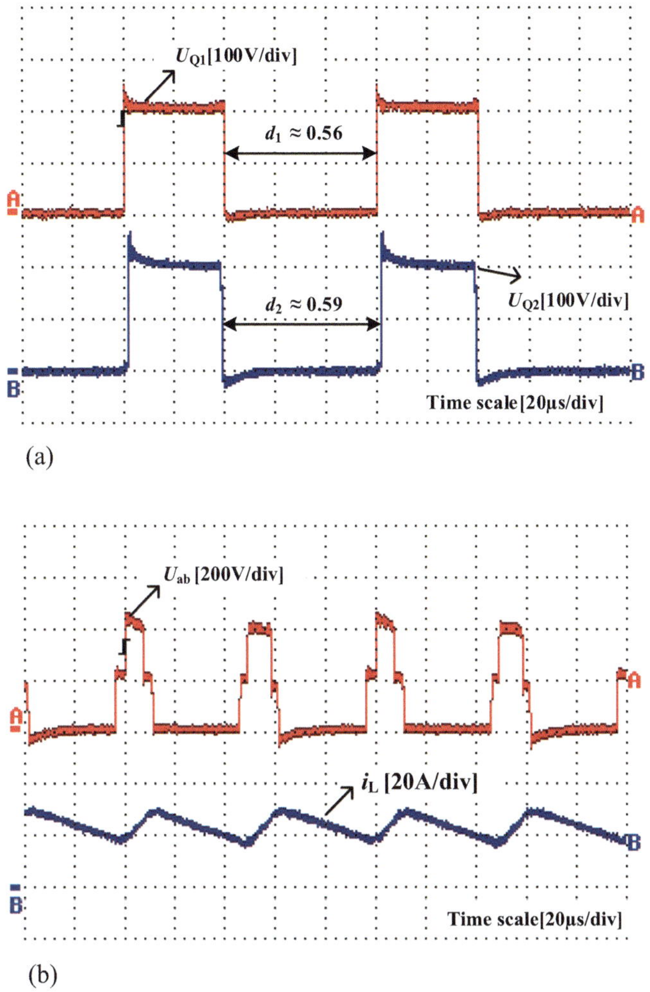

Fig. 6.10 **a** Voltages U_{Q1} and u_{Q2} across Q_1 and Q_2 and **b** pulse voltage U_{ab} and inductor current i_L in step-down mode with $k_{Buck} = 0.1$

Comparing the experimental results in Fig. 6.11 to that in Fig. 6.10, it is evident that the frequency of the inductor current ripple has doubled, and the fluctuation has been reduced by approximately 46.2%. Meanwhile, it should be acknowledged that the duty cycles are now closer to the permitted range. These findings emphasize the trade-offs involved in selecting duty cycles and their impact on inductor current fluctuations. Finding the optimal balance between these factors is crucial for achieving stable and efficient operation of the converter.

To assess the effectiveness of the duty cycle disturbance control method, an experiment without duty cycle disturbance was conducted for comparison. The experiment

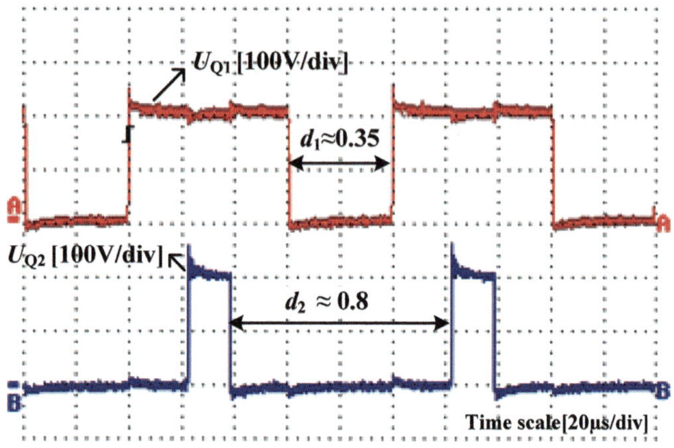

(a)

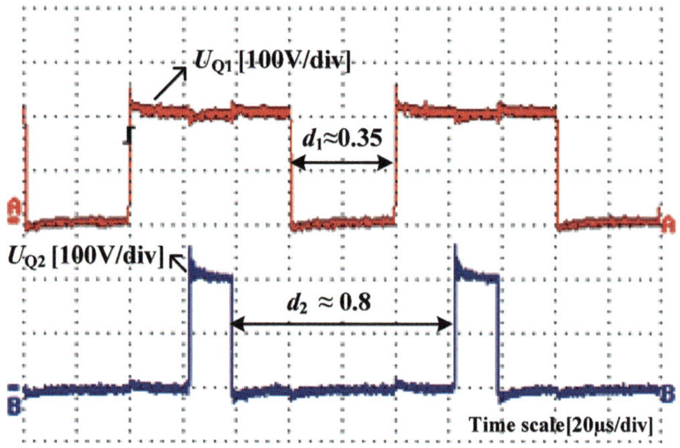

(b)

Fig. 6.11 **a** Voltages U_{Q1} and U_{Q2} across Q_1 and Q_2 and **b** pulse voltage U_{ab} and inductor current i_L in step-down mode with $k_{Buck} = 1.5$

was performed with U_{high} at 400 V and k_{Buck} at 1.5. The experimental waveforms of voltages across Q_1, Q_2, C_1, and C_2 in the steady state are presented in Fig. 6.12. It is evident that C_1 is charged to a higher voltage than C_2, resulting in an imbalance between U_{C1} and U_{C2}. It is notable that the voltage across Q_1 is approximately 240 V, while the voltage across Q_2 is nearly 160 V. Comparing these results with those shown in Fig. 6.11a, the voltage stresses across power switches are significantly higher due to the unbalanced capacitor voltages. Consequently, the power switches are at a higher risk of being damaged under these circumstances.

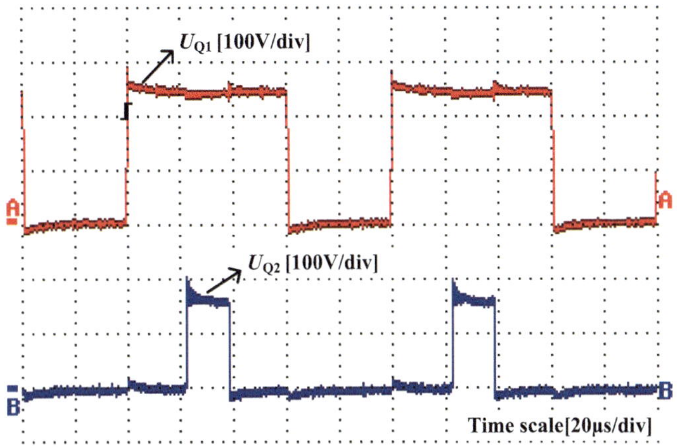

Fig. 6.12 Voltages across Q_1 and Q_2 in Buck mode without duty cycle disturbance control

In another experimental scenario, U_{high} is held constant at 400 V, while U_{low} varies continuously between 60 and 160 V. The efficiencies of the prototype in the Buck mode at different power levels are depicted in Fig. 6.13. The maximum efficiency achieved is approximately 94.9%, while the minimum efficiency is about 83.4%. The experimental results underscore the importance of the duty cycle disturbance control method in addressing capacitor voltage imbalances and ensuring the integrity and efficiency of the converter operation.

The experimental results of the prototype operating under step-up voltage gain M_{Boost} at 6.67 with the duty cycle disturbance control method are presented in Figs. 6.14 and 6.15, illustrating the blocking voltages across Q_3, Q_4, U_{ab}, and the inductor current i_L. Notably, the capacitor voltages are well balanced, and the

Fig. 6.13 Efficiency of the prototype in Buck mode versus output voltage with $k_{\text{Buck}} = 1.6$

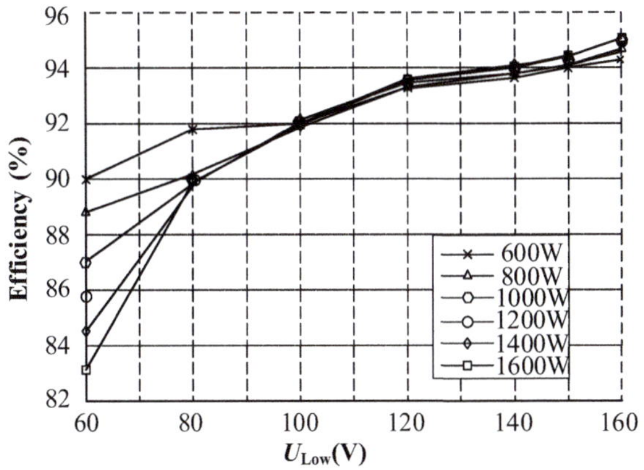

blocking voltages across Q_1 and Q_2 are equal to half of U_{high}, which is consistent to the previous theoretical analysis.

By selecting the restriction function k_{Boost} as 0.1, duty cycles close to 0.5 can be obtained, as depicted in Fig. 6.14a, with d_3 at 0.44 and d_4 at 0.41. In this configuration, power flow in the prototype occurs from the low voltage side to the high voltage side. Furthermore, Fig. 6.14b demonstrates that the inductor current i_L exhibits a ripple between – 16 A and – 29 A, with a ripple frequency twice that of the switching frequency.

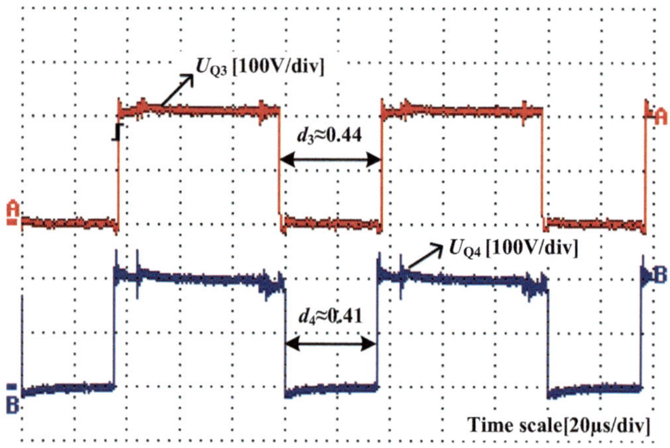

(a)

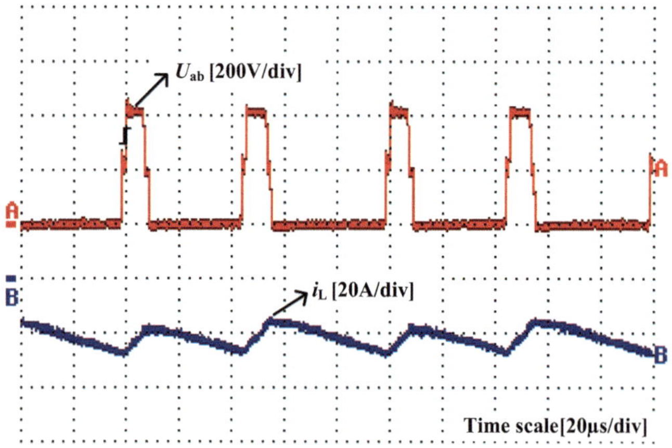

(b)

Fig. 6.14 **a** Voltages U_{Q1} and U_{Q2} across Q_1 and Q_2 and **b** pulse voltage U_{ab} and inductor current i_L in step-up mode with $k_{Boost} = 0.1$

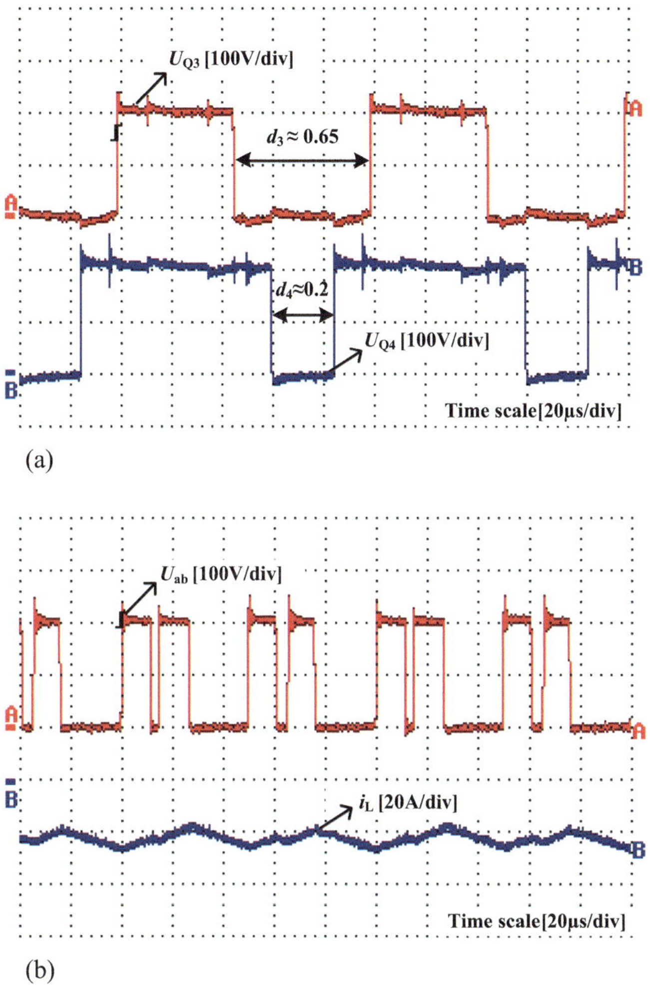

Fig. 6.15 **a** Voltages U_{Q1} and U_{Q2} across Q_1 and Q_2 and **b** pulse voltage U_{ab} and inductor current i_L in step-up mode with $k_{Boost} = 1.5$

Alternatively, when k_{Boost} is chosen as 1.5, the duty cycles approach the non-extreme range of 0.2 to 0.8. Theoretically, this results in d_3 at 0.65 and d_4 at 0.2, as shown in Fig. 6.15a. Additionally, Fig. 6.15b reveals that the inductor current varies from – 18 A to – 27 A, with a ripple frequency four times that of the switching frequency. Consequently, the current fluctuation is reduced by 30.8% compared to the previous condition.

To assess the effectiveness of the duty cycle disturbance control method in the Boost mode, an experimental investigation is conducted without implementing this method. The experimental setup involves a configuration with U_{high} at 400 V and

k_{Boost} at 0.1. The blocking voltages across Q_1 and Q_2 in the steady state are depicted in Fig. 6.16. The results clearly demonstrate that capacitor C_1 is charged to a significantly higher voltage than C_2, with Q_1 reaching approximately 245 V and Q_2 only reaching 155 V. This imbalance in voltage poses a significant risk to the power switches, as they may be subjected to excessive stress.

Furthermore, the measured efficiencies of the prototype in the Boost mode are evaluated at different power levels, as illustrated in Fig. 6.17. The efficiency varies with the input voltage and the output power level, with the lowest efficiency occurring at the highest voltage gain and highest power level. It can be observed that the maximum efficiency achieved is approximately 94.5%, while the minimum efficiency is approximately 81.2%.

In order to validate the bidirectional operation performance of the prototype, an experiment was carried out to charge and discharge batteries, and the results are shown in Fig. 6.18. The low voltage side of the converter is connected to battery stacks with a nominal voltage of 48 V, while the high voltage side is connected to the DC link. The converter controls the power flow as a current-source converter.

In Fig. 6.18a, the converter operates in the Boost mode with the reference inductor current I_L set at – 3 A, and the energy flows from the low voltage side to the high voltage side. After approximately 3 s, the power flow direction is controlled to be reversed. It then flows from the high voltage side to the low voltage side, and the converter operates in the Buck mode with the reference inductor current I_L set at 3 A.

Figure 6.18b provides a zoomed-in view of the transient process from the Buck mode to the Boost mode. The dynamic process is completed within 4 ms, where the inductor current quickly drops from 3 A to zero, and then stabilizes at – 3 A after 2 ms. Figure 6.18c shows the opposite process, where the inductor current changes from – 3 A to 3 A within 10 ms. This represents the power flow direction switching to the low voltage side from the high voltage side. It can be observed that the transition

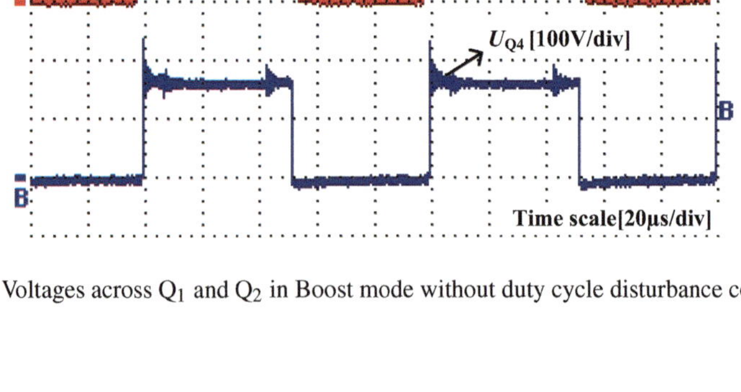

Fig. 6.16 Voltages across Q_1 and Q_2 in Boost mode without duty cycle disturbance control

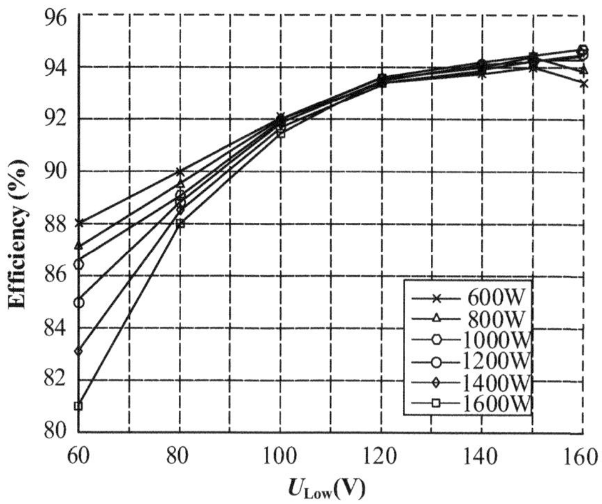

Fig. 6.17 Efficiency of the prototype in Boost mode versus output voltage with $k_{Boost} = 3$

from the Buck mode to the Boost mode is smooth and rapid. These experimental results demonstrate the bidirectional operation capability of the converter, allowing for efficient charging and discharging of batteries with smooth and quick power flow direction changes.

The power loss distribution of the prototype was calculated for an experiment with a low voltage side voltage U_{low} of 160 V, a high voltage side voltage U_{high} of 400 V, and a power level P of 1200 W, with both Buck and Boost modes having restriction factors k_{Buck} and k_{Boost} of 0.25. As presented in Fig. 6.19, the total losses of the converter in the Buck mode were found to be 53.71 W, with the majority of losses coming from the power switches Q_1-Q_8, which accounted for 71.50% of the total losses. This was due to turn-on, turn-off, and conduction losses of the power switches. The conduction losses of diodes D_{c1}-D_{c4} accounted for 5.57% of the total losses, while the conduction losses of the capacitors and the inductor accounted for 19.48%. The remaining 3.45% of the total losses were attributed to the core losses of the inductor. In the Boost mode, the total losses were found to be 57.98 W, with the majority of losses also coming from the power switches Q_1-Q_8, accounting for 72.79% of the total losses. The conduction losses of diodes D_{c1}-D_{c4} accounted for 5.16% of the total losses, while the conduction losses of the capacitors and the inductor accounted for 18.86%. The remaining 3.19% of the total losses were attributed to the core losses of the inductor. Overall, these results indicate that the power switches Q_1-Q_8 were the primary source of power loss in both modes. It is worth noting that the core losses of the inductor were relatively small, but still worth considering for future optimization.

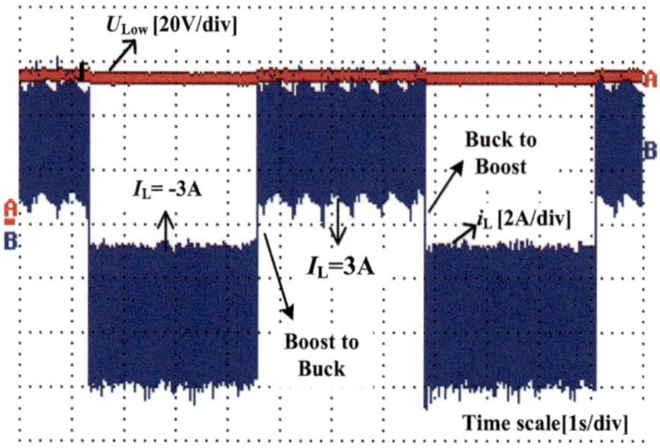

(a)

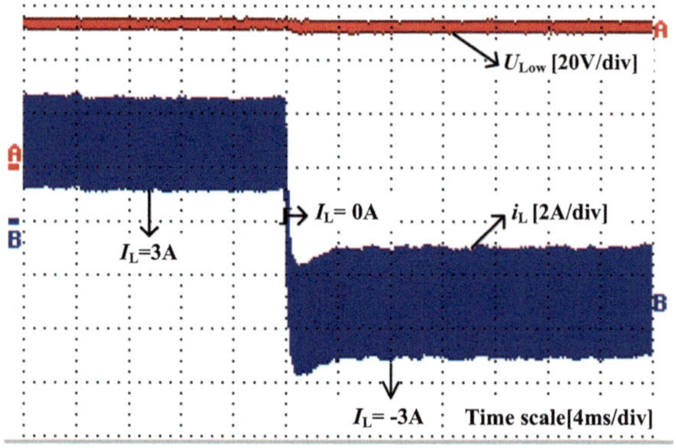

(b)

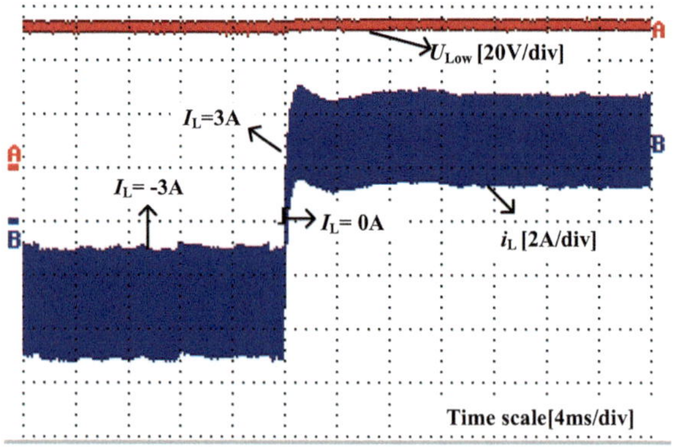

(c)

Fig. 6.18 Transient process during **a** bidirectional power flow, **b** step-down to step-up mode, and **c** step-up to step-down mode

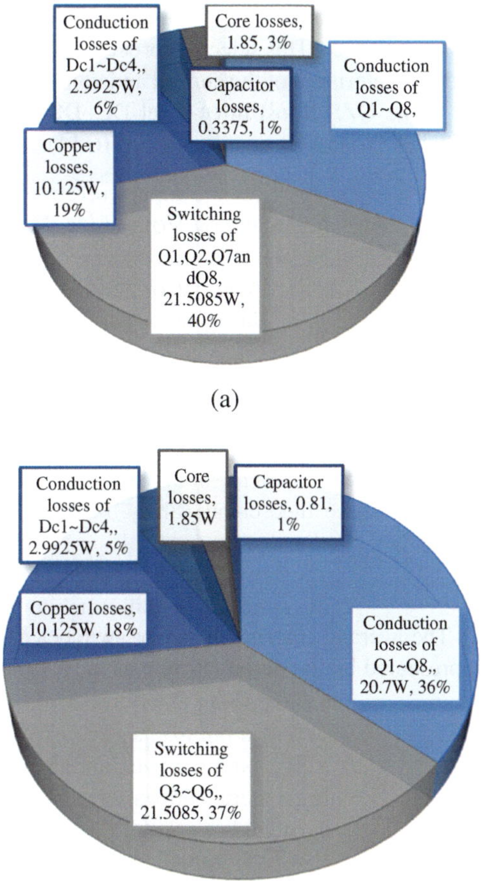

Fig. 6.19 Power loss distributions for the prototype in **a** step-down and **b** step-up modes when $U_{\text{low}} = 160$ V, $U_{\text{high}} = 400$ V, $P = 1200$ W

6.4 Example of the Quasi-Z-Source Boost Three-Level DC–DC Converter

6.4.1 Parameter Design of the Quasi-Z-Source Boost Three-Level DC–DC Converter

In terms of parameter design for the QZS Boost three-level DC–DC converter, the analysis of switching states reveals that the voltage U_{Cfly} across the flying capacitor C_{fly} is determined as half of the output voltage U_{out}, along with the voltage sum of capacitors C_1 and C_2. By examining the energy flow paths among the voltage source, inductors, and capacitors during their effective switching states, the voltage stresses on the power switches and diodes employed in this topology can be deduced.

The choice of power components is determined by evaluating their current and voltage stresses. This selection process involves carefully considering the magnitude and characteristics of the current and voltage that these components will be subjected to. In the QZS Boost three-level DC–DC converter, the voltage stresses of power switches Q_1 and Q_2 are obtained as (6.14).

$$\begin{cases} U_{Q1} = U_{Cfly} = \dfrac{U_{out}}{2} \\ U_{Q2} = U_{out} - U_{Cfly} = \dfrac{U_{out}}{2} \end{cases} \tag{6.14}$$

Meanwhile, the voltage stresses of diodes D_1-D_3 are obtained as below.

$$\begin{cases} U_{D1} = U_{C1} - U_{C2} = \dfrac{U_{out}}{2} \\ U_{D2} = U_{Cfly} = \dfrac{U_{out}}{2} \\ U_{D3} = U_{out} - U_{Cfly} = \dfrac{U_{out}}{2} \end{cases} \tag{6.15}$$

The average currents of power switches and diodes in their ON state are can be represented by the output current as in (6.16) and (6.17), respectively.

$$\begin{cases} I_{Q1} = \dfrac{4}{3-4d} I_{out} \\ I_{Q2} = \dfrac{4}{3-4d} I_{out} \end{cases} \tag{6.16}$$

$$\begin{cases} I_{D1} = \left(\dfrac{4}{3-4d} - \dfrac{1}{1-d}\right) I_{out} \\ I_{D2} = \dfrac{1}{1-d} I_{out} \\ I_{D3} = \dfrac{1}{1-d} I_{out} \end{cases} \tag{6.17}$$

When S_1S_2 is set to 10 or 01, Q_1 and Q_2 experience lower current stresses, as do D_3 and D_2. However, when S_1S_2 is set to 11, the current stresses on Q_1 and Q_2 can be as much as twice the average currents of the inductors. Therefore, it is important to consider the specific switching states and refer to (6.14)–(6.17) to ensure the appropriate selection of semiconductors Q_1, Q_2, D_1-D_3 in the QZS Boost three-level DC–DC converter.

On the other hand, it can be observed that L_1 and L_2 are in the charging state when S_1S_2 is set to 11. Thus, the inductances of L_1 and L_2 can be determined by (6.18), where ΔI_{L1} and ΔI_{L2} represent the current fluctuations of L_1 and L_2, respectively, and f_s denotes the switching frequency.

$$\begin{cases} L_1 = (2d-1)\dfrac{U_{C1}+U_{in}}{2\Delta i_{L1} f_s} \\ L_2 = (2d-1)\dfrac{U_{C2}}{2\Delta i_{L2} f_s} \end{cases} \tag{6.18}$$

By combining (6.18) with the voltage gain and capacitor voltages presented in the steady-state analysis, the inductances of L_1 and L_2 can be calculated, as expressed in (6.19), which establishes a relationship between the output voltage U_{out}, the current fluctuations ΔI_{L1} and ΔI_{L2}, the switching frequency f_s, and the duty cycle d.

$$\begin{cases} L_1 = (2d-1)(1-d)\dfrac{U_{out}}{2\Delta i_{L1} f_s} \\ L_2 = (2d-1)(1-d)\dfrac{U_{out}}{2\Delta i_{L2} f_s} \end{cases} \tag{6.19}$$

During the design considerations for capacitances, it is observed that when S_1S_2 is set to 11, C_1 and C_2 are in the discharging state. The capacitances of C_1 and C_2 can be determined by (6.20), where ΔU_{C1} and ΔU_{C2} represent the capacitor voltage fluctuations of C_1 and C_2, respectively.

$$\begin{cases} C_1 = \dfrac{(2d-1)I_{out}}{(3-4d)\Delta U_{C1} f_s} \\ C_2 = \dfrac{(2d-1)I_{out}}{(3-4d)\Delta U_{C2} f_s} \end{cases} \tag{6.20}$$

On the other hand, when S_1S_2 is set to 10, the flying capacitor C_{fly} is discharged. Its capacitance can be obtained via (6.21), where ΔU_{Cfly} denotes the capacitor voltage fluctuation of C_{fly}, which is independent of the duty cycle d of the power switches.

$$C_{fly} = \frac{I_{out}}{\Delta U_{Cfly} f_s} \tag{6.21}$$

Regarding the output capacitor C_{out}, it is only charged when S_1S_2 is set to 10. The capacitance of C_{out} can be deduced as below, where the capacitor voltage fluctuation of C_{out} is denoted as ΔU_{out}.

$$C_{out} = \frac{dI_{out}}{\Delta U_{out} f_s} \tag{6.22}$$

6.4.2 Experimental Results of the Quasi-Z-Source Boost Three-Level DC–DC Converter in Fuel Cell Vehicle

A downscaled prototype of the QZS Boost three-level DC–DC converter with a power rating of 1.2 kW was constructed to evaluate its feasibility and effectiveness for fuel cell vehicles. In the experiment, the fuel cell stack source U_{FC}, namely the input voltage U_{in} was replaced by an adjustable DC voltage source, offering a range of 60 to 150 V DC input source. Control of the converter voltage loop was achieved using a TMS320F28335 DSP, while the power circuit utilized IXTK102N30P MOSFETs, rated at 300 V and 102 A, along with DSEC60-03A Schottky Barrier Diodes. Additional specifications included a switching frequency of f_s at 10 kHz, a dead time of t_d set to1 μs, and inductances that $L_1 = 228$ μH and $L_2 = 225$ μH. The load resistor R_L ranged from 133 to 400 Ω, and the desired reference output voltage was set at 400 V. To provide a comprehensive overview, the main experimental parameters of the prototype are presented in Table 6.4.

Figure 6.20 illustrates the experimental pulse voltage U_{pn} for an input voltage of U_{in} at 40 V. The frequency of U_{pn} is twice that of the switching frequency. Notably, under the influence of the voltage control loop, the actual duty cycles are observed to be approximately 0.7, deviating from the typical Boost converter where the duty cycle often exceeds 0.9. Furthermore, the amplitude of U_{pn} is measured at 200 V, varying in conjunction with the flying-capacitor voltage U_{Cfly} and $U_{out} - U_{Cfly}$. Importantly, this amplitude corresponds precisely to half of the output voltage. Hence, it can be concluded that in the steady state, the flying-capacitor voltage U_{Cfly} self-balances effectively without the need for any additional control measures.

The demonstration of synchronous rectification ZVS operation can be observed based on the driven signal and voltage across the synchronous rectifier Q_{SR} shown in Fig. 6.21. Due to the presence of a dead time t_d, Q_{SR} experiences a delay in turning

Table 6.4 Parameters of the prototype

Parameter/components	Value
Rated power P_n	1.2 kW
Input DC voltage U_{in}	60–150 V
Output DC voltage U_{out}	400 V
Switching frequency f_s	10 kHz
Dead time t_d	1 μs
Inductor L_1	228 μH
Inductor L_2	225 μH
Capacitors C_1, C_2, C_{fly}	660 μF
Capacitor C_{out}	440 μF
Load R_{load}	133–400 Ω
MOSFETs Q_1, Q_2, Q_{SR}	IXTK102N30P
Diodes D_2, D_3, D_{FC}	DSEC60-03A

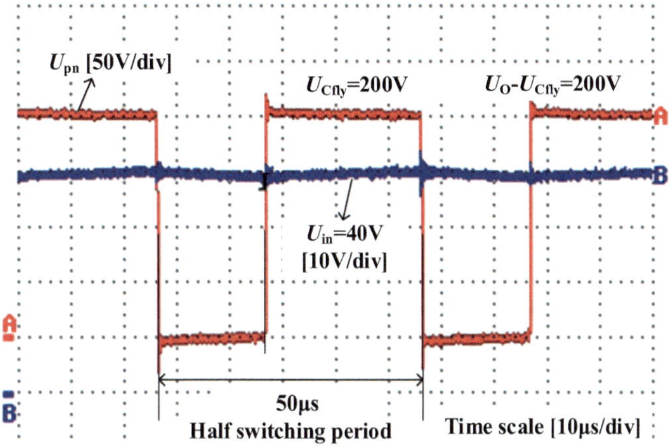

Fig. 6.20 Output pulse voltage U_{pn} with $U_{in} = 40$ V and $M = 10$

ON, meanwhile the anti-parallel body diode D_{SR} to conduct during this dead time. The voltage across Q_{SR} is clamped at the on-state voltage drop of D_{SR}, achiving ZVS ON operation. Due to the presence of parasitic capacitor, the voltage across Q_{SR} rises from the on-state voltage drop of D_{SR} to $U_{high}/2$ after the gate signal U_{GS} is set low. This ZVS operation reduces switching losses and enhances the overall performance of the synchronous rectification process.

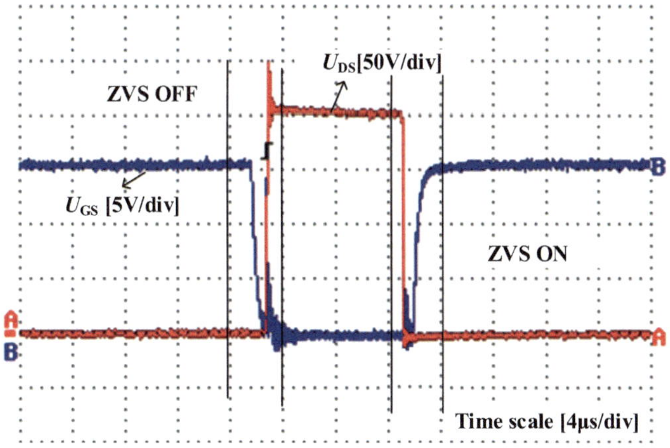

Fig. 6.21 Driven signal U_{gs} and voltage U_{ds} across Q_{SR}

An experimental demonstration of the suitability of the converter to fuel cell vehicles is shown in Fig. 6.22a, during which the input voltage U_{in} is gradually reduced from 120 to 40 V over a period of several seconds. By using the voltage control loop, it can be seen that the output voltage UO closes in on the reference voltage 400 V and the wide step-up voltage-gain range varies between 3.3 and 10. As a matter of fact, due to the losses compensation of the converter's operation, the ratio of currents on the input and output side is more than 3.3 to 10. As shown in Fig. 6.22b, the input current i_{L1} increases gradually with the wide-range change from 120 to 40 V in input voltage.

In the open loop, based on the output voltage U_{out} that varies between 200 and 400 V, the voltage of the flying capacitor U_{Cfly} is changed accordingly, as shown in Fig. 6.23. Even during dynamic states I and II, the flying-capacitor voltage U_{Cfly} remains half of the output voltage U_{out}. Due to the fact that the voltage across the flying capacitor is clamped by the total voltage from the QZ source capacitors whose voltage relates to the corresponding real-time duty cycle and output voltage.

Since the converter utilizes a voltage control loop by adjusting the duty cycle to maintain the output voltage U_{out} at 400 V, the output power P_{out} is regulated to 1.2 kW. Figure 6.24a shows the output pulse voltage U_{pn} and the current flowing through the inductor L_1. When the instantaneous voltage of U_{pn} is zero while S_1S_2 is set to 11, the inductor L_1 is charged. After U_{pn} has reached $U_{out}/2$ while S_1S_2 is set to 01 or 10, the inductor L_1 is discharged. Moreover, Fig. 6.24b shows that the current of inductor L_2 is also affected by the duty cycle and is nearly equal to the current of L_1. During each switching period, the inductors will be charged and discharged twice.

An experiment was conducted to verify the dynamic behavior of the converter. Figure 6.25 displays the output voltage and inductor current when the load is stepped between 133 Ω and 200 Ω. The output voltage U_{out} can be maintained around the constant reference voltage at 400 V under the voltage loop control. Meanwhile, the inductor currents have corresponding responses between 8 and 12 A. The load step-change from 200 Ω to 133 Ω lead to i_{L1} to rises from 8 to 12 A over a 20 ms period, and the load step-change from 133 Ω to 200 Ω causes it to drops from 12 to 8 A over the same period.

In order to measure the conversion efficiency of the prototype over a wide input-voltage range and the different output powers, a Power Analyzer (Yokogawa-WT3000) is used. As shown in Fig. 6.26, the efficiency of the system, the variable input voltages, and the different output powers are related. As can be seen in the figure, the maximum efficiency measured in SR operation is approximately 95.66%. Furthermore, decreasing input voltage with constant output power results in corresponding decreases in efficiency due to the increasing losses caused by growing input current. The efficiency curve under DR operation is also assessed under the same conditions above, and SR operation efficiency is greater than DR operation efficiency. A minimum efficiency difference area is located near the medium input voltage ($U_{in} = 120$ V), and its average efficiency difference is approximately 0.6%. In this case, the maximum efficiency difference area can be observed around input voltages of 80 V and 150 V, with an average value of approximately 0.85%.

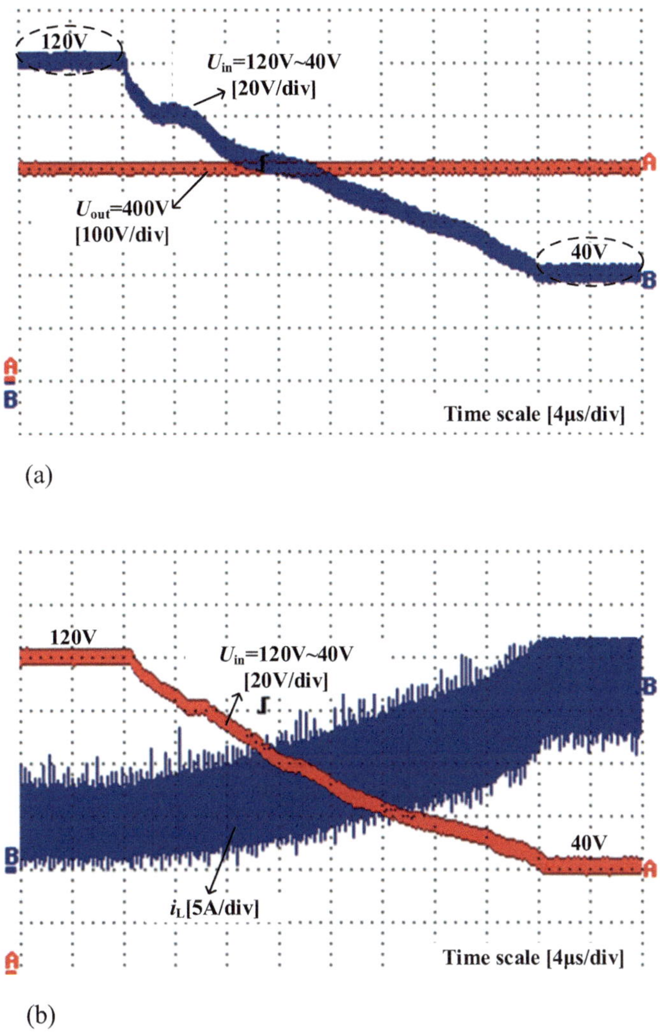

Fig. 6.22 **a** Input voltage U_{in} and output voltage U_{out} and **b** input voltage U_{in} and inductor current i_L with input voltage varying from 120 to 40 V

Figure 6.27 displays the computed loss distributions for the experiment with U_{in} at 150 V and P_{out} at 1200 W. The converter's total losses under DR operation are 57.06 W, which is displayed in Fig. 6.27a. As can be seen from the illustration, 39.87% of the overall losses under DR operation were made up of turn-on and turn-off losses and conduction loss in Q_1 and Q_2. Specifically, due to the greater conduction loss of D_1, the total conduction losses of diodes account for 41.57% of the overall losses, which is somewhat more than the switching and conduction losses of Q_1 and Q_2. However, under SR operation, the converter's overall losses are decreased to 49.26W.

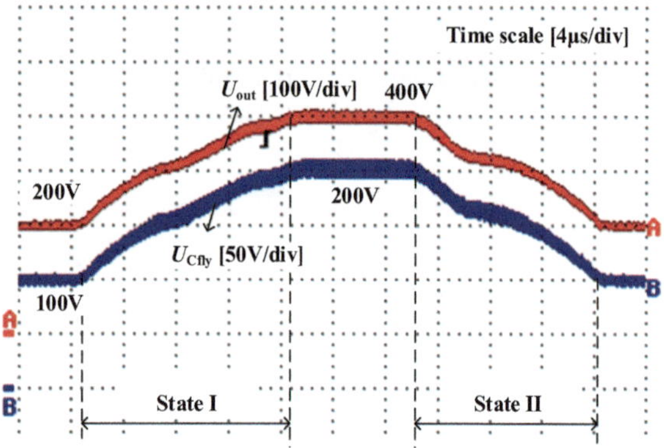

Fig. 6.23 Flying-capacitor voltage u_{Cfly} with output voltage U_{out} varying between 200 and 400 V

As can be observed in Fig. 6.27b, as a result of the SR operation of Q_{SR}, the switching and conduction losses of Q_1 and Q_2 account for 46.19% of the total losses, while the conduction losses of D_2, D_3, D_{FC}, and Q_{SR} are reduced to 32.32% of the total losses.

6.5 Example of the Switched-Quasi-Z-Source Bidirectional DC–DC Converter

6.5.1 Parameter Design of the Switched-Quasi-Z-Source Bidirectional DC–DC Converter

Based on the previous analysis of the switched-quasi-Z-source bidirectional DC–DC converter, it is possible to achieve a soft-switching operation within the rated operating range of the converter by appropriately designing the values of the inductors. This soft-switching operation refers to the ability of the converter to facilitate ZVS ON for all power switches and quasi-ZCS OFF for synchronous rectification switches in both Boost and Buck modes. The specific conditions for achieving soft-switching in the switched-quasi-Z-source bidirectional DC–DC converter are described by

$$\frac{U_{\text{low}} D_{\text{boost}} T_s}{2L \cdot \left[P\left(\frac{1}{U_{\text{high}}} + \frac{1}{U_{\text{low}}}\right) + \delta\right]} = 1 \tag{6.23}$$

It is revealed in (6.23) that the soft-switching operation of the converter is influenced by various factors, such as the input and output voltage, output power, voltage gain, as well as the values of inductors L_1 and L_2. The interdependence of these parameters highlights the significance of carefully selecting the inductance values to optimize the performance of the converter. Furthermore, it's worth noting that the choice of inductance values is a trade-off between meeting the soft-switching

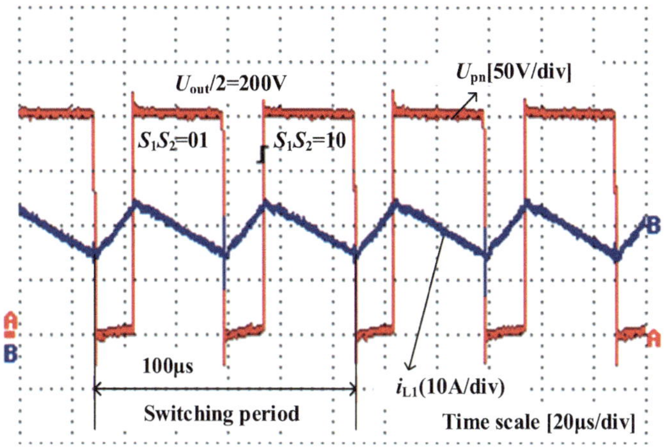

(a)

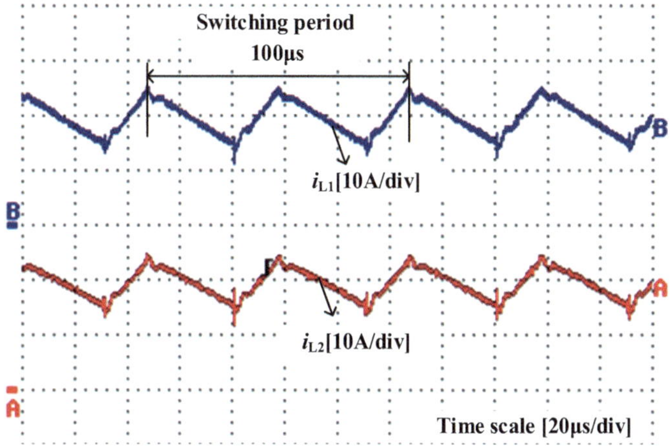

(b)

Fig. 6.24 **a** Output pulse voltage U_{pn} and **b** inductor currents i_{L1} and i_{L2} of the prototype

conditions and managing the current ripple through the inductors. Smaller inductance values allow for a wider range of soft-switching conditions to be satisfied, but they also lead to larger current ripple. Therefore, a careful balance must be struck to determine the optimal inductance values that satisfy the soft-switching requirements while maintaining desirable performance characteristics.

To facilitate the design process, the following section elaborates on the determination of the inductance values for the quasi-Z-source inverter based on the system operating parameters provided in Table 6.5. This detailed analysis will guide the

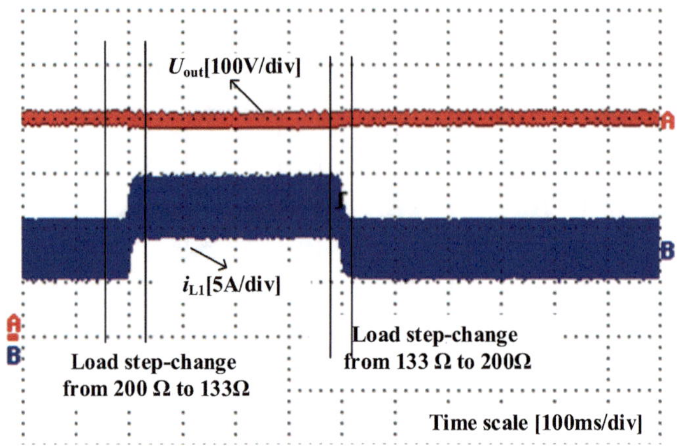

Fig. 6.25 Output voltage U_{out} and inductor current i_{L1} during load transient

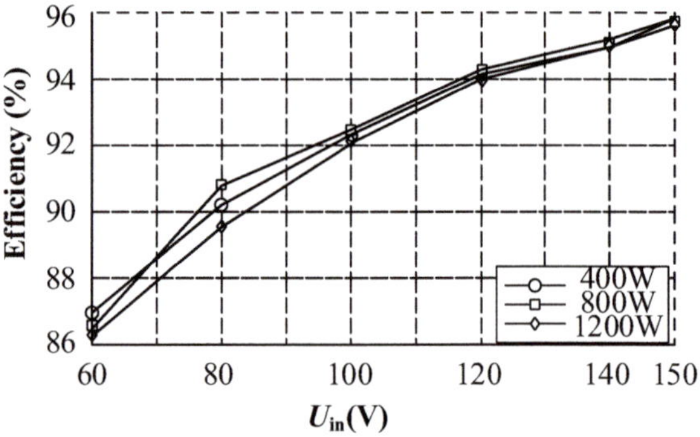

Fig. 6.26 Operation efficiency of the prototype under SR operation

selection of suitable inductance values that meet the desired soft-switching conditions and overall system performance.

Figure 6.28 illustrates the threshold value of the equivalent inductance, denoted as L, in (6.23) for different operational conditions when U_{high} is maintained at 300 V with a switching frequency fs of 36 kHz. The green region in the figure represents the range of operation conditions that enable ZVS operation of main power switches. It can be observed that as the value of L decreases, the ZVS range expands.

To ensure ZVS operation throughout the entire range and avoid excessive current ripple in the inductor, the inductance value is designed based on the minimum value of L within the ZVS region, denoted as L^*. By substituting the boundary operation conditions ($P = 300$ W and $U_{low} = 50$ V as displayed in Table 6.5) into Fig. 6.28, the value of L^* is determined to be 70.86 μH.

In other words, as long as the values of L_1 and L_2 satisfy the condition described by (6.23), the ZVS operation can be achieved for all the operation conditions listed in

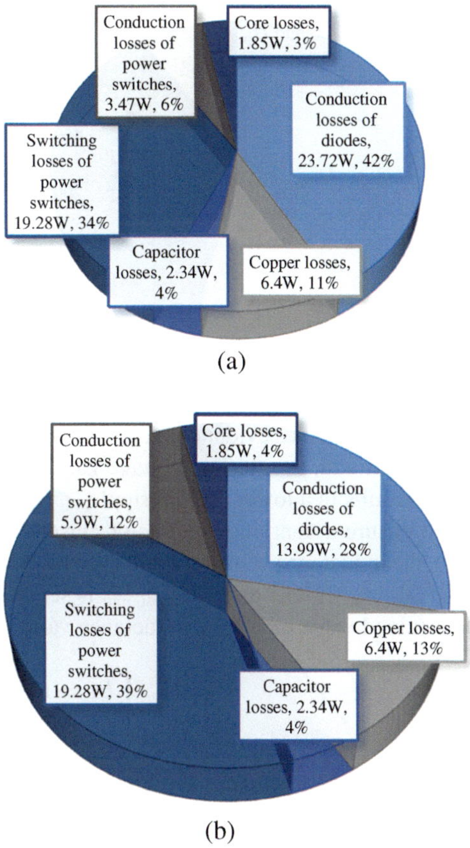

Fig. 6.27 Power loss distribution of the prototype under **a** DR and **b** SR operation with U_{in} at 150 V and P_{out} at 1200 W

Table 6.5 Parameters of operating condition

Parameter	Value
Rated power: P_n	300 W
Low-side voltage: U_{low}	50–150 V
High-side voltage: U_{high}	300 V
Minimum switching frequency f_s	35 kHz

Table 6.5. This design approach ensures efficient and reliable ZVS operation while considering the specific operational parameters and constraints of the switched-quasi-Z-source bidirectional DC–DC converter.

At the boundary condition specified in (6.23), the current ripple θ_1 of inductor L_1 can be calculated by

$$\theta_1 = \frac{\Delta i_{L1}}{I_{L1}} = \frac{2}{1 + {L_1}/{L_2}}(1 + \frac{1+\delta}{M_{boost}}) \tag{6.24}$$

Fig. 6.28 Threshold of L when $f_s = 36$ kHz, and $U_{high} = 300$ V

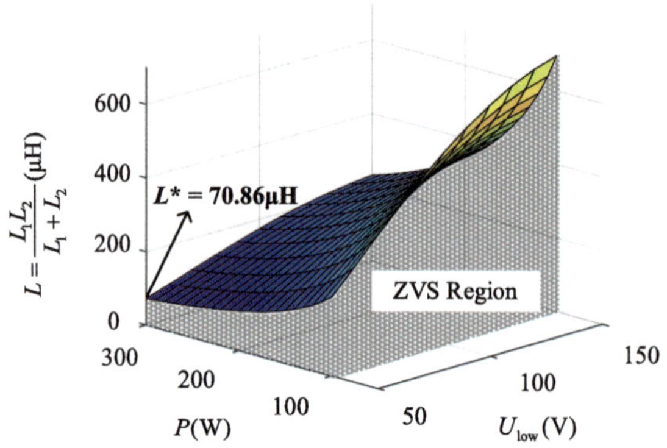

Thus, it becomes evident that, as the inductance values are fixed, θ_1 has a negative correlation with the step-up voltage gain M_{boost}. This indicates that a higher Boost factor results in a lower current ripple.

To ensure a continuous inductor current throughout the entire range of voltage gain, the inductance values are designed based on the minimum voltage gain, denoted as M_{boost_min}. In this case, M_{boost_min} is set to 2. Taking into account the previously determined value of L^*, the inductance values can be obtained as follows:

$$\begin{cases} L_1 = 147\ \mu\text{H} \\ L_2 = 95\ \mu H \end{cases} \tag{6.25}$$

By selecting these specific inductance values, the converter can maintain a continuous inductor current while achieving soft operation within the desired full voltage gain range.

6.5.2 Experimental Results of the Switched-Quasi-Z-Source Bidirectional DC–DC Converter in Energy Storage System

Within renewable energy generation systems, the switched-quasi-Z-source bidirectional DC–DC converter assumes the crucial role of a charge–discharge controller for energy storage devices. Its primary function is to regulate the charging and discharging operations of various devices such as batteries and super-capacitors. This control capability is achieved through the implementation of bidirectional operation and precise voltage-current control.

In order to validate the accuracy and effectiveness of the variable switching-frequency control strategy of switched-quasi-Z-source bidirectional DC–DC converter, an experimental prototype has been constructed using the parameters listed in Table 6.6. The MOSFETs in the prototype are driven by Si8238 chips, which

Table 6.6 Experimental parameters of the switched-quasi-Z-source bidirectional DC–DC converter

Parameter	Value
Rated power: P_n	300 W
Low-side voltage: U_{low}	50–150 V
High-side voltage: U_{high}	300 V
Power MOSFETs: Q_1, Q_2, Q_3	IRFP4868PBF
Low-side filtering capacitor: C_{low}	50 μF
High-side filtering capacitor: C_{high}	30 μF
Switched capacitors: C_1, C_2	30 μF
Inductor L_1	147 μH
Inductor L_2	95 μH

provide the necessary control signals. The inductor core employed in the prototype is of type PC95-PQ50/50, chosen based on its suitability for the application.

Experimental results in step-up mode with the proposed soft-switching operation method are presented in Fig. 6.29. The experiments were conducted at a high-side voltage U_{high} of 300 V and an output power P of 300 W. The inductor current waveforms for two different low-side voltages, $U_{low} = 50$ V and $U_{low} = 150$ V, are shown in Fig. 6.29a, b, respectively. It can be observed that the inductor current i_{L1} remains in continuous conduction mode under both operating conditions.

Figure 6.29c–h present the voltage and current waveforms of the power switches utilizing the variable switching-frequency control method at $U_{low} = 50$ V and $U_{low} = 150$ V, respectively. In these waveforms, the switching frequency f_s is adjusted to 36 kHz and 82 kHz, correspondingly. From the experimental results, it is evident that all power switches achieve ZVS ON. Moreover, for power switches Q_2 and Q_3, quasi-ZCS with a small positive turn-off current is achieved.

Consequently, in step-up mode, by employing the variable switching frequency control strategy, the converter operates close to the critical soft-switching condition. This allows for effective reduction of switching losses per switching cycle. The experimental results highlight the effectiveness of the proposed soft-switching operation method in achieving efficient and low-loss operation of the converter in step-up mode.

Figure 6.30 illustrates the experimental results of the converter operating in step-down mode using the soft-switching operation method. The inductor current waveforms for two different low-side voltages, $U_{low} = 50$ V and $U_{low} = 150$ V, are displayed in Fig. 6.30a, b, respectively. It is evident from the waveforms that the inductor current i_{L1} remains in continuous conduction mode as U_{low} varies.

By employing the variable switching-frequency control method, the voltage and current waveforms of power switches Q_1-Q_3 are observed at $U_{low} = 50$ V and $U_{low} = 150$ V, as shown in Fig. 6.30c–h, respectively. The switching frequency f_s is adjusted to 36 kHz and 77 kHz, respectively, for these operating conditions. In both cases, all power switches achieve ZVS ON operations.

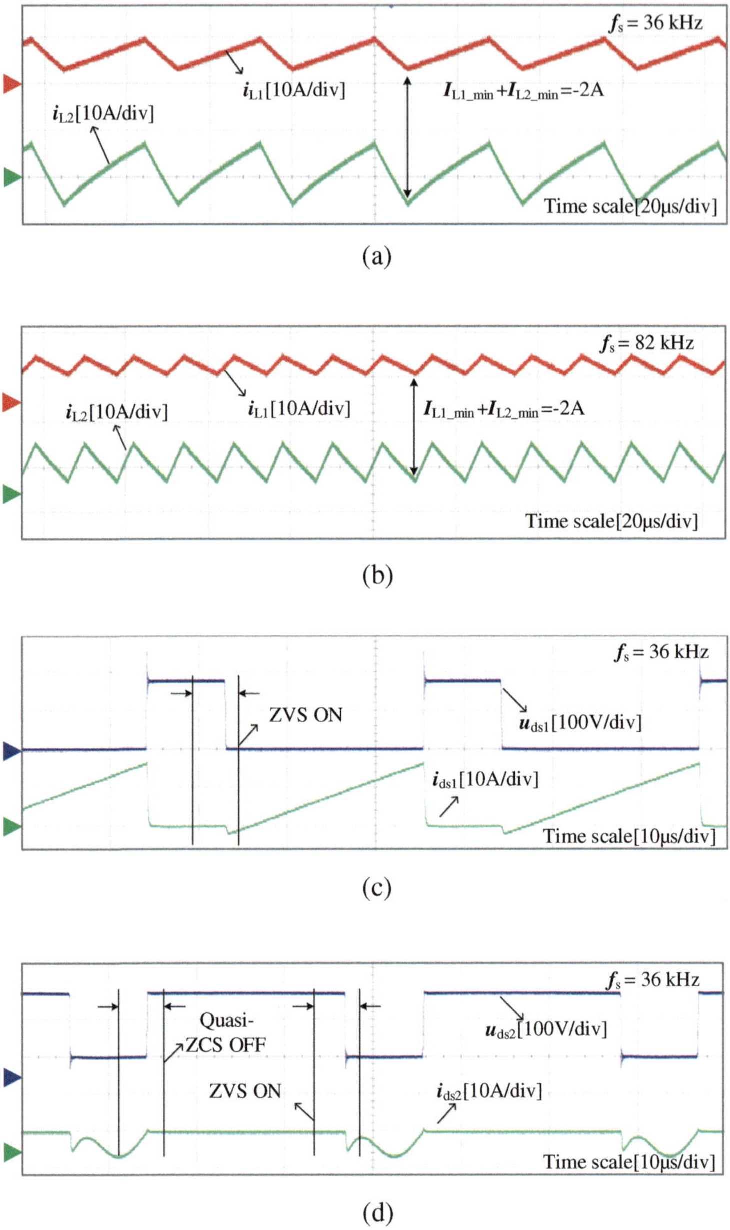

Fig. 6.29 Experimental results of in step-up mode at $U_{\text{high}} = 300$ V, $P = 300$ W. **a** Inductor currents at $U_{\text{low}} = 50$ V. **b** Inductor current at $U_{\text{low}} = 150$ V. **c** Voltage and current stresses of Q_1 at $U_{\text{low}} = 50$ V. **d** Voltage and current stresses of Q_2 at $U_{\text{low}} = 50$ V. **e** Voltage and current stresses of Q_3 at $U_{\text{low}} = 50$ V. **f** Voltage and current stresses of Q_1 at $U_{\text{low}} = 150$ V. **g** Voltage and current stresses of Q_2 at $U_{\text{low}} = 150$ V. **h** Voltage and current stresses of Q_3 at $U_{\text{low}} = 150$ V

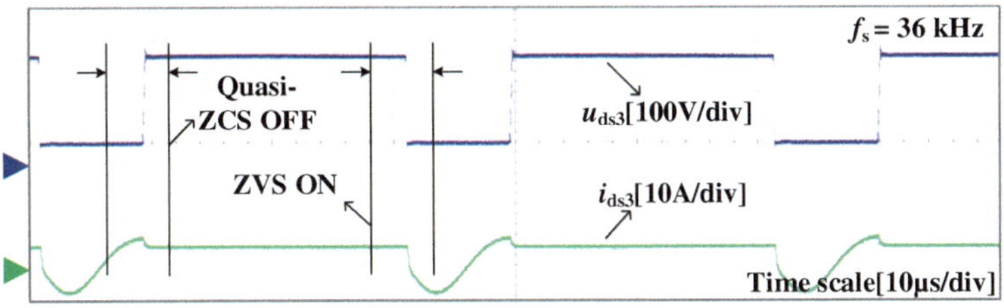

(e)

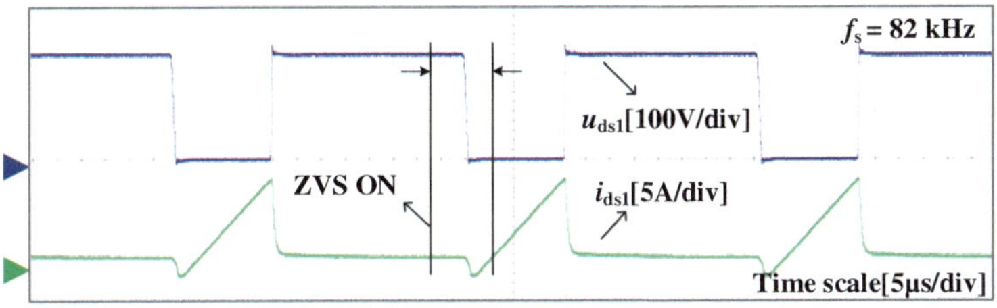

(f)

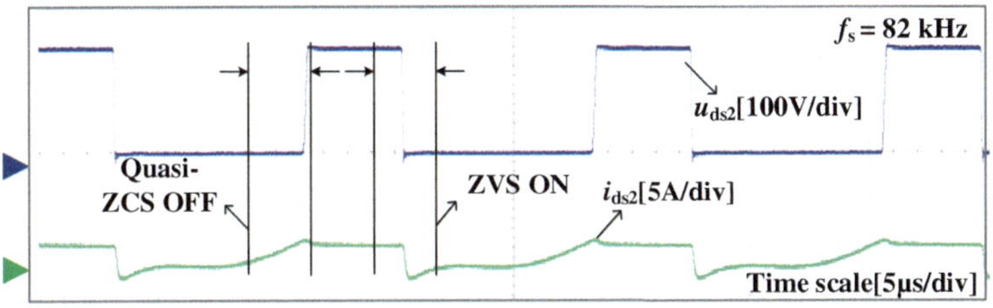

(g)

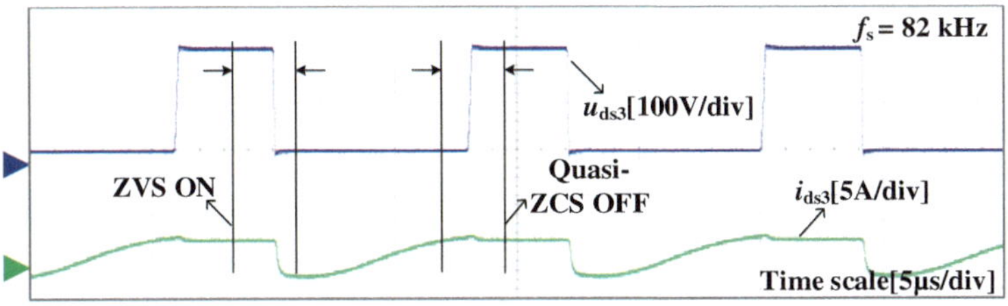

(h)

Fig. 6.29 (continued)

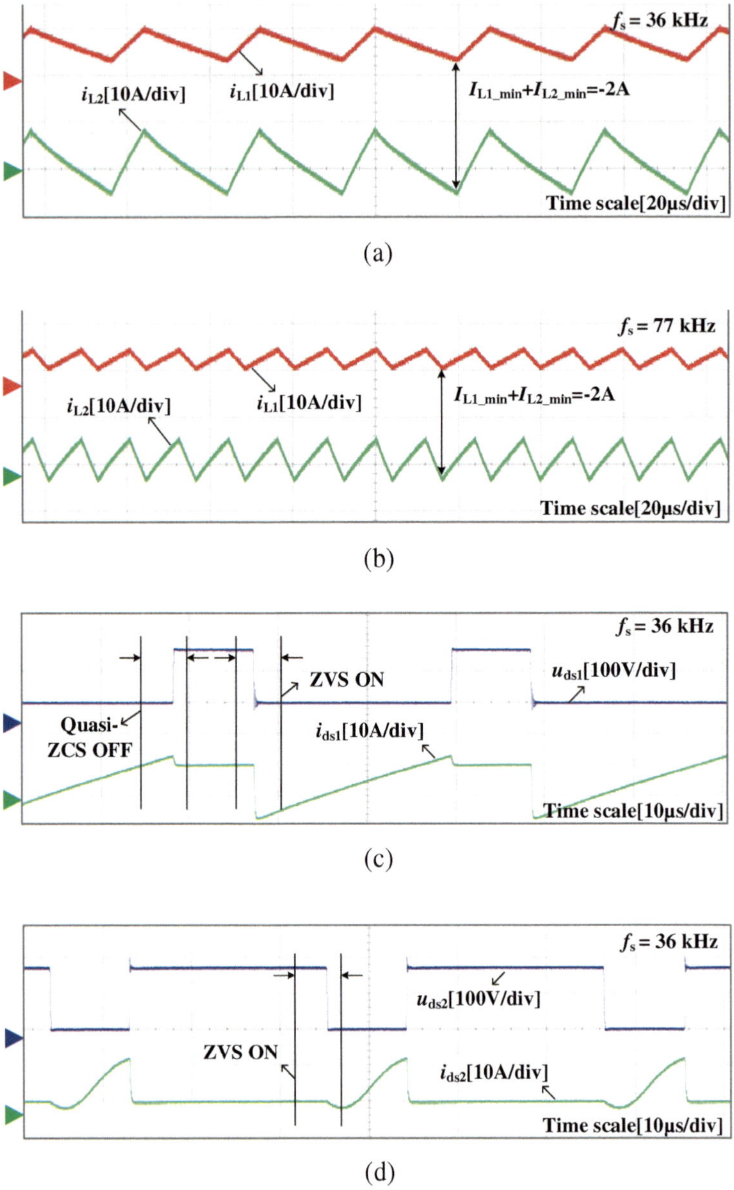

Fig. 6.30 Experimental results of in step-down mode at $U_{\text{high}} = 300$ V, $P = 300$ W. **a** Inductor currents at $U_{\text{low}} = 50$ V. **b** Inductor current at $U_{\text{low}} = 150$ V. **c** Voltage and current stresses of Q_1 at $U_{\text{low}} = 50$ V. **d** Voltage and current stresses of Q_2 at $U_{\text{low}} = 50$ V. **e** Voltage and current stresses of Q_3 at $U_{\text{low}} = 50$ V. **f** Voltage and current stresses of Q_1 at $U_{\text{low}} = 150$ V. **g** Voltage and current stresses of Q_2 at $U_{\text{low}} = 150$ V. **h** Voltage and current stresses of Q_3 at $U_{\text{low}} = 150$ V

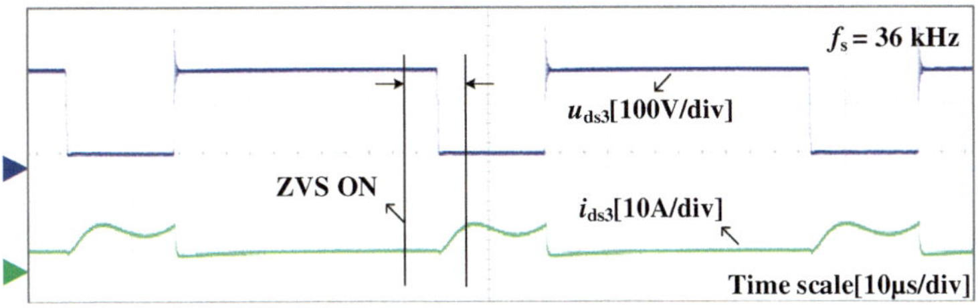

(e)

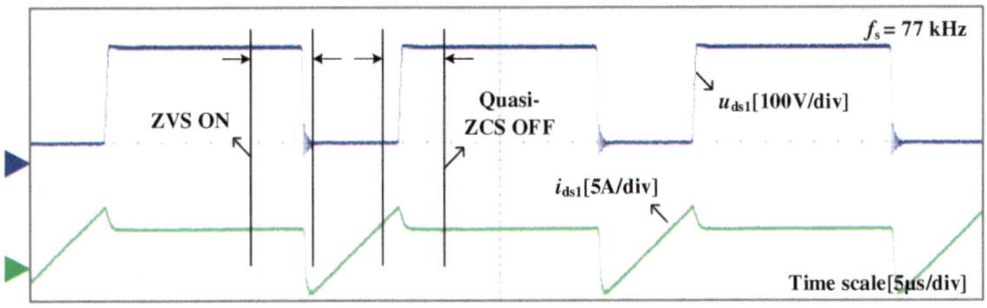

(f)

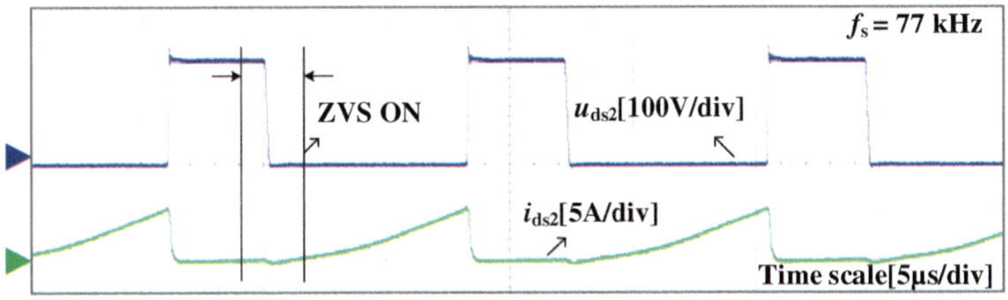

(g)

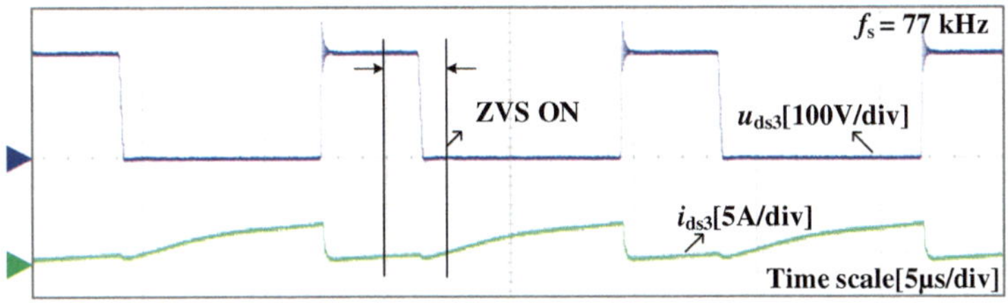

(h)

Fig. 6.30 (continued)

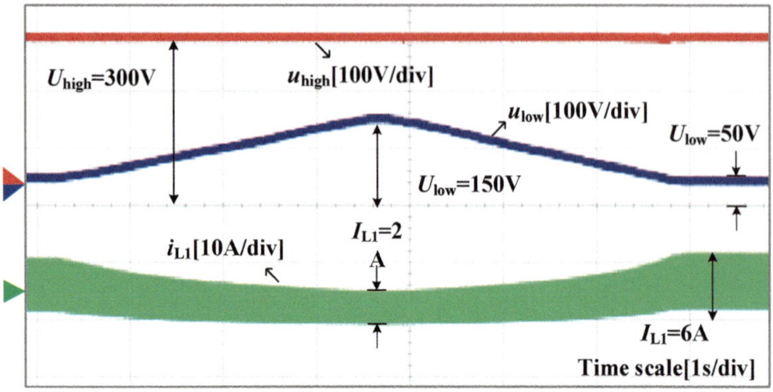

Fig. 6.31 High-side voltage U_{low} with variable low-side voltage U_{low} between 50 and 150 V

Furthermore, for power switch Q_1, it operates with quasi-ZCS during turn-off. The turn-off current is small and positive. This validation indicates that the soft-switching condition in step-down mode aligns with that in step-up mode under the same operating parameter conditions.

The experimental results presented in Figs. 6.29 and 6.30 demonstrate the effectiveness and consistency of the soft-switching operation method for both step-up and step-down modes. The converter achieves efficient and low-loss operation while maintaining soft-switching characteristics across different operating conditions.

In Fig. 6.31, the behavior of U_{low} varying from 50 to 150 V and then returning to 50 V at a rate of 33 V/s is presented. It can be observed that U_{high}, the high-side voltage, remains relatively stable at 300 V throughout the process. This indicates that the converter is capable of operating within a wide voltage gain range of 2 to 6. The ability to maintain a stable U_{high} demonstrates the converter's flexibility and suitability for different operating conditions.

To evaluate the dynamic response during load transients, Figs. 6.32 and 6.33 illustrate the performance of the converter in step-up and step-down modes, respectively, when transitioning between full load and 10% of full load. Following a drop or rise in the inductor current, the output voltage quickly recovers to 300 V. This rapid response indicates the converter's capability to maintain stable output voltage during load variations, ensuring reliable and consistent power delivery.

Additionally, the current ripple of i_{L1} provides insights into the adjustment of the switching frequency f_{s} based on the output power level. It can be observed that fs is increased when the output power is at 30 W compared to when it is at 300 W. This adaptive frequency adjustment enhances the dynamic performance and robustness of the control strategy, allowing the converter to effectively respond to changes in load conditions.

The experimental results presented in Figs. 6.31, 6.32 and 6.33 validate the dynamic performance, stability, and robustness of the system. The converter demonstrates its ability to handle load transients, maintaining stable output voltage and ensuring efficient operation across a wide voltage gain range. These findings provide

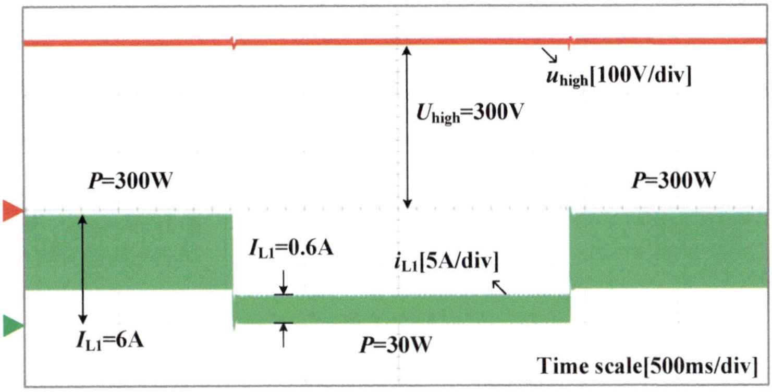

Fig. 6.32 High-side voltage U_{low} and inductor current i_{L1} during load transient in step-up mode

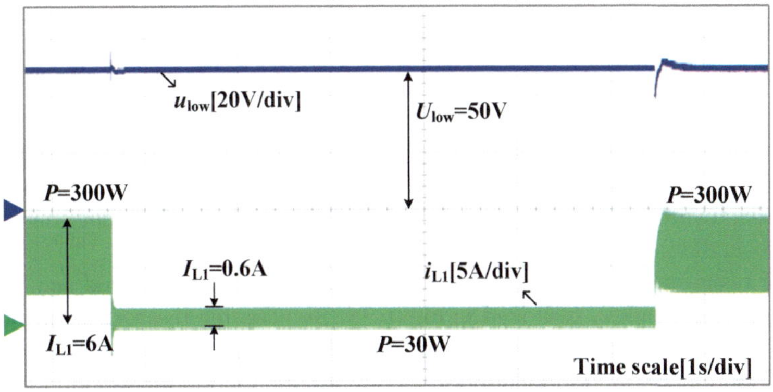

Fig. 6.33 Low-side voltage U_{low} and inductor current i_{L1} during load transient in step-down mode

further evidence of the effectiveness and reliability of the soft operation control strategy for the converter's practical implementation.

Figure 6.34 presents the experimental results of the prototype during bidirectional power flow. The inductor current of L_1 undergoes a reversal, transitioning from 6 A to – 6 A as the operation mode switches. The low voltage side of the converter is connected to a DC voltage source, which can either supply power or absorb power based on the mode of operation. Throughout the switching processes, the high voltage side voltage is controlled at a constant value of 300 V, ensuring stable operation. The switching frequency f_s is automatically adjusted, adapting to the specific operating conditions of the system.

In a multisource system, the converter serves as an energy transfer interface between the super-capacitors and the DC bus. During step-up mode, the converter transfers energy from the super-capacitors to the DC bus, effectively increasing the voltage level. Conversely, during step-down mode, the converter absorbs energy from the DC bus, reducing the voltage level. This bidirectional power flow capability

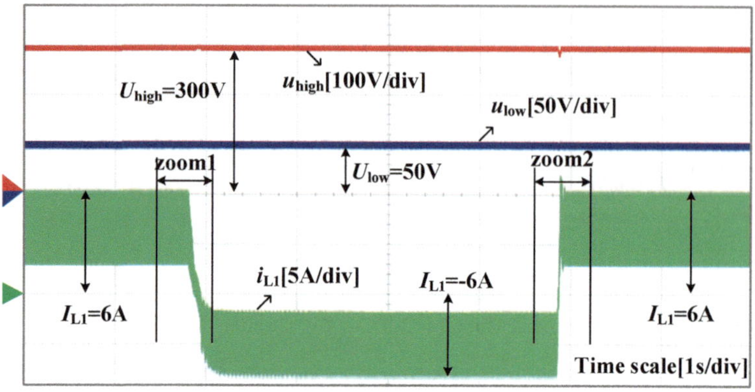

Fig. 6.34 High-side and low-side voltages and inductor current i_{L1} during bidirectional power flow processes

makes the converter well-suited for applications where it is necessary to transfer power between super-capacitors and a DC bus. The experimental results presented in Fig. 6.34 demonstrate the converter's ability to effectively handle bidirectional power flow and serve as a reliable power interface between the super-capacitors and the DC bus. This functionality expands the range of applications where the converter can be deployed, offering enhanced energy management capabilities and contributing to the overall efficiency and stability of the system.

The operating efficiency of the prototype, considering both fixed and variable switching frequencies, is plotted in Fig. 6.35. By adopting the variable switching-frequency control strategy and reducing the inductances, the converter achieves ZVS operation, resulting in significantly reduced switching losses. Consequently, the operating efficiency surpasses that of the reference [13] across different voltage gain values.

When the converter operates at a fixed frequency of 36 kHz, the maximum operating efficiencies in step-up and step-down modes are 96.76% and 97.83%, respectively. It's worth noting that due to the smaller inductance values of L_1 and L_2 compared to those in [13], the operation efficiency slightly decreases as the low-side voltage increases, primarily due to the backflow currents in the power switches. In step-up mode, the minimum operating efficiency reaches 95.7%, while in step-down mode, it reaches 95.94%.

On the other hand, when the variable switching-frequency control strategy is employed, the switching frequency increases as the voltage gain decreases. However, the switching losses per switching cycle are reduced. As a result, the efficiency improves compared to that at a lower fixed frequency. The maximum operating efficiencies achieved with variable switching-frequency control are 97.78% in step-up mode and 98.28% in step-down mode.

The experimental results in Fig. 6.35 demonstrate that the variable switching-frequency control strategy, combined with reduced inductance values, contributes to higher operating efficiencies in both step-up and step-down modes. These findings

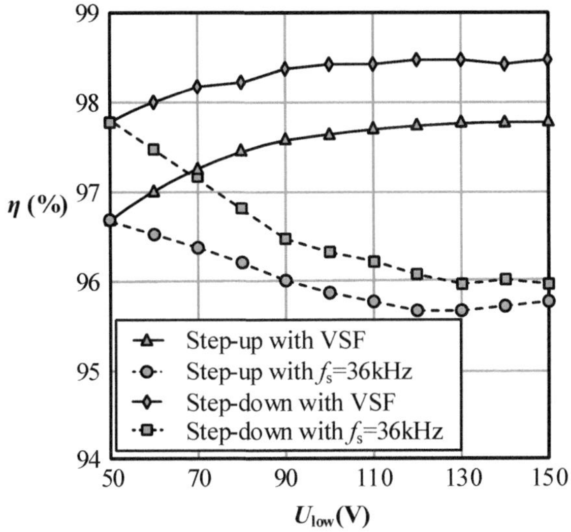

Fig. 6.35 Operation efficiency of the prototype with variable and fixed switching frequency

emphasize the effectiveness of the proposed approach in achieving efficient and low-loss operation of the converter across a range of voltage gain values.

The decrease in efficiency can be attributed to several factors. Firstly, as the voltage gain and output power decrease, the switching losses per switching cycle are reduced due to the variable switching-frequency control strategy. However, the overall switching losses increase because of the dramatic increase in switching frequency. This increase in total switching losses offsets the benefits of reduced losses per cycle. Additionally, the equivalent series resistance of the inductor windings and core losses also play a role in reducing efficiency. As the voltage gain and output power decline, these losses increase, negatively impacting the overall efficiency of the converter.

It is important to note that the efficiency reduction at lighter loads and higher voltage gains is a common characteristic in power converters. These operating conditions pose challenges due to the increased losses and reduced power transfer efficiency. However, despite the relatively lower efficiencies in these extreme operating regions, the converter still maintains high efficiency levels throughout a wide range of voltage gain and output power values. Thereby, the experimental results emphasize the trade-off between efficiency and operating conditions in the converter, and highlight the importance of optimizing the system design and control strategies to achieve a balance between efficiency and performance, considering factors such as switching losses, ESR, and core losses.

Figure 6.36 displays the power losses distribution, categorized into different components: switching losses, core and copper losses, conduction losses, and capacitor losses. In step-up mode, the switching losses primarily arise from the turn-off loss in Q_1, while in step-down mode, they stem from the turn-off losses in Q_2 and Q_3.

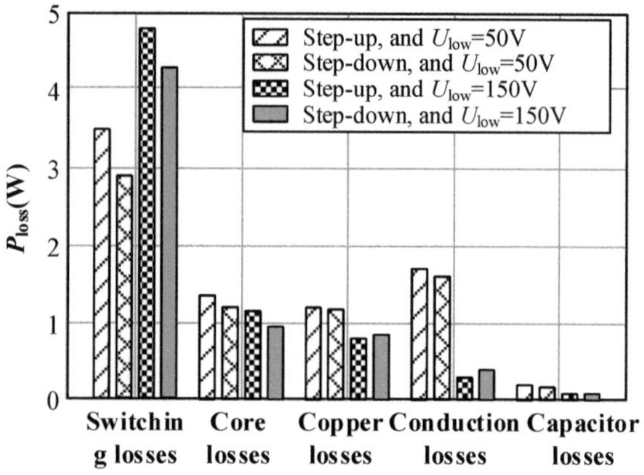

Fig. 6.36 Calculated power losses distribution of the prototype with variable switching frequency

The figure reveals that switching losses account for the largest proportion of the total losses. With the implementation of the variable switching-frequency control strategy, an increase in U_{low} from 50 to 150 V leads to a rise in switching frequency. Consequently, although the switching losses per cycle at $U_{\text{low}} = 150$ V are lower, the total switching losses are higher compared to those at $U_{\text{low}} = 50$ V in both step-up and step-down modes.

Furthermore, the total inductance values are relatively small due to the utilization of the soft-switching operation scheme. As a result, the core and copper losses, which are associated with the inductor, are relatively low. It is evident that optimizing switching losses is crucial for enhancing the overall efficiency of the converter. The variable switching-frequency control strategy effectively manages switching losses, leading to improved performance. Additionally, the reduced total inductance values contribute to lower core and copper losses, further enhancing the converter's efficiency.

6.6 Example of the Input-Parallel Output-Series Boost DC–DC Converter

6.6.1 Parameter Design of the Input-Parallel Output-Series Boost DC–DC Converter

For the input-parallel output-series Boost DC–DC converter with an output voltage of $U_{\text{out}} = 400$ V, the voltage stress on each power semiconductor can be determined as 200 V based on the previous analysis on steady-state characteristic.

When the voltage gain is 8, where the duty cycle is 0.75, and the load resistance is 100 Ω, the current stress, which represents the average currents in the ON state, for the power semiconductors can be calculated as follows:

$$\begin{cases} I_{\mathrm{Q1}} = 16\,\mathrm{A} \\ I_{\mathrm{Q2}} = 21.3\,\mathrm{A} \\ I_{\mathrm{D1}} = I_{\mathrm{D2}} = 16\,\mathrm{A} \\ I_{\mathrm{D3}} = 5.3\,\mathrm{A} \end{cases} \tag{6.26}$$

With these referenced voltage and current stresses, appropriate power switches and diodes can be selected for the converter. It is important to choose components that can handle the specified voltage and current levels to ensure reliable and efficient operation of the converter.

Furthermore, it is important to accurately calculate the values of inductances and capacitance to ensure proper operation and performance of the converter. Based on (6.26), the values of the inductances L_1 and L_2 can be calculated as follows:

$$\begin{cases} L_1 = \dfrac{d \times U_{\mathrm{in}} \times T_{\mathrm{s}}}{\Delta i_{\mathrm{L1}}} \\ L_2 = \dfrac{d \times U_{\mathrm{in}} \times T_{\mathrm{s}}}{\Delta i_{\mathrm{L2}}} \end{cases} \tag{6.27}$$

Additionally, considering the switching state analysis, when the power switch Q_2 is turned off, capacitor C_1 is charging from inductor L_2 and the DC source. The capacitance of C_1 can be determined as (6.28), where ΔU_{C1} represents the capacitor voltage fluctuation.

$$C_1 = \frac{(1-d)I_{L2}T_s}{\Delta U_{\mathrm{C1}}} \tag{6.28}$$

When power switch Q_1 is turned on, capacitor C_2 is discharging. Similarly, when power switch Q_2 is turned off, capacitor C_3 is discharging. As a result, the capacitances of C_2 and C_3 can be determined by (6.29), where ΔU_{C2} and ΔU_{C3} represent the capacitor voltage fluctuations of C_1 and C_2, respectively.

$$\begin{cases} C_2 = \dfrac{d \times I_{\mathrm{o}} \times T_{\mathrm{s}}}{\Delta U_{\mathrm{C2}}} \\ C_3 = \dfrac{(1-d) \times I_o \times T_s}{\Delta U_{\mathrm{C3}}} \end{cases} \tag{6.29}$$

By combining Eqs. (6.26)–(6.29), the values of inductances L_1 and L_2, as well as the capacitances of C_1, C_2, and C_3, can be designed in this research paper. These design parameters play a crucial role in achieving desired performance and functionality of the circuit. Careful consideration and calculation of these values ensure proper operation and optimize the overall system performance.

6.6.2 Experimental Results of the Input-Parallel Output-Series Boost DC–DC Converter in Photovoltaic System

To verify the feasibility and effectiveness of the input-parallel output-series Boost DC–DC converter, an experimental setup for photovoltaic system was constructed. Instead of using photovoltaic panel, an adjustable DC source with a voltage range of 50 V to 120 V was employed. The voltage loop of the converter was controlled by a TMS320F28335 DSP, providing precise control over the output voltage. The power circuit of the converter utilizes IXTK102N30P MOSFETs, which have a rated voltage of 300 V and a rated current of 102 A. These MOSFETs were chosen to handle the desired output voltage of 400 V. Additionally, DPG60C300HB Schottky Barrier Diodes were used in the power circuit.

The converter was operated at a switching frequency of 20 kHz, providing the necessary switching action for the power conversion process. The inductors L_1 and L_2 were designed with values of 227 μH and 225 μH, respectively. These inductors play a crucial role in energy storage and current regulation within the converter. The capacitances C_1, C_2, and C_3 were set at 470 μF each, contributing to energy storage and voltage stability in the system. The reference output voltage of the converter was set to 400 V, and the load resistance connected to the output was maintained at 100 Ω. This load resistance determines the current flowing through the circuit and allows for testing under specific operating conditions.

By implementing this experimental setup with the specified components and parameters, the performance of the prototype could be evaluated, ensuring its feasibility and effectiveness in achieving the desired voltage conversion and power regulation.

Figure 6.37 illustrates the voltage distribution across the power semiconductors in the converter when the input voltage U_{in} is 50 V. In Fig. 6.37a, the voltages of power switches Q_1 and Q_2 are displayed. It can be observed that both Q_1 and Q_2 operate with a voltage of 200 V, which is half of the output voltage of the converter. Figure 6.37b shows the voltage across diodes D_1 and D_2. Similar to the power switches, the voltage across D_1 and D_2 is also 200 V, as indicated in the figure. This balanced voltage distribution ensures proper operation and voltage stress management of the power semiconductors in the converter. Furthermore, Fig. 6.37c presents the voltage across diodes D_2 and D_3. Similar to the previous observations, the voltage across D_2 and D_3 is also maintained at 200 V, which corresponds to half of the output voltage of 400 V.

In Fig. 6.38, the voltages of capacitors C_2 and C_3 are presented when the input voltage is $U_{in} = 50$ V. It can be observed that both C_2 and C_3 have a voltage of 200 V, which corresponds to half of the output voltage of the converter. The voltage across capacitor C_3 represents the potential difference between the output and input sides of the converter. As shown in the figure, the voltage across C_3 exhibits minimal ripple, indicating effective voltage filtering and a stable output voltage.

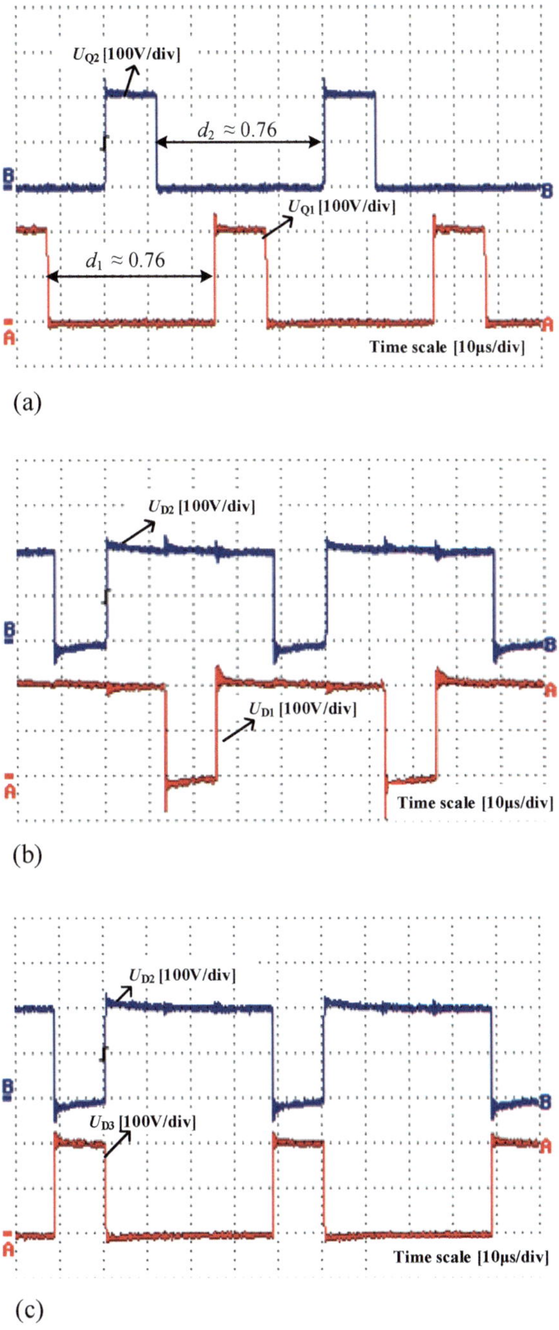

Fig. 6.37 **a** Voltages of power switches Q_1 and Q_2, **b** voltages of diodes D_1 and D_2, and **c** voltages of D_2 and D_3 with U_{in} at 50 V

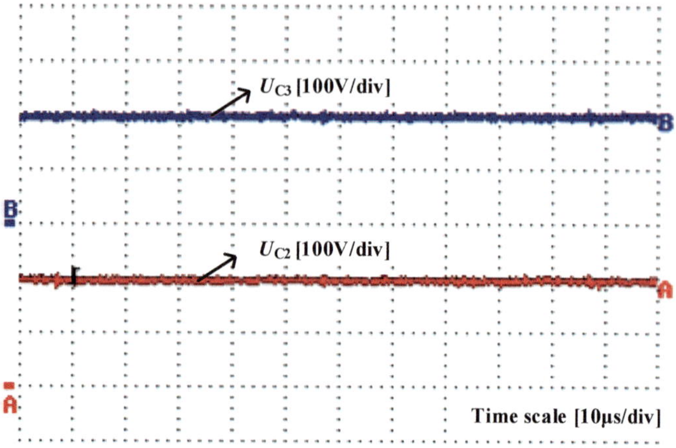

Fig. 6.38 Capacitor voltages of C_2 and C_3 with U_{in} at 50 V

Figure 6.39 displays the input current i_{in}, as well as the inductor currents i_{L1} and i_{L2}, when the input voltage is $U_{in} = 50$ V. Analyzing the inductor currents i_{L1} and i_{L2} in Fig. 6.39a, it can be observed that i_{L1} exhibits a ripple rate of 53.13%, while i_{L2} exhibits a ripple rate of 56.25%. On the other hand, Fig. 6.39b depicts the input current i_{in} and the inductor current i_{L1}. The ripple rate of the input current is calculated to be 17.65%.

Moreover, it is notable that the current ripple of i_{in} is significantly lower than the current ripples of i_{L1} and i_{L2}, as indicated by the waveforms in Fig. 6.39. Refer to theoretical analysis, the ripple rates of i_{L1} and i_{L2} should be approximately 51.81%, while the ripple rate of i_{in} should be around 17.28%. The experimental results closely resemble these theoretical values, providing further validation of the theoretical models used in the design and analysis of the converter.

The waveform shown in Fig. 6.40 represents the output voltage of a prototype operating in maximum power point tracking (MPPT) mode. In this setup, the input terminal of the photovoltaic generator controller is connected to a photovoltaic simulator, which emulates the DC voltage generated by the photovoltaic cells. The output terminal is connected to an electronic load, and the photovoltaic simulator is activated. By setting the output load as a constant resistance, the output voltage directly indicates the level of output power, and the perturbation and observation algorithm is utilized. Within a mere 0.45 s after starting the system, the operating point of the photovoltaic cells quickly tracks and settles at the maximum power point. This signifies that the prototype is capable of effectively and rapidly adjusting to the optimal operating conditions for maximum power extraction.

Based on these results, it can be concluded that the prototype demonstrates its viability for deployment in photovoltaic systems as a photovoltaic generator controller. By accurately tracking the maximum power point of the photovoltaic cells, it facilitates the efficient utilization and effective management of photovoltaic energy. The voltage control loop effectively maintains the output voltage in close

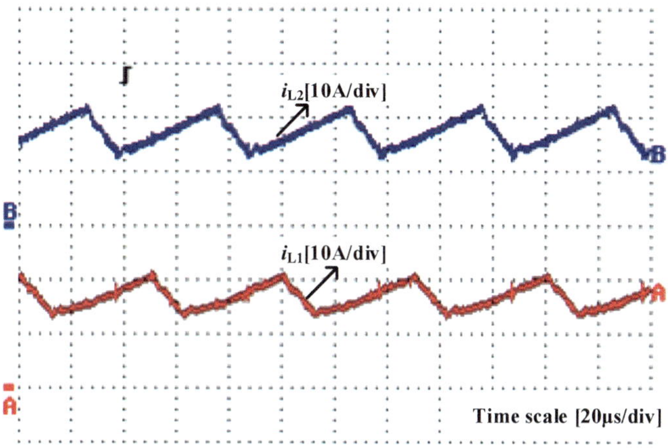

(a)

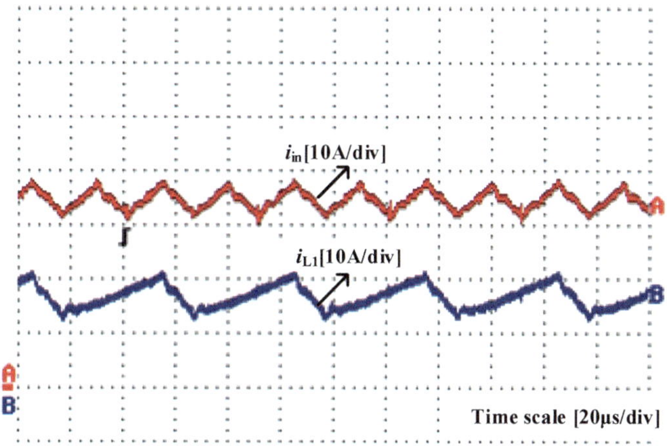

(b)

Fig. 6.39 **a** Inductor currents i_{L1} and i_{L2}. **b** Input current i_{in} and the inductor current i_{L1} with U_{in} at 50 V

proximity to the reference voltage of 400 V. The dynamic response of the system, as depicted in Fig. 6.41, provides insight into the behavior of the output voltage and input current when the input voltage is varied from 120 to 50 V.

Figure 6.41a presents the plotted output and input voltages. As per the graph, when the input voltage, denoted as U_{in}, is gradually adjusted from 120 to 50 V, the output voltage remains remarkably stable around the 400 V mark. This remarkable consistency in the output voltage implies that the converter possesses a wide voltage gain range, exhibiting variability from 3.3 to 8. Therefore, irrespective of the substantial reduction in the input voltage, the converter effectively maintains a

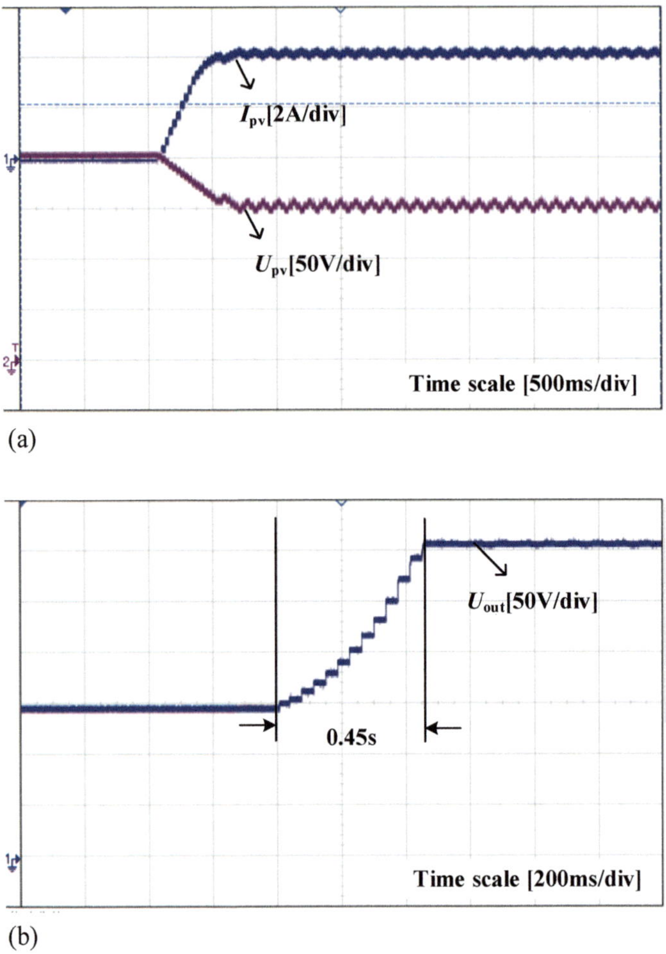

Fig. 6.40 **a** Photovoltaic voltage and current. **b** Output voltage during MPPT operation

steady output voltage level of approximately 400 V. Figure 6.41b showcases the relationship between the input current and voltage. As the input voltage is decreased from 120 to 50 V, the input current exhibits a gradual increment, rising from 13 to 34 A. This observed behavior aligns with expectations, as the converter necessitates drawing more current to compensate for the diminishing input voltage while concurrently upholding a consistent output voltage level. In summary, the findings from the figures demonstrate that the proposed converter exhibits a wide voltage gain range, enabling it to sustain the output voltage in the vicinity of 400 V even when the input voltage undergoes substantial variations.

Figure 6.42 illustrates the ripple rate of the input current within the wide input voltage range of 50 V to 120 V when the output voltage is maintained at 400 V

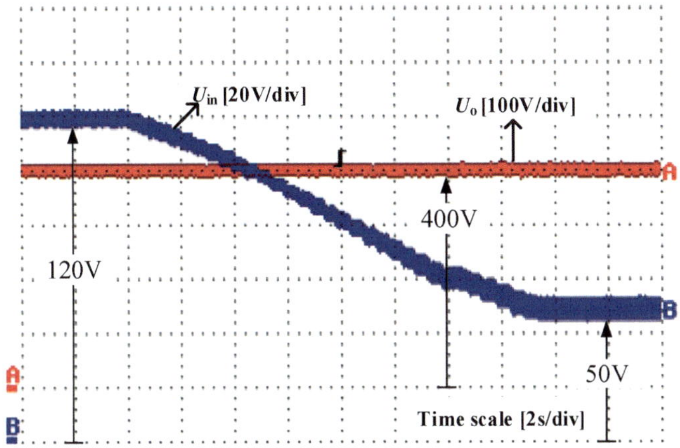

(a) Output voltage U_{out} and input voltage U_{in}

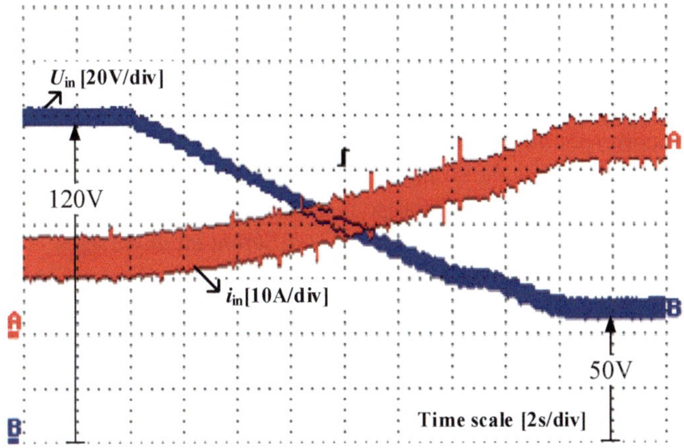

(b) Input current i_{in} and input voltage U_{in}

Fig. 6.41 Output voltage and input current with varied input voltage from 120 to 50 V

and the load resistance is 100 Ω. In the graph, $\Delta i_{in1}/I_{in1}$ represents the experimental ripple rate of the input current, while $\Delta i_{in2}/I_{in2}$ denotes the ripple rate of the input current. Based on the information provided, it is observed that the experimental and theoretical results exhibit similarity. The theoretical calculations align closely with the experimental measurements, validating the accuracy and effectiveness of the theoretical analysis referring to the ripple rate of the input current.

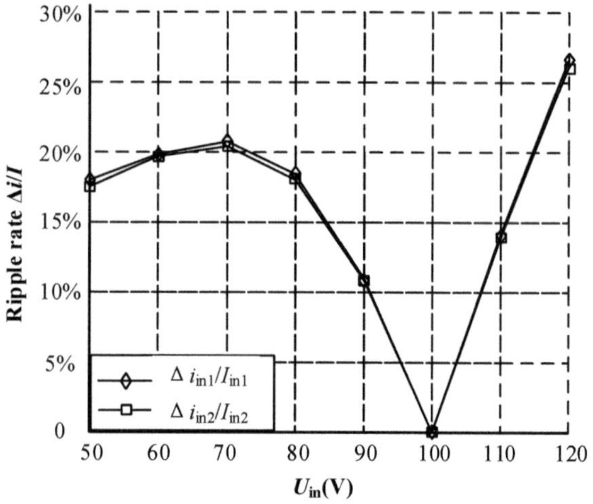

Fig. 6.42 Ripple rate of input current of the prototype

In Fig. 6.43, the efficiency η_e of the prototype is measured using a Power Analyzer (Yokogawa-WT3000) for different input voltages, ranging from 120 to 50 V, while maintaining the output voltage U_{out} at 400 V and the load resistance R at 100 Ω. According to the graph, the maximum efficiency achieved is 96.62% when the voltage gain is 3.3, corresponding to an input voltage of 120 V. Conversely, the minimum efficiency is 94.14% when the voltage gain reaches 8, which occurs at an input voltage of 50 V. The efficiency of the prototype is influenced by the change in voltage gain, as demonstrated by the varying efficiencies at different input voltages. As the voltage gain increases from 3.3 to 8, the input current also increases. This rise in input current leads to a reduction in overall efficiency.

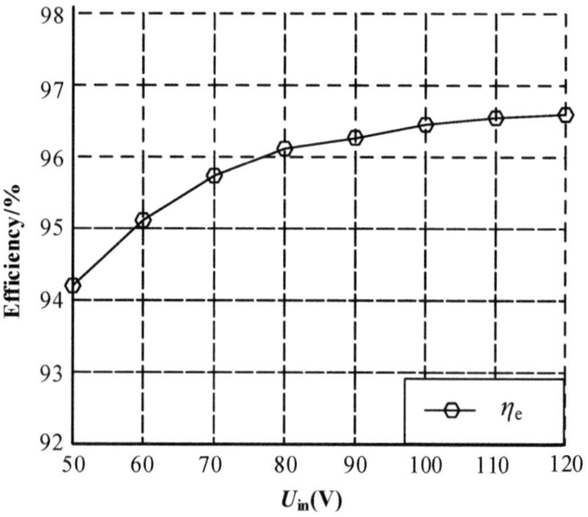

Fig. 6.43 Operation efficiency of the prototype

References

1. E. Du, N. Zhang, B. Hodge, Q. Wang, C. Kang, B. Kroposki, and Q. Xia, The role of concentrating solar power toward high renewable energy penetrated power systems. IEEE Trans. Power Sys. 33(6), 6630–6641 (2018).
2. M. G. Villalva, J. R. Gazoli and E. R. Filho, Comprehensive approach to modeling and simulation of photovoltaic arrays. IEEE Trans. Power Electron. 24(5), 1198–1208 (2009).
3. L. O. Santos, P. C. M. de Carvalho and C. O. C. Filho, Photovoltaic cell operating temperature models: a review of correlations and parameters. IEEE J. Photovolt. 12(1), 179–190 (2022).
4. C. Jin, X. Sheng, and P. Ghosh, Optimized electric vehicle charging with intermittent renewable energy sources. IEEE journal of selected topics in signal processing, 8(6), 1063–1072 (2014).
5. B. Zeng, J. Zhang, X. Yang, J. Wang, J. Dong, and Y. Zhang, Integrated planning for transition to low-carbon distribution system with renewable energy generation and demand response. IEEE Trans. Power Syst. 29(3), 1153–1165 (2014).
6. A. Soroudi, R. Caire, N. Hadjsaid, and M. Ehsan, Probabilistic dynamic multi-objective model for renewable and non-renewable distributed generation planning. IET Gener. Transm. Distrib. 5(11), 1173–1182 (2011).
7. G. Fontes, C. Turpin, and S. Astier, A large-signal and dynamic circuit model of a H_2/O_2 PEM fuel cell: description, parameter identification, and exploitation. IEEE Trans. Ind. Electron. 57(6), 1874–1881 (2010).
8. A. Askarzadeh, and A. Rezazadeh, An innovative global harmony search algorithm for parameter identification of a PEM fuel cell model. IEEE Trans. Ind. Electron. 59(9), 3473–3480 (2012).
9. J. Morales-Morales, I. Cervantes, and U. Cano-Castillo, On the design of robust energy management strategies for FCHEV. IEEE Trans. Veh. Technol. 64(5), 1716–1728 (2015).
10. G. Su, L. Tang, A reduced-part, triple-voltage DC–DC converter for EV/HEV power management. IEEE Trans. Power Electron. 24(10), 2406–2410 (2009).
11. U. R. Prasanna, A. K. Rathore, Dual three-pulse modulation-based high-frequency pulsating DC link two-stage three-phase inverter for electric/hybrid/fuel cell vehicles applications. IEEE J. Emerging Sel. Top. Power Electron. 2(3), 477–486, (2014).
12. J. Jia, G. Wang, Y. T. Cham, Y. Wang, and M. Han, Electrical characteristic study of a hybrid PEMFC and ultracapacitor system. IEEE Trans. Ind. Electron. 57(6), 1945–1953 (2010).
13. C. C. Chan, A. Bouscayrol and K. Chen, Electric, hybrid, and fuel-cell vehicles: Architectures and modeling. IEEE Trans. Veh. Technol. 59(2), 589–598 (2010).

Y. Zhang and S. Gao, *High Efficiency Non-isolated DC-DC Converters with Wide Voltage Gain Range for Renewable Energies*, CPSS Power Electronics Series,
https://doi.org/10.1007/978-981-97-0648-8

14. A. Tavakoli, M. Negnevitsky, D. T. Nguyen and K. M. Muttaqi, Energy exchange between electric vehicle load and wind generating utilities. IEEE Trans. Power Syst. 31(2), 1248–1258 (2016).
15. J. Shen, A. Khaligh, A supervisory energy management control strategy in a battery/ultracapacitor hybrid energy storage system. IEEE Trans. Transp. Electrif. 1(3), 223–231 (2015).
16. Y. Tang, D. Fu, T. Wang, and Z. Xu, Hybrid switched-inductor converters for high step-up conversion. IEEE Trans. Ind. Electron. 62(3), 1480–1491 (2015).
17. Y. Tang, T. Wang, and D. Fu, Multicell switched-inductor/switched-capacitor combined active-network converters. IEEE Trans. Power Electron. 30(4), 2063–2073 (2015).
18. B. Axelrod, Y. Berkovich, and A. Ioinovici, Switched-capacitor/switched-inductor structures for getting transformerless hybrid DC–DC PWM converters. IEEE Trans. Circuits Syst. I, Reg. Papers, 55(2) 687–697 (2008).
19. J. Liu, J. Hu, and L. Xu, Dynamic modeling and analysis of Z source converter—derivation of AC small signal model and design-oriented analysis. IEEE Trans. Power Electron. 22(5), 1786–1797 (2007).
20. M. M. Haji-Esmaeili, E. Babaei, and M. Sabahi, High step-up quasi-Z source DC–DC converter. IEEE Trans. Power Electron. 33(12), 10563–10572 (2018).
21. J. Fang, F. Blaabjerg, S. Liu, and S. M. Goetz, A review of multilevel converters with parallel connectivity. IEEE Trans. Power Electron. 36(11), 12468–12489 (2021).
22. Z. Guo, Y. Zhu, and D. Sha, Zero-voltage-switching asymmetrical PWM full-bridge DC–DC converter with reduced circulating current. IEEE Trans. Ind. Electron. 68(5), 3840–3944 (2021).
23. S. Dusmez, A. Hasanzadeh, and A. Khaligh, Comparative analysis of bidirectional three-level DC–DC converter for automotive applications. IEEE Trans. Ind. Electron. 62(5), 3305–3316 (2015).
24. A. Alkhaldi, A. Elkhateb, and D. Laverty, Voltage lifting techniques for non-isolated DC/DC converters. Electronics, 12(3) (2023).
25. F. Yi, F. Wang, Review of voltage-bucking/boosting techniques, topologies, and applications," Energies, 16(2) (2023).
26. Y. Babazadeh, M. Sabahi, E. Babaei, and S. Kai, A new continuous input current nonisolated bidirectional interleaved Buck-Boost DC–DC converter. Int. Trans. Electr. Energy Syst. 2022 (2022).
27. K. Wu, C. W. Silva, and W. G. Dunford, Stability analysis of isolated bidirectional dual active full-bridge DC–DC converter with triple phase-shift control. IEEE Trans. Power Electron. 27(4) (2012).
28. N. Denniston, A. M. Massoud, S. Ahmed, and P. N. Enjeti, Multiple-module high-gain high-voltage DC–DC transformers for offshore wind energy systems. IEEE Trans. Ind. Electron. 58(5), 1877–1886 (2011).
29. A. Rodriguez, A. Vazquez, D. G. Lamar, M. M. Hernando, and J. Sebastian, Different purpose design strategies and techniques to improve the performance of a dual active bridge with phase-shift control. IEEE Trans. Power Electron. 30(2), 790–804 (2015).
30. A. Ajami, H. Ardi, and A. Farakhor, A novel high step-up DC/DC converter based on integrating coupled inductor and switched-capacitor techniques for renewable energy applications. IEEE Trans. Power Electron. 30(8), 4255–4263 (2015).
31. R. Wai and R. Duan, High-efficiency bidirectional converter for power sources with great voltage diversity. IEEE Trans. Power Electron. 22(5), 1986–1996 (2007).
32. C. Lin, L. Yang, and G. Wu, Study of a non-isolated bidirectional DC–DC converter. IET Power Electron. 6(1), 30–37 (2013).
33. M. Kwon, S. Oh, and S. Choi, High gain soft-switching bidirectional dc-dc converter for eco-friendly vehicles. IEEE Trans. Power Electron. 29(4), 1659–1666 (2014).
34. P. Jose, N. Mohan, A novel ZVS bidirectional Cuk converter for dual voltage systems in automobiles. Proc. 29th Annu. Conf. IEEE Ind. Electron. Soc. 117–122 (2003).

35. I. Kim, S. Paeng, J. Ahn, E. Nho, and J. Ko, New bidirectional ZVS PWM Sepic/Zeta DC–DC converter. Proc. IEEE ISIE Conf. Rec. 555–560 (2007).
36. P. U R, A. K. Rathore, Extended range ZVS active-clamped current-fed full-bridge isolated DC/DC converter for fuel cell applications: analysis, design, and experimental results. IEEE Trans. Ind. Electron. 60(7), 2661–2672 (2013).
37. Y. Wang, L. Xue, C. Wang, P. Wang, and W. Li, Interleaved high-conversion-ratio bidirectional DC–DC converter for distributed energy-storage systems—circuit generation, analysis, and design. IEEE Trans. Power Electron. 31(8), 5547–5561 (2016).
38. S. Tan, S. Kiratipongvoot, S. Bronstein, A. Ioinovici, Y. M. Lai, and C. K. Tse, Adaptive mixed on-time and switching frequency control of a system of interleaved switched-capacitor converters. IEEE Trans. Power Electron. 26(2), 364–380 (2011).
39. B. Wu, K. Smedley, and S. Sigmond, A new 3X interleaved bidirectional switched capacitor converter. Proc. IEEE Appl. Power Electron. Conf. Expo. 1433–1439 (2014).
40. O. Cornea, G. Andreescu, N. Muntean and D. Hulea, Bidirectional power flow control in a DC microgrid through a switched-capacitor cell hybrid DC–DC converter. IEEE Trans. Ind. Electron. 64(4), 3012–3022 (2017).
41. J. Zhang, J. Ge, Analysis of Z-source DC-DC converter in discontinuous current mode. Proc. Power Energy Eng. Conf. 1–4 (2010).
42. Y. Zhang, Y. Gao, J. Li, M. Sumner, P. Wang, and L. Zhou, High ratio bidirectional DC–DC converter with synchronous rectification H-bridge for hybrid energy sources electric vehicles. J. Power Electron. 16(6), 2035–2044 (2016).
43. Y. Zhang, Y. Gao, J. Li, and M. Summer, A wide voltage-gain range asymmetric H-bridge bidirectional DC–DC converter with a common ground for energy storage system. J. Power Electron. 18(2), 343–355 (2018).
44. B. Wang, J. Xu, R. J. Wai, and B. Cao, Adaptive sliding-mode with hysteresis control strategy for simple multimode hybrid energy storage system in electric vehicles. IEEE Trans. Ind. Electron. 64(2),1404–1414 (2017).
45. W. Li, W. Li, X. He, D. Xu, and B. Wu, General derivation law of nonisolated high-step-up interleaved converters with built-in transformer. IEEE Trans. Ind. Electron. 59(3), 1650–1661 (2012).
46. D. S. Gautam, F. Musavi, W. Eberle, and W. G. Dunford, A zero-voltage switching full-bridge DC–DC converter with capacitive output filter for plug-in hybrid electric vehicle battery charging. IEEE Trans. Power Electron. 28(12), 5728–5735 (2013).
47. X. Ruan, B. Li, Q. Chen, S. Tan, and C. K. Tse, Fundamental considerations of three-level DC–DC converters: topologies, analyses, and control. IEEE Trans. Circuits and Systems—I: Regular Papers, 55(11), 3733–3743 (2008).
48. N. D. Benavides, P. L. Chapman, Mass-optimal design methodology for DC–DC converters in low-power portable fuel cell applications. IEEE Trans. Power Electron. 23(3), 1545–1555 (2008).
49. G. Dotelli, R. Ferrero, P. G. Stampino, S. Latorrata, and S. Toscani, PEM fuel cell drying and flooding diagnosis with signals injected by a power converter. IEEE Trans. Instrum. Meas. 64(8), 2064–2071 (2015).
50. S. Dusmez, A. Hasanzadeh, and A. Khaligh, "Comparative analysis of bidirectional three-level DC–DC converter for automotive applications," IEEE Trans. Ind. Electron., vol. 62, no. 5, pp. 1063–1072, May 2015.
51. H. Chen, J. Liao, Design and implementation of sensorless capacitor voltage balancing control for three-level Boosting PFC. IEEE Trans. Power Electron. 29(7), 3808–3817 (2014).
52. L. Shi, B. P. Baddipadiga, M. Ferdowsi, and M. L. Crow, Improving the dynamic response of a flying-capacitor three-level Buck converter. IEEE Trans. Power Electron. 28(5), 2356–2365 (2013).
53. Y. J. A. Alcazar, D. S. Oliveira, Jr., F. L. Tofoli, and R. P. Torrico-Bascopé, DC–DC nonisolated Boost converter based on the three-state switching cell and voltage multiplier cells. IEEE Trans. Ind. Electron. 60(10), 4438–4449 (2013).

54. Y. Tang, D. Fu, T. Wang, and Z. Xu, Hybrid switched-inductor converters for high step-up conversion. IEEE Trans. Ind. Electron. 62(3), 1480–1490 (2015).
55. W. Qian, D. Cao, J. G. Cintrón-Rivera, M. Gebben, D. Wey, and F. Z. Peng, A switched-capacitor DC–DC converter with high voltage gain and reduced component rating and count. IEEE Trans. Ind. Appl. 48(4), 1397–1406 (2012).
56. A. M. Jorge, L. Rodrigo, P. Elvia, and L. G. Jorge, Modelling and control of a DC–DC quadratic boost converter with R2P2. IET Power Electron. 7(1), 11–22 (2014).
57. R. Gules, L. L. Pfitscher, and L. C. Franco, An interleaved Boost DC-DC converter with large conversion ratio, 2003 IEEE Int. Symp. Ind. Electron. 411–416 (2003).
58. Q. Zhao, F. C. Lee, High-efficiency, high step-up DC–DC converters. IEEE Trans. Power Electron. 18(1), 65–73 (2003).
59. G. A. L. Henn, R. N. A. L. Silva, P. P. Praça, L. H. S. C. Barreto, and D. S. Oliveira, "Interleaved-Boost converter with high voltage gain," IEEE Trans. Power Electron., vol. 25, no. 11, pp. 2753–2761 (2010).
60. P. P. Praça, G. A. L. Henn, D. S. Oliveira, L. H. S. C. Barreto, and R. N. A. L. Silva, High voltage gain single stage DC–DC converter based on three-state commutation cell. Proc. 28th Annu. IEEE Appl. Power Electron. Conf. Expo. 2882–2887 (2013).
61. V. P. Galigekere, M. K. Kazimierczuk, Analysis of PWM Z-source DC–DC converter in CCM for steady state. IEEE Trans. Circuits Syst. I, Reg. Papers, 59(4), 854–86 (2012).
62. S. Yang, F. Z. Peng, Q. Lei, R. Inoshita, and Z. Qian, Current-fed quasi-Z-source inverter with voltage Buck–Boost and regeneration capability. IEEE Trans. Ind. Appl. 47(2), 882–892 (2011).
63. O. Ellabban, H. Abu-Rub, and B. Ge, A quasi-Z-source direct matrix converter feeding a vector controlled induction motor drive. IEEE J. Emerging Sel. Top. Power Electron. 3(2), 1063–1072 (2015).
64. K. Patidar, A. C. Umarikar, High step-up pulse-width modulation DC–DC converter based on quasi-Z-source topology. IET Power Electron. 8(4), 477–488 (2015).
65. B. Poorali, A. Torkan, and E. Adib, High step-up Z-source DC–DC converter with coupled inductors and switched capacitor cell. IET Power Electron. 8(8), 1394–1402 (2015).
66. H. Shen, B. Zhang, D. Qiu, and L. Zhou, A common grounded Z-source DC–DC converter with high voltage gain," IEEE Trans. Ind. Electron. 63(5), 2925–2935 (2016).
67. Y. Zhang, X. Sun, Y. Wang, and H. Shao, Transformerless three-level DC–DC Buck converter with a high step-down conversion ratio. J. Power Electron. 13(1), 70–76 (2013).
68. Y. Zhang, J. Sun, Y. Wang, Hybrid Boost three-level DC–DC converter with high voltage gain for photovoltaic generation systems. IEEE Trans. Power Electron. 28(8), 3659–3664 (2013).
69. P. Wang, C. Zhao, Y. Zhang, J. Li, and Y. Gao, A bidirectional three-level DC–DC converter with a wide voltage conversion range for hybrid energy source electric vehicles. J. Power Electron. 17(2), 334–345 (2017).
70. Y. Zhang, J. Shi, L. Zhou, J. Li, M. Summer, P. Wang, and C. Xia, Wide input-voltage range Boost three-level DC–DC converter with quasi-Z source for fuel cell vehicles. IEEE Trans. Power Electron. 32(9), 6728–6738 (2017).
71. K. Jin, M. Yang, X. Ruan, and M. Xu, Three-level bidirectional converter for fuel-cell/battery hybrid power system. IEEE Trans. Ind. Electron. 57(6), 1976–1986 (2010).
72. P. J. Grbovic, P. Delarue, P. Le Moigne, and P. Bartholomeus, A bi-directional three-level DC–DC converter for the ultra-capacitor applications. IEEE Trans. Ind. Electron. 57(10), 3415–3430 (2010).
73. S. Busquets-Monge, S. Alepuz, and J. Bordonau, A novel bidirectional multilevel boost-buck DC–DC converter," IEEE Trans. Power Electron. 26(8), 2172–2183 (2011).
74. C. Li, L. Herrera, J. Jia, L. Fu, A. Isurin, A. Cook, Y. Huang and J. Wang, Design and implementation of a bidirectional isolated Cuk converter for low-voltage and high-current automotive DC source applications. IEEE Trans. Veh. Technol. (63(6), 2567–2577 (2014).
75. H.-Y. Lee, T.-J. Liang, J.-F. Chen, and K.-H. Chen, "Design and implementation of a bidirectional SEPIC-Zeta DC–DC converter, Proc. IEEE Int. Symp. Circuits Syst. 101–104 (2014).

76. Y. Hsieh, J. Chen, L. Yang, C. Wu, and W. Liu, High conversion-ratio bidirectional DC–DC converter with coupled inductor. IEEE Trans. Ind. Electron. 61(1), 210–222 (2014).
77. L. Jiang, X. Zhang, C. Yin, C. Mi, S. Li, and M. Zhang, A novel soft-switching bidirectional DC–DC converter with coupled inductors. IEEE Trans. Ind. Appl. 49(6), 2730–2740 (2013).
78. H. Wu, K. Sun, L. Chen, and L. Zhu, High step-up/step-down soft-switching bidirectional DC–DC converter with coupled-inductor and voltage matching control for energy storage systems. IEEE Trans. Ind. Electron. 63(5), 2892–2903 (2016).
79. H. Shen, B. Zhang, D. Qiu, and L. Zhou, A common grounded Z-source DC–DC converter with high voltage gain. IEEE Trans. Ind. Electron. 63(5), 2925–2935 (2016).
80. Y. Tang, T. Wang, and D. Fu, Multicell switched-inductor/switched-capacitor combined active-network converters. IEEE Trans. Power Electron. 30(4), 2063–2072 (2015).
81. H. Ardi, A. Ajami, F. Kardan, and S. N. Avilagh, Analysis and implementation of a nonisolated bidirectional DC–DC converter with high voltage gain. IEEE Trans. Power Electron. 63(8), 4878–4888 (2016).
82. Y. S. Lee, Y. P. Ko, Switched-capacitor bi-directional converter performance comparison with and without quasi-resonant zero-current switching. IET Power Electron. 3(2), 269–278 (2010).
83. L. S. Yang, T. J. Liang, Analysis and implementation of a novel bidirectional dc–dc converter. IEEE Trans. Ind. Electron. 59(1), 422–434 (2012).
84. M. K. Nguyen, T. D. Duong, and Y. C. Lim, Switched-capacitor-based dual-switch high-boost DC–DC converter. IEEE Trans. Power Electron. 33(5), 4181–4189 (2018).
85. H. Bi, P. Wang, and Y. Che, A capacitor clamped H-type boost DC–DC converter with wide voltage-gain range for fuel cell vehicles. IEEE Trans. Veh. Technol. 68(1), 276–290 (2019).
86. N. Elsayad, H. Moradisizkoohi, and O. A. Mohammed, Design and implementation of a new transformerless bidirectional DC–DC converter with wide conversion ratios. IEEE Trans. Ind. Electron. 66(9), 7067–7077 (2019).
87. M. Hasan Babayi Nozadian, E. Babaei, S. H. Hosseini and E. Shokati Asl, Switched Z-source networks: A review. IET Power Electron. 12(7), 1616–1633 (2019).
88. M. R. Mohammadi, H. Farzanehfard, A new family of zero-voltage-transition nonisolated bidirectional converters with simple auxiliary circuit. IEEE Trans. Ind. Electron. 63(3), 1519–1527 (2016).
89. X. Zhang, L. Sun, Y. Guan, S. Han, H. Cai, Y. Wang, and D. Xu, Novel high step-up soft-switching DC–DC converter based on switched capacitor and coupled inductor. IEEE Trans. Power Electron. 35(9), 9471–9481 (2020).
90. P. Das, S. A. Mousavi and G. Moschopoulos, Analysis and design of a nonisolated bidirectional ZVS-PWM DC–DC converter with coupled inductors. IEEE Trans. Power Electron. 25(10), 2630–2641 (2010).
91. Y. Yang, J. Ma, C. N. Ho and Y. Zou, A new coupled-inductor structure for interleaving bidirectional DC–DC converters. IEEE J. Emerg. Sel. Top. Power Electron. 3(3), 841–849 (2015).
92. M. Aamir, S. Mekhilef and H. Kim, High-gain zero-voltage switching bidirectional converter with a reduced number of switches. IEEE Trans. Circuits Syst. II-Express Briefs. 62(8), 816–820 (2015).
93. J. Yang, H. Do, High-efficiency bidirectional DC–DC converter with low circulating current and ZVS characteristic throughout a full range of loads. IEEE Trans. Ind. Electron. 61(7), 3248–3256 (2014).
94. J. K. Reed G. Venkataramanan, Bidirectional high conversion ratio DC–DC converter. Proc. IEEE Power Energy Conf. Illinois. 1–5 (2012).
95. L. Sun, F. Zhuo, F. Wang, and T. Zhu, A novel topology of high voltage and high power bidirectional ZCS DC–DC converter based on serial capacitors. Proc. IEEE Appl. Power Electron. Conf. Expo. 810–815 (2016).
96. B. Axelrod, Y. Berkovich, and A. Ioinovici, Transformerless DC–DC converters with a very high DC line-to-load voltage ratio. J. Circuits, Syst. Comput. 13(3), 467–475 (2004).

97. B. Axelrod, Y. Berkovich, and A. Ioinovici, Switched-capacitor/switched-inductor structures for getting transformerless hybrid DC–DC PWM converters. IEEE Trans. Circuits Syst. I, Reg. Papers. 55(2), 687–696 (2008).
98. G. Zhang, Z. Li, B. Zhang, D. Qiu, W. Xiao, and W. Halang, A Z-source half-bridge converter. IEEE Trans. Ind. Electron. 61(3), 1269–1279 (2014).
99. T. Takiguchi, H. Koizumi, Quasi-Z-source DC–DC converter with voltage-lift technique. Proc. IEEE 39th Annu. Conf. Ind. Electron. Soc. 1191–1196 (2013).
100. X. Zhu, B. Zhang, Z. Li, H. Li, and L. Ran, Extended switched-boost DC–DC converters adopting switched-capacitor/switched-inductor cells for high step-up conversion. IEEE J. Emerg. Sel. Topics Power Electron. 5(3), 1020–1030 (2017).
101. Y. Zhang, Q. Liu, J. Li and M. Sumner, A common ground switched-quasi-Z-Source bidirectional DC–DC converter with wide-voltage-gain range for EVs with hybrid energy sources. IEEE Trans. Ind. Electron. 65(6), 5188–5200 (2018).
102. Y. Zhang, Q. Liu, Y. Gao, J. Li, and M. Summer, Hybrid Switched-capacitor/switched-quasiz-source bidirectional DC–DC converter with wide-voltage-gain range for hybrid energy sources EVs. IEEE Trans. Ind. Electron. 66(4), 2680–2690 (2019).
103. Z. Wang, H. Li, A soft switching three-phase current-fed bidirectional DC–DC converter with high efficiency over a wide input voltage range. IEEE Trans. Power Electron. 27(2), 669–684 (2012).
104. G. Wu, X. Ruan, and Z. Ye, Nonisolated high step-up DC–DC converters adopting switched-capacitor cell," IEEE Trans. Ind. Electron. 61(1), 383–393 (2015).
105. P. Thummala, D. Maksimovic, Z. Zhang, and M. A. E. Andersen, Digital control of a high-voltage (2.5 kV) bidirectional DC–DC flyback converter for driving a capacitive incremental actuator. IEEE Trans. Power Electron. 31(12) 8500–8516 (2016).
106. C. C. Lin, L. S. Yang, and G. W. Wu, Study of a non-isolated bidirectional DC–DC converter. IET Power Electron. 6(1), 30–37 (2013).
107. J. Yang, H. Do, Soft-switching bidirectional DC–DC converter using a lossless active snubber. IEEE Trans. Circuits Syst. I, Reg. Papers, 61(5), 1588–1596 (2014).
108. S. Li, K. M. Smedley, D. R. Caldas, and Y. W. Martins, Hybrid bidirectional DC–DC converter with low component counts. IEEE Trans. Ind. Appl. 54(2), 1573–1582 (2018).
109. R. J. Wai, R. Y. Duan, and K. H. Jheng, High-efficiency bidirectional DC–DC converter with high-voltage gain. IET Power Electron. 5(2), 173–184 (2012).
110. H. Liu, L. Wang, Y. Ji, and F. Li, A novel reversal coupled inductor high-conversion-ratio bidirectional DC–DC converter. IEEE Trans. Power Electron. 33(6), 4968–4979 (2018).
111. C. Hong, L. Yang, T. Liang, and J. Chen, Novel bidirectional DC–DC converter with high step-up/down voltage gain. Proc. Energy Convers. Congr. Expo. 60–66 (2009).
112. J. Yao, A. Abramovitz, and K. M. Smedley, Steep-gain bidirectional converter with a regenerative snubber. IEEE Trans. Power Electron. 30(12), 6845–6856 (2015).
113. N. Molavi, E. Adib, and H. Farzanehfard, Soft-switching bidirectional DC–DC converter with high voltage conversion ratio. IET Power Electron. 11(1), 33–42 (2018).
114. K. Jin, X. Ruan, M. Yang, and M. Xu, A hybrid fuel cell power system. IEEE Trans. Ind. Electron. 56(4), 1212–1222 (2009).
115. F. Z. Peng, F. Zhang, and Z. Qian, A magnetic-less DC–DC converter fordual-voltage automotive systems. IEEE Trans. Ind. Appl. 39(2), 511–518 (2003).
116. S. J. Amodeo, H. G. Chiacchiarini, and A. R. Oliva, High-performance control of a DC–DC Z-source converter used for an excitation field driver. IEEE Trans. Power Electron. 27(6), 2947–2957 (2012).
117. R. Wai, R. Duan, High step-up converter with coupled-inductor. IEEE Trans. Power Electron. 20(5), 1025–1035 (2005).
118. W. Li, X. He, Review of nonisolated high-step-up DC/DC converters in photovoltaic grid-connected applications. IEEE Trans. Ind. Electron. 58(4), 1239–1250 (2011).
119. H. Liu, H. Hu, H. Wu, Y. Xing and I. Batarseh, Overview of high-step-up coupled-inductor boost converters. IEEE J. Emerging Sel. Top. Power Electron. 4(2), 689–704 (2016).

120. F. L. Tofoli, D. d. C. Pereira, W. Josias de Paula and D. d. S. Oliveira Júnior, Survey on non-isolated high-voltage step-up DC–DC topologies based on the boost converter. IET Power Electron. 8(10), 2044–2057 (2015).
121. Y. Zhang, Y. Gao, L. Zhou, and M. Sumner, A switched-capacitor bidirectional DC–DC converter with wide voltage gain range for electric vehicles with hybrid energy sources. IEEE Trans. Power Electron. 33(11), 9459–9469 (2018).
122. Y. Zhang, W. Zhang, F. Gao, S. Gao and D. J. Rogers, A switched-capacitor interleaved bidirectional converter with wide voltage-gain range for super capacitors in EVs. IEEE Trans. Ind. Electron. 35(2), 1536–1547 (2020).
123. R. H. Ashique, Z. Salam, A high-gain, high-efficiency nonisolated bidirectional DC–DC converter with sustained ZVS Operation. IEEE Trans. Ind. Electron. 65(10), 7829–7840 (2018).
124. Y. Hsieh, J. Chen, L. Yang, C. Wu, and W. Liu, High-conversion-ratio bidirectional DC–DC converter with coupled inductor. IEEE Trans. Ind. Electron. 61(1), 210–222 (2014).
125. S. Liao, J. Teng, and S. Chen, Bidirectional DC–DC converter with high step-down and step-up voltage conversion ratio. IEEE Annual Southern Power Electron. Conf. 1–6 (2016).
126. H. Ardi, A. Ajami, F. Kardan, and S. N. Avilagh, Analysis and implementation of a nonisolated bidirectional DC–DC converter with high voltage gain. IEEE Trans. Ind. Electron. 63(8), 4878–4888 (2016).
127. Y. Zhang, Y. Gao, J. Li, and M. Sumner, Interleaved switched-capacitor bidirectional DC–DC converter with wide voltage-gain range for energy storage systems. IEEE Trans. Power Electron. 33(5), 3852–3869 (2018).
128. L. Yang, T. Liang, and J. Chen, Transformerless DC–DC converters with high step-up voltage gain. IEEE Trans. Ind. Electron. 56(8), 3144–3152 (2009).
129. X. Hu, C. Gong, A high gain input-parallel output-series DC/DC converter with dual coupled inductors. IEEE Trans. Power Electron. 30(3), 1306–1317 (2015).
130. X. Hu, L. Li, Y. Li and G. Wu, Input-parallel output-series DC–DC converter for non-isolated high step-up applications. Electron. Lett. 52(20), 1715–1717 (2016).
131. Y. Zhang, L. Zhou, M. Summer, and P. Wang, Single-switch, wide voltage-gain range, Boost DC–DC converter for fuel cell vehicles. IEEE T. Veh. Technol. 67(1), 134–145 (2018).
132. V. F. Piresa, D. Foitoa, and J. F. Silvab, A single switch hybrid DC/DC converter with extended static gain for photovoltaic applications. Elect. Power Syst. Res. 146, 228–235 (2017).
133. E. Jamshidpour, P. Poure, E. Gholipour, and S. Saadate, Single-switch DC–DC converter with fault-tolerant capability under open- and short circuit switch failures. IEEE Trans. Power Electron. 30(5), 2703–2712 (2015).
134. Y. Zhang, C. Fu, M. Sumner and P. Wang, A wide input-voltage range quasi-Z-source boost DC–DC converter with high-voltage gain for fuel cell vehicles. IEEE Trans. Ind. Electron. 65(6), 5201–5212 (2018).
135. M. Prudente, L. L. Pfitscher, G. Emmendoerfer, E. F. Romaneli, and R. Gules, Voltage multiplier cells applied to non-isolated DC–DC converters. IEEE Trans. Power Electron. 23(2), 871–887 (2008).
136. Y. Shindo, M. Yamanaka, and H. Koizumi, Z-source DC–DC converter with cascade switched capacitor. Proc. 37th Annu. Conf. IEEE Ind. Electron. Soc. 1665–1670 (2011).
137. L. W. Zhou, B. X. Zhu, Q. M. Luo, and S. Chen, Interleaved non-isolated high step-up DC/DC converter based on the diode–capacitor multiplier. IET Power Electron. 7(2), 390–397 (2014).
138. Y. Zhang, H. Liu, J. Li, M. Summer, and C. Xia, DC–DC Boost Converter with a wide input range and high voltage gain for fuel cell vehicles. IEEE. Trans. Power Electron. 34(5), 4100–4111 (2019).
139. Y. R. Novaes, A. Rufer, and I. Barbi, A new quadratic, three-level, DC/DC converter suitable for fuel cell applications. Proc. Power Convers. Conf. 601–607 (2007).
140. C. T. Pan, C. F. Chuang, and C. C. Chu, A novel transformer-less adaptable voltage quadrupler DC converter with low switch voltage stress. IEEE Trans. Power Electron. 29(9), 4787–4796 (2014).

141. Y. Zhang, H. Liu, J. Li, and M. Summer, A low-current ripple and wide voltage-gain range bidirectional DC–DC converter with coupled inductor. IEEE Trans. Power Electron. 35(2), 1525–1535 (2020).
142. E. Babaei, Z. Saadatizadeh, and C. Cecati, High step-up high step-downbidirectional DC/DC converter. IET Power Electron. 10(12), 1556–1571 (2017).
143. H. Wu, K. Sun, L. Chen, L. Zhu, and Y. Xing, High step-up/step-downsoft-switching bidirectional DC–DC converter with coupled-inductor andvoltage matching control for energy storage systems. IEEE Trans. Ind. Electron. 63(5), 2892–2903 (2016).

GPSR Compliance

The European Union's (EU) General Product Safety Regulation (GPSR) is a set of rules that requires consumer products to be safe and our obligations to ensure this.

If you have any concerns about our products, you can contact us on ProductSafety@springernature.com

In case Publisher is established outside the EU, the EU authorized representative is:

Springer Nature Customer Service Center GmbH
Europaplatz 3
69115 Heidelberg, Germany

Batch number: 10732090

Printed by Printforce, the Netherlands

CPSS Power Electronics Series

This series comprises advanced textbooks, research monographs, professional books, and reference works covering different aspects of power electronics, such as Variable Frequency Power Supply, DC Power Supply, Magnetic Technology, New Energy Power Conversion, Electromagnetic Compatibility as well as Wireless Power Transfer Technology and Equipment. The series features leading Chinese scholars and researchers and publishes authored books as well as edited compilations. It aims to provide critical reviews of important subjects in the field, publish new discoveries and significant progress that has been made in development of applications and the advancement of principles, theories and designs, and report cutting-edge research and relevant technologies. The CPSS Power Electronics series has an editorial board with members from the China Power Supply Society and a consulting editor from Springer.

Readership: Research scientists in universities, research institutions and the industry, graduate students, and senior undergraduates.

Indexed by EI Compendex